L'Organisation de Coopération et de Développement Économiques (OCDE), qui a été instituée par une Convention signée le 14 décembre 1960, à Paris, a pour objectif de promouvoir des politiques visant :

— à réaliser la plus forte expansion possible de l'économie et de l'emploi et une progression du niveau de vie dans les pays Membres, tout en maintenant la stabilité financière, et contribuer ainsi au développement de l'économie mondiale ;

— à contribuer à une saine expansion économique dans les pays Membres, ainsi que non membres, en voie de développement économique ;

— à contribuer à l'expansion du commerce mondial sur une base multilatérale et non discriminatoire, conformément aux obligations internationales.

Les Membres de l'OCDE sont : la République Fédérale d'Allemagne, l'Australie, l'Autriche, la Belgique, le Canada, le Danemark, l'Espagne, les États-Unis, la Finlande, la France, la Grèce, l'Irlande, l'Islande, l'Italie, le Japon, le Luxembourg, la Norvège, la Nouvelle Zélande, les Pays-Bas, le Portugal, le Royaume-Uni, la Suède, la Suisse et la Turquie.

L'Agence de l'OCDE pour l'Énergie Nucléaire (AEN) a été créée le 20 avril 1972, en remplacement de l'Agence Européenne pour l'Énergie Nucléaire de l'OCDE (ENEA) lors de l'adhésion du Japon à titre de Membre de plein exercice.

L'AEN groupe désormais tous les pays Membres européens de l'OCDE ainsi que l'Australie, le Canada, les États-Unis et le Japon. La Commission des Communautés Européennes participe à ses travaux.

L'AEN a pour principaux objectifs de promouvoir, entre les gouvernements qui en sont Membres, la coopération dans le domaine de la sécurité et de la réglementation nucléaires, ainsi que l'évaluation de la contribution de l'énergie nucléaire au progrès économique.

Pour atteindre ces objectifs, l'AEN :

— *encourage l'harmonisation des politiques et pratiques réglementaires dans le domaine nucléaire, en ce qui concerne notamment la sûreté des installations nucléaires, la protection de l'homme contre les radiations ionisantes et la préservation de l'environnement, la gestion des déchets radioactifs, ainsi que la responsabilité civile et les assurances en matière nucléaire ;*

— *examine régulièrement les aspects économiques et techniques de la croissance de l'énergie nucléaire et du cycle du combustible nucléaire, et évalue la demande et les capacités disponibles pour les différentes phases du cycle du combustible nucléaire, ainsi que le rôle que l'énergie nucléaire jouera dans l'avenir pour satisfaire la demande énergétique totale ;*

— *développe les échanges d'informations scientifiques et techniques concernant l'énergie nucléaire, notamment par l'intermédiaire de services communs ;*

— *met sur pied des programmes internationaux de recherche et développement, ainsi que des activités organisées et gérées en commun par les pays de l'OCDE.*

Pour ces activités, ainsi que pour d'autres travaux connexes, l'AEN collabore étroitement avec l'Agence Internationale de l'Énergie Atomique de Vienne, avec laquelle elle a conclu un Accord de coopération, ainsi qu'avec d'autres organisations internationales opérant dans le domaine nucléaire.

FOREWORD

The disposal of radioactive waste in underground repositories may itself create various phenomena in the local host rock, and it is essential to check that such phenomena do not affect the ability of the geologic formation to isolate the radioactive material for very long periods of time. Changes in the local geological environment of a waste repository, or *near-field*, associated with the heat generation and radiation field of radioactive waste are the subject of the contributions to this NEA Workshop.

Prediction of the behaviour of radioactive waste and the geologic host medium is a complex problem, involving an understanding of many chemical and physical phenomena. Topics covered by this Workshop include rock mechanics in stressed and heated conditions ; thermally induced groundwater flow in fractured rock ; chemical changes to rock surfaces associated with groundwater and changes in the thermal and chemical environment ; the chemical solubilities and sorption properties of radionuclides ; and the long-term integrity of containers and packaging for radioactive waste.

An understanding of these phenomena and the interaction between them is essential to reliable predictive modeling of the long-term safety of geologic disposal of radioactive waste. The objectives of this Workshop were :

- to review experimental work in NEA Member countries designed to provide information on near-field phenomena in various possible host rocks ;

- to review analytical and modelling activities relevant to the near-field behaviour of radionuclides and the host geology ;

- to identify research needs and areas of current interest ;

- to promote co-ordination of activities, in the interest of effective use of research facilities in NEA Member countries.

These proceedings reproduce the papers contributed to the meeting and the views expressed are the responsibility of the authors. A record of discussions is also given, together with brief overviews of research and development prepared by working groups convened during the Workshop. The overviews concern granite formations, salt formations, clay and tuff, and engineered barriers to radionuclide dispersion.

AVANT-PROPOS

Le fait d'évacuer des déchets radioactifs dans des dépôts souterrains est susceptible d'engendrer des phénomènes divers dans les roches situées à proximité et il est essentiel de veiller à ce que ceux-ci n'affectent pas la capacité des formations géologiques à isoler les substances radioactives au cours de très longues périodes. Ce sont ces modifications, intervenant dans le milieu géologique situé à proximité des dépôts de déchets radioactifs, ou *champ proche*, liées au dégagement de chaleur et de rayonnements par les déchets radioactifs qui ont fait l'objet des exposés présentés à cette réunion de travail de l'AEN.

La prévision du comportement des déchets radioactifs et du milieu géologique hôte est un problème complexe, impliquant la compréhension de nombreux phénomènes chimiques et physiques. Les sujets couverts par cette réunion de travail comprenaient la mécanique des roches dans des conditions de dégagement de chaleur et sous contraintes ; la circulation d'eau souterraine dans les roches fissurées due au gradient thermique ; les modifications chimiques de la surface des roches liées à la présence d'eaux souterraines et aux altérations thermiques et chimiques du milieu ; la solubilité chimique et les propriétés de sorption des radionucléides ; et l'intégrité à long terme des conteneurs et des emballages des déchets radioactifs.

Une compréhension de ces phénomènes et leur interaction est essentielle pour l'établissement de modèles prévisionnels fiables relatifs à la sûreté à long terme de l'évacuation des déchets radioactifs dans des formations géologiques. Les objectifs de cette réunion de travail ont donc été les suivants :

- examiner les travaux expérimentaux dans les pays Membres de l'AEN afin de fournir des informations sur les phénomènes dans le champ proche en ce qui concerne les diverses variétés de roches envisagées pour le dépôt des déchets radioactifs ;

- examiner les activités analytiques et de modélisation relatives au comportement dans le champ proche des radionucléides et des roches avoisinantes ;

- définir les recherches nécessaires et les domaines d'intérêt ;

- promouvoir la coordination des activités en vue d'une utilisation efficace des moyens de recherche disponibles dans les pays Membres de l'AEN.

Le compte rendu comporte les communications qui ont été présentées à la réunion, dont le contenu n'engage que leurs auteurs, les discussions qui ont suivies, ainsi que des résumés établis par des groupes de travail au cours de la réunion. Ces résumés portent sur des travaux de recherche et de développement relatifs aux formations granitiques, salines, argileuses et aux dépôts de tuf ainsi qu'aux barrières artificielles destinées à contenir la dispersion des radionucléides.

TABLE OF CONTENTS
TABLE DES MATIERES

Session 3 - Séance 3

WATER MOVEMENT
ECOULEMENT DES EAUX

Chairman - Président : N.A. CHAPMAN (United Kingdom)

Session 4 - Séance 4

CLAY AND TUFF
FORMATIONS ARGILEUSES ET TUF VOLCANIQUE

Chairman - Président : N.A. CHAPMAN (United Kingdom)

Session 5 - Séance 5

SALT FORMATIONS
FORMATIONS SALINES

Chairman - Président : R.H. KOSTER (Federal Republic of Germany)

Session 6 - Séance 6

BARRIER EFFECTS
ACTION DES BARRIERES

Chairman - Président : R.H. HEREMANS (Belgium)

ACTINIDE PROPERTIES AND SOURCE TERMS

Chairman - Président

H. BURKHOLDER

(United States)

Séance 1

PROPRIETES DES ACTINIDES ET TERMES-SOURCES

ACTINIDE SOLUBILITIES IN THE NEAR-FIELD
OF A NUCLEAR WASTE REPOSITORY

Dhanpat Rai, R. G. Strickert, and J. L. Swanson
Pacific Northwest Laboratory
P.O. Box 999
Richland, Washington, U.S.A. 99352

ABSTRACT

Information on the solubilities of actinide compounds present in
nuclear waste forms is needed to predict the potential hazard of actinide
disposal in geologic repositories. Because solubilities are independent
of release scenarios, hydrologic properties, and sorption coefficients,
they provide a means of predicting maximum radionuclide concentrations that
would be available for transport. Due to the lack of available data on
experimentally-determined solubilities of actinide compounds, studies
were conducted on the solubilities of some compounds of Pu, Np, and Am,
including the Np and Pu compounds present in borosilicate glass. The results
of several years research effort are briefly summarized.

INTRODUCTION

Several of the actinide radioisotopes, present in high level radio-active wastes, have long half-lives. The solid compounds of these elements that are present in the solidified high-level waste forms have specific solubilities at equilibrium in a given weathering environment, which could control the maximum possible concentration of actinides in groundwater and thus the amount of element available for transport to the biosphere. Therefore, knowledge of actinide compounds present in the wastes and their solubilities is needed to predict maximum possible actinide concentrations in ground-waters. Although sorption reactions may be occurring in and near the waste material, as long as the actinide compound is in contact with the groundwater, the equilibrium concentration would be determined by the solubility of that compound. We have determined the solubility of several plutonium, neptunium, and americium solids which may be important for near-field repository conditions:

1. Np-doped crushed borosilicate glass, similar in compo-sition to proposed glass waste forms

2. PuO_2 (c, c = crystalline)

3. Pu(IV) polymer

4. $Pu(OH)_4$ (a, a = amorphous)

5. Am solid phase present in weathered sediments contaminated by reprocessing wastes during the 1950's.

Experimental

The experimental details are described in earlier publications [1-5]. Briefly, the solubility was determined in a batch system in which a small amount of the solid material (a few mg of pure solids $^{237}NpO_2$(c), $^{239}PuO_2$(c), $^{239}Pu(OH)_4$(a), or ^{239}Pu(IV) polymer to a gram of crushed Pu/Np-doped borosilicate glass and ^{241}Am-contaminated sediments) was suspensed in 20 to 30 ml of dilute salt solution (0.0015 $\underline{M}$ $CaCl_2$). Quinhydrone, a redox buffer (pe + pH = 11.82), was employed in the experiments involving the crushed borosilicate glasses (PNL 76-68). All the suspensions were equilibrated in air for different times. In all cases, less than one percent of the total actinide element present was in the solution phase. Periodically the pH and redox potential of the suspensions were measured with glass and platinum electrodes, respectively. A small aliquot of the suspension was also filtered through a 0.015 µm or smaller membrane filter. The oxidation states of Pu and Np in filtered solutions were determined by spectrophoto-metric and solvent extraction techniques [2, 4, 6, 7]. Actinide concentrations were determined by alpha and gamma (in the case of Am) counting techniques. Neutron activation [8] and inductively coupled plasma techniques were used to determine the concentrations of nonradioactive elements in solutions.

RESULTS AND DISCUSSION

Thermodynamic data [9, 10] have been used to predict the stability of Pu minerals. These predictions indicate that crystalline PuO_2 is more stable than the hydrous Pu oxides; in addition, for oxidizing or reducing conditions, PuO_2 is predicted to be the most stable Pu compound among the simple oxides, hydroxides, carbonates, and phosphates for which information is available [11]. Thermodynamic considerations also dictate that compounds with high free energy would eventually convert to compounds with low free energy. This means that with time $Pu(OH)_4$(a) may be expected to convert to PuO_2(c).

The steady state experimental solubility of different Pu solids in air-equilibrated solutions (Figure 1) does indeed show that $^{239}PuO_2$ is

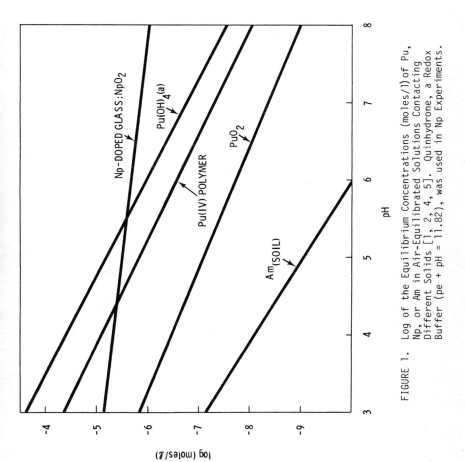

FIGURE 1. Log of the Equilibrium Concentrations (moles/l) of Pu, Np, or Am in Air-Equilibrated Solutions Contacting Different Solids [1, 2, 4, 5]. Quinhydrone, a Redox Buffer (pe + pH = 11.82), was used in Np Experiments.

more stable than the ^{239}Pu(IV) hydroxide or polymer. However, a recent study [12] has shown that regardless of whether one starts with ^{239}PuO$_2$(c) or ^{239}Pu(OH)$_4$(a), the aging process in which dehydration and crystallization are opposed by radiolytic effects, will convert these solids to a material having steady state properties between those of the true crystalline PuO$_2$ and those of fresh hydrated oxide. Crystalline ^{238}PuO$_2$ was found to convert to a solid that had steady state properties, including solubility, similar to that of ^{239}Pu(IV) polymer. Both the ^{239}PuO$_2$(c) and ^{239}Pu(OH)$_4$(a) were shown by X-ray measurements to convert to PuO$_2$(lc, lc = less crystalline) in approximately 1300 days. The PuO$_2$(lc) solubility was similar to that represented for PuO$_2$(c) in Figure 1. Alpha radiolysis, in addition to its effect on the crystallinity of solids, also produces nitric acid in air-equilibrated solutions [13]. However, nitric acid primarily decreases pH and the effect of its production is predictable from pH-concentration relationships. The Pu solution species in equilibrium with the Pu solid phases were found to be primarily Pu(V) at pH values greater than about 3.5 (Table I).

The solubility of the solid phase that may be controlling the Pu concentration (see Figure 1) must be determined to ascertain the potential amount of soluble Pu that could leach from a repository by a given amount of water. For total understanding, the concentrations of ligands and the formation constants of their complexes with Pu should also be known in order to assess their effect on solubility.

In studies employing ^{239}Pu-doped glasses (a simulated high-level waste form), Rai and Strickert [3] observed that under controlled redox conditions (pe + pH 12) the concentration of Pu in solution was solubility-limited at levels similar to those observed for PuO$_2$. These results show that the soluble Pu concentration that can result from leaching this waste form can be predicted from the solubility behavior of PuO$_2$ and that knowledge of PuO$_2$ solubility in a given environment would allow placing a limit on the maximum expected Pu concentration in that environment.

To help predict concentrations of Np leached from nuclear waste repositories in geologic environments, the solubility of a Np-doped boro-silicate glass was also investigated [4]. The results with Np present are the same as those with Pu; that is, the concentrations of Np in solutions contacting the crushed doped glass were found to be controlled by a Np solid phase similar to crystalline NpO$_2$ in solubility (Figure 1). Thus, the maximum concentration of the Np leached from this waste form can be predicted from the solubility of NpO$_2$(c). In addition, predictions [11] based on the available thermodynamic data show that NpO$_2$(c) is the most stable compound among the Np oxides and hydroxides, at all pH and Eh values encountered in groundwaters [14]. Tetravalent U, Np, Puand Am have similar ionic radii, and tetravalent oxides of these elements can exist as solid solutions in unprocessed wastes such as spent fuel. Thus NpO$_2$ (and PuO$_2$ and AmO$_2$) would be expected to be present in this potential waste as well. Solvent extraction techniques showed Np(V) + Np(VI) to be the primary solution species (Table I) in the solutions contacting Np-doped glass and NpO$_2$(c). Neptunium (V) would be expected to be the primary oxidation state because the redox buffer (quinhydrone) used in this study maintains potentials (pe + pH $\approx$ 12) at which thermodynamic data [15] predict Np(V) to be stable and at which the Np(VI) stability field is far removed from the Np(V) stability field.

Work with Am has involved a different type of solid phase than the work with Pu and Np. The Am-containing solid (Am(soil)) employed in the study was a contaminated sediment resulting from disposal of Pu and Am-contaminated solutions onto the ground. Many years elapsed between this disposal and the study of the contaminated sediment. The Am compound present in these sediments has not yet been identified, but it does not appear to be Am(OH)$_3$. The steady state Am concentrations in solutions contacting ^{Am}(soil) solid were found to decrease approximately 10-fold with on unit increase in pH (Figure 1). In many Am sorption experiments

Table I. Concentrations and Oxidation State Distributions of Pu and Np
in Filtered Solutions Contacting Different Solids

Solid	Approximate Contact time (days)	pH	log Concentration (moles/l)	Oxidation State (V + VI)* % of total soluble
		Plutonium		
239 PuO$_2$	250	4.00	-5.94	82
239PuO$_2$	250	5.21	-6.83	92
239Pu(OH)$_4$(a)	250	3.50	-4.77	93
239Pu(OH)$_4$(a)	250	4.90	-5.39	91
239Pu(OH)$_4$(a)	250	7.73	-7.30	73
239Pu(IV) polymer	20	3.87	-4.97	86
239Pu(IV) polymer	20	6.48	-7.20	87
		Neptunium		
237NpO$_2$(c)	27	5.4	-5.85	99
237NpO$_2$(c)	103	4.4	-5.67	99
237Np-doped glass beads	82	4.5	-5.48	99
237Np-doped glass beads	82	6.3	-5.73	96

* Data based on solvent extraction techniques [2, 4, 6]; Pu(V) was identified to be the primary species in several solutions where concentrations were high enough that spectrophotometric techniques could be used [6, 7]; Np(V) is expected to be the only Np oxidation state because of the presence of redox buffer (pe + pH 12) and because the Np(VI) stability field is far removed from that of Np(V) at these redox potentials.

with a large number of rocks, soils, and minerals in which initial Am solution concentrations exceeded the solubility-limited concentration (Figure 1), the final Am solution concentrations were similar to Am concentrations in solutions contacting ^{Am}(soil) [5]. Predictions based on thermodynamic data suggest that Am solution species in the pH range of 3 to 8 is likely to be Am(OH)$_2^+$. Although the type of the Am solids that may be present in the high-level wastes is not known, the study of ^{Am}(soil) solubility suggest that the maximum Am concentrations in solutions at a short distance away from the waste packages may be controlled by the ^{Am}(soil) solubility. However, it should again be emphasized that the actual Am concentrations in equilibrium with a given solid would depend upon the nature of the dominant solution species, which is a function of groundwater chemistry.

The results discussed so far (Figure 1 and Table 1) were obtained under oxidizing conditions and in low ionic strength solutions. Wood and Rai [16] have attempted to estimate, using available thermodynamic data, changes that would result in actinide solubilities and solution species in repository groundwaters that are reducing and that contain complexing ligands. Their estimates showed that oxidizing groundwaters containing unusually high concentrations of different ligands (Cl$^-$, NO$_3^-$, SO$_4^{2-}$, H$_2$PO$_4^-$) would not significantly affect the solubilities shown in Figure 1. Although extrapolation of solubility to reducing conditions (pe + pH $\approx$ 2.4) required several assumptions because of the uncertain quality of available thermodynamic data, Wood and Rai [16] estimated that the solubilities of Np and Pu compounds would be either close to or lower than the solubilities under oxidizing conditions. Comparison of these solubilities with the Maximum Permissible Concentrations (MPC) established by the U.S. Nuclear Regulatory Commission (10 CFR-20) showed that the solubility limited concentrations (under both the oxidizing and reducing conditions) in near-neutral solutions for most of actinide radioisotopes present in spent fuel would be either close to or lower than the MPC's.

BIBLIOGRAPHIC REFERENCES

1. Rai, Dhanpat, R. J. Serne and D. A. Moore: "Solubility of Plutonium Compounds and Their Behavior in Soils," Soil Sci. Soc. Am. J. 44, 490-495 (1980).

2. Rai, Dhanpat and J. L. Swanson: "Properties of Plutonium (IV) Polymer of Environmental Importance," Nucl. Tech. 54, 107-112 (1981).

3. Rai, Dhanpat and R. G. Strickert: "Maximum Concentrations of Actinides in Geologic Media," Trans. Am. Nucl. Soc. 35, 185-186 (1980).

4. Rai, Dhanpat, R. G. Strickert and G. L. McVay: "Neptunium Concentrations in Solutions Contacting Actinide-Doped Glass," U.S. Dept. of Energy Rep. PNL-SA-9699 (1981). Submitted to Nucl. Tech.

5. Rai, Dhanpat, R. G. Strickert, D. A. Moore and R. J. Serne: "Influence of an Americium Solid Phase on Americium Concentrations in Solutions," Geochim. Cosmochim. Acta, in press (1981).

6. Rai, Dhanpat, R. J. Serne and J. L. Swanson: "Solution Species of Plutonium in the Environment," J. Environ. Qual. 9, 417-420 (1980).

7. Swanson, J. L. and Dhanpat Rai: "Spectrophotometric Measurements of Ionic Plutonium Species in Solution at Low Concentrations," Radiochem. Radioanal. Lett., in press (1981).

8. Laul, J. C.: "Neutron Activation of Geological Materials," At. Energy Rev. 3, 603-693 (1979).

9. Rai, Dhanpat and R. J. Serne: "Plutonium Activities in Soil Solutions and the Stability and Formation of Selected Plutonium Minerals," J. Environ. Qual. 6, 89-95 (1977).

10. Polzer, W. L.: "Solubility of Plutonium in Soil/Water Environments," Proc. of Rocky Flats Symp. on Safety in Plutonium Handling Facilities, U.S. Atomic Energy Comm. Symp. Ser. CONF-710401, pp 411-430, Dow Chem. Co., Rocky Flats Div., Golden, Colorado, (1971).

11. Rai, Dhanpat and R. J. Serne: "Solid Phases and Solution Species of Different Elements in Geologic Environments," U.S. Dept. of Energy Rep. PNL-2651, (1978).

12. Rai, Dhanpat and J. L. Ryan: "Crystallinity and Solubility of Pu(IV) Oxide and Hydroxide in Aged Aqueous Suspensions," U.S. Dept. of Energy Rep. PNL-SA-9722, (1981). Submitted to Radiochim. Acta.

13. Rai, Dhanpat, R. G. Strickert and J. L. Ryan: "Alpha Radiation Induced Production of HNO_3 During Dissolution of Pu Compounds," Inorg. Nucl. Chem. Lett. 16 551-555, (1981).

14. Baas Becking, I.G.M., I. R. Kaplan and D. A. Moore: "Limits of the Natural Environment in Terms of pH and Oxidation-Reduction Potentials," J. Geol. 68, 243-284, (1961).

15. Allard, B., H. Kipatsi and J. O. Liljenzin: "Expected Species of Uranium, Neptunium, and Plutonium in Neutral Aqueous Solutions," J. Inorg. Nucl. Chem. 42, 1015-1027 (1980).

16. Wood, B. J. and Dhanpat Rai: "Nuclear Waste Isolation: Actinide Containment in Geologic Repositories," U.S. Dept. of Energy Rep. PNL-SA-9546/RHO-BWI-SA-143, (1981).

DISCUSSION

A.T. JAKUBICK, Canada

Your presentation suggests no differences in solubilities of Pu under different redox conditions. This seems to me somewhat surprising.

Could you please give me more details on the source of your data ?

D. RAI, United States

Our experimental measurements of PuO_2 solubility were done under oxidizing conditions (air saturation) where we found Pu(V) species in solution. With every 0.1 V decrease in Eh, Pu concentrations will roughly decrease by 50 fold so long as we stay in the Pu(V) stability field. Once we enter the Pu(IV) field the redox potential would not effect the concentration. If we enter the Pu(III) field, then further reduction in Eh would increase the Pu concentration. Thus, to extrapolate the results (obtained under oxidizing conditions) to lower redox potentials requires the knowledge of Eh and pH at which transition from Pu(V) to Pu(IV) and then to Pu(III) occurs. Accurate knowledge about these transitions is not available. Therefore, our results under reducing conditions represents our best estimate which is the same as under oxidizing conditions.

R.H. KOSTER, Federal Republic of Germany

You mentioned that release of actinides is independent of release scenarios. I think the second equation you mentioned shows that the total amount of released radionuclides depends on the flow rate, which is of course dependent on the postulated release scenario.

D. RAI, United States

I was referring to the solubility of the actinide compounds which of course are independent of the release scenario. The solubility of the compounds determines the concentration in solution which in turn determines the potential hazard. An example of a release scenario was provided to emphasise only that for most radionuclides only a small fraction will be released to the environment in 100,000 yrs.

I.R. GRENTHE, Sweden

What is the composition of the aqueous phase under your environmental conditions ? Will the solubility not change as a result of changes in the water composition ?

D. RAI, United States

We used 0.0015 M $CaCl_2$ solutions. We have taken a look at the available thermodynamic data for the Cl^-, SO_4^-, HCO_3^-, NO_3^-, $H_2PO_4^-$ and OH^- complexes with Pu(V), Np(V) and Am(III). Calculations based on these data show that the solubility of NpO_2, PuO_2, $Am_{(soil)}$ under oxidizing conditions would not be affected even if the solubility were measured in ground waters containing unusually high concentrations of different ligands (Cl^-, NO_3^-, SO_4^{2-}, $H_2PO_4^-$, HCO_3^-).

ACTINIDE SPECIES IN GROUND WATER SYSTEMS

I Grenthe and D Ferri
Department of Inorganic Chemistry, the Royal Institute of
Technology
Stockholm, Sweden

ABSTRACT

Radioactive products may be released from a nuclear waste
repository to the biosphere through dissolution in, and transport
by ground water. In order to predict the amounts of harmful substances which may reach man, it is necessary to know how the various
constituents of the repository interacts with one another and ground
water. The first stage in such an analysis is usually a description
of the equilibrium properties of the system by using thermodynamic
data. The following subjects will be discussed :

- The chemical composition of soluble actinide species
  in ground water,

- redox properties of actinides in ground water,

- solubilities of actinide species in ground water.

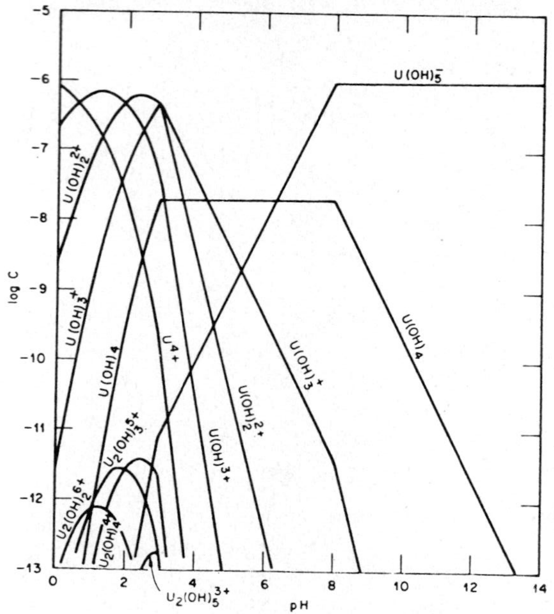

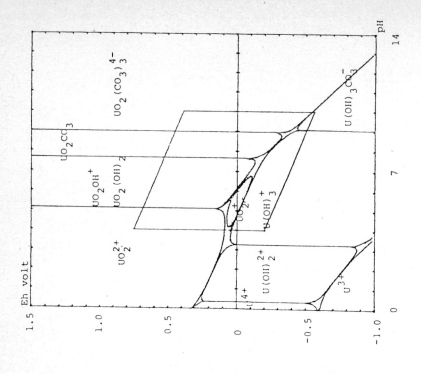

FIGURE 2. PREDOMINANCE AREA, E_H-PH DIAGRAM FOR THE
URANIUM - H_2O SYSTEM (FROM REF.2).

FIGURE 1. LOGARITMIC DIAGRAM DEPICTING THE CON-
CENTRATION OF URANIUM(IV)HYDROXIDE SPECIES AS A
FUNCTION OF PH (FROM REF.1).

1. INTRODUCTION

Radioactive products may be released from a nuclear waste repository to the biosphere through dissolution and transport by ground water. It is necessary to know how ground water and the various constituents of the repository interacts with one another in order to be able to predict the amounts of harmful substances which may reach the biosphere and eventually man. The first stage of such an analysis will be to describe the equilibrium properties of the system by using thermodynamic data. The description may then be rendered more complete by adding kinetic information. This communication deals only with equilibrium properties of hydroxide and carbonate equilibria under "ground water" conditions.

Thermodynamic data will be used to discuss the following subjects :

- the chemical composition of soluble actinide species in ground water,

- redox properties of actinides in ground water,

- solubility of actinide species in ground water.

2. THERMODYNAMIC DATA

Thermodynamic models are no better than the quality of the experimentally determined data base used. Thermodynamic data are often determined under conditions differing from those in ground water and one should always ensure that the numerical values of the thermodynamic quantities are applicable also under ground water conditions. In general, this is not much of a problem, since thermodynamic quantities do not change much for moderate changes of pressure, temperature and the composition of the ionic medium. Much more important is that all relevant equilibria must be included in the model, otherwise totally erroneous conclusions may be drawn about the system. This point is illustrated in Figures 1 and 2 which depict the concentrations of U(IV) hydroxide complexes as a function of pH [1] and a potential - pH diagram of the uranium - H_2O system [2], respectively. When constructing Figure 2, the author has not taken the formation of $U(OH)_4$ and $U(OH)_5^-$ into account, species which are present in much larger concentrations than $U(OH)_3^+$ in most ground waters, hence, the diagram does not give a correct representation of the predominating species.

2.1 General chemistry of the actinide elements

The actinide elements form in all oxidation states very strong soluble complexes and a number of different solid phases containing hydroxide or carbonate, in some cases, possibly both anions, and these species are expected to dominate the ground water chemistry of these elements (vide infra). The experimental studies of the actinide carbonate and hydroxide species are complicated by the redox properties of the elements and by the possible occurence of polynuclear and ternary complexes $M_p(OH)_q(CO_3)_r$. There is a pronounced scarcity of data on actinide carbonate complexes. An important part of our experimental program has thus been devoted to a study of equilibria in actinide(VI), actinide(IV) and actinide(III) $H_2O - CO_2(g) - CO_3^{2-}$ systems. Some of our results have been published [3-5], a summary of the findings is presented in the following section.

All actinides in the oxidation state +4 form very insoluble oxide hydrates ($M(OH)_4(s)$ or $MO_2(s)$). The solubility of these solids depends on pH and pCO_3^{2-} and also on the degree of crystallinity of the solid phase. It is not always easy to reproduce this, a fact which may explain the fairly large variations in their solubility products. The possible occurence of phase changes in the solid actinide(IV) oxide hydrates has important consequences for their migration as discussed in sections 2.2 and 4.

2.2 Data base used

An excellent critical compilation of thermodynamic data for the solid actinide oxide hydrates and for soluble actinide hydroxide complexes is given by Baes and Mesmer [6].

Redox data are taken from Stability Constants [7] which also contains some information on stability constant data for actinide carbonate complexes. However, as important data were missing or of questionable accuracy, we found it necessary to start some new experimental studies, the first part of which consists of investigations of carbonate complexes of U(VI), U(IV), La(III) and Ce(III). The experimental studies have been made in a medium of constant ionic strength 3 M (Na)ClO$_4$ by using emf, spectrophotometric and solubility methods. The main results are summarized in Table 1.

Table I. Thermodynamic data for U(VI), U(IV) and some lanthanoid-(III) equilibria in $M - H_2O - CO_2(g) - CO_3^{2-}$ systems. All data refer to 25 °C and a 3 M(Na)ClO$_4$ medium.

Reaction	log K
$CO_2(g) + H_2O \rightleftharpoons HCO_3^- + H^+$	-7.99 ± 0.01
$HCO_3^- \rightleftharpoons CO_3^{2-} + H^+$	-9.63 ± 0.01
$3UO_2(CO_3)_3^{4-} + 3H_2O \rightleftharpoons (UO_2)_3(CO_3)_6^{6-} + 6HCO_3^-$	-6.43 ± 0.08
$UO_2(CO_3)(s) + 2H^+ \rightleftharpoons UO_2^{2+} + CO_2(g) + H_2O$	3.8 ± 0.2
$3UO_2CO_3(s) + 6HCO_3^- \rightleftharpoons (UO_2)_3(CO_3)_6^{6-} + 3CO_2(g) + 3H_2O$	13.7 ± 0.2

From the above experimental data one can calculate the constants for the following two equilibria :

$UO_2^{2+} + 3CO_3^{2-} \rightleftharpoons UO_2(CO_3)_3^{4-}$	23.8 ± 0.4
$3UO_2^{2+} + 6CO_3^{2-} \rightleftharpoons (UO_2)_3(CO_3)_6^{6-}$	60.5 ± 0.1
$U^{4+} + 2H_2O \rightleftharpoons UO_2(s) + 4H^+$	-0.3 ± 0.2
$U^{4+} + 5CO_3^{2-} \rightleftharpoons U(CO_3)_5^{6-}$	38.2

$UO_2^{2+} + 4H^+ + 2e^- \rightleftharpoons U^{4+} + 2H_2O$	$E^O = 0.318$ V
$UO_2(CO_3)_3^{4-} + 2CO_2(g) + 2e^- \rightleftharpoons U(CO_3)_5^{6-}$	$E^O = -0.279$ V

$La_2(CO_3)_3(s) \rightleftharpoons 2La^{3+} + 3CO_3^{2-}$	-31.83 ± 0.05
$2La^{3+} + CO_3^{2-} \rightleftharpoons La_2CO_3^{4+}$	7.81 ± 0.03
$La^{3+} + HCO_3^- \rightleftharpoons LaHCO_3^{2+}$	1.40 ± 0.05
$La^{3+} + CO_3^{2-} \rightleftharpoons LaCO_3^+$	5.7 ± 0.2

$Ce_2(CO_3)_3(s) \rightleftharpoons 2Ce^{3+} + 3CO_3^{2-}$	-30.9 ± 0.2
$Ce^{3+} + CO_3^{2-} \rightleftharpoons CeCO_3^+$	$6.5_0 \pm 0.2$
$Ce^{3+} + 2CO_3^{2-} \rightleftharpoons Ce(CO_3)_2$	$10.9_5 \pm 0.2$
$Ce^{3+} + 3CO_3^{2-} \rightleftharpoons Ce(CO_3)_3^{3-}$	~ 14.5

The species $(UO_2)_3(CO_3)_6^{6-}$ and $UO_2(CO_3)_3^{4-}$ are dominating under ground water conditions. No other complexes, e.g. $(UO_2)_p(OH)_q(CO_3)_r$ are present in the pH range $5 < pH < 12$. An experimental study of the complexes formed at higher acidities has been made by Ciavatta et al. [8].

The solubility product of $UO_2(s)$ as given in the literature varies considerably, Baes and Mesmer quote a value of $\log K = -1.8$ for the reaction

$$UO_2(s,cryst) + 4H^+ \rightleftharpoons U^{4+} + 2H_2O$$

The solubility of amorphous uraninite $U(OH)_4(s)$ "may be a factor of 10^6 (and perhaps $10^8 - 10^9$) times greater than that of wellcrystallized uraninite" [9]. The secondary phase transformation in the solid oxidehydrates are important because dissolution is usually governed by the solubility of the thermodynamically most stable, i.e. the crystalline, phase while precipitation is determined by the solubility of the amorphous phase (c.f. Figure 3).

The normal potentials given in Table 1 shows that complex formation with carbonate changes the stability of U(VI) and U(IV) very much as compared to the uncomplexed state.

The stability constant for the formation of $U(CO_3)_5^{6-}$ has been determined in the same pH-range as found in many ground waters, i.e. pH around 8. However, the total concentration of bicarbonate, 0.1-0.2 M, was larger than the values, $10^{-3} - 10^{-2}$ M, usually found in water of these kinds. Further experimental work is presently carried out in order to decide whether $U(CO_3)_5^{6-}$ is the dominating U(IV) species in reducing ground water, or not.

Only small amounts of the total uranium is present as U(V) in reducing ground waters. From some preliminary experimental data in carbonate solutions, where U(VI) was reduced to U(IV), we can estimate that the conditional redox potential U(V)/U(IV) is lower than -400 mV in the pH range 8-9. Hence, U(V) accounts for less than 1 % of the total uranium concentration. However, the percentage of U(V) increases with increasing pH and U(V) complexes are the predominating species at pH > 10.9.

The preliminary measurements indicate that the limiting complex of U(V) has the composition $UO_2(CO_3)_3^{5-}$.

The dominating lanthanoid(III) species present in ground water are $LnCO_3^+$ and $Ln(CO_3)_2^-$. The concentrations of $Ln_2CO_3^{4+}$ and $LnHCO_3^{2+}$ are negligible and need not to be taken into account when making a chemical transport model.

3. THERMODYNAMIC DATA AND THE COMPOSITION OF MULTICOMPONENT SYSTEMS

A convenient way to present the composition of multicomponent systems is to use various types of logaritmic diagrams which describe how the composition of the system changes as a function of one or more master variables. The most commonly used variables are the redox potential and the pH. Several examples of so called potential-pH diagrams are presented in the following section. Diagrams of this type tends to be rather complicated when the number of components increase. In these cases, it is often more convenient to study sections of the potential-pH diagrams, e.g. sections at constant E_h or constant pH. Diagrams of the latter type will be used when surveying the dominant actinide species in ground waters.

The principle is illustrated by the following set of equations:

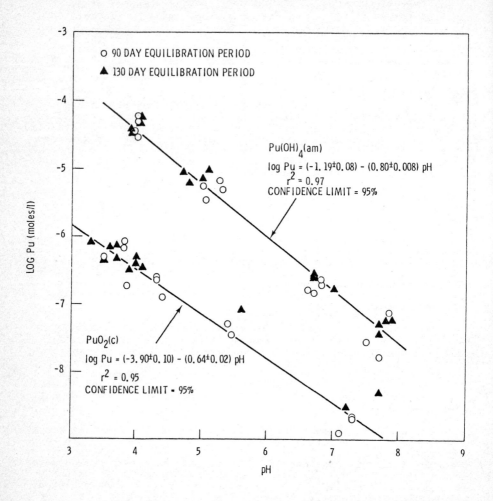

FIGURE 3. EXPERIMENTAL SOLUBILITIES OF PU(OH)$_4$(S,AMORF) AND
PUO$_2$(S,CRYST) AS A FUNCTION OF PH (FROM REF.10).

$$UO_2^{2+} + 4H^+ + 2e^- \rightleftharpoons U^{4+} + 2H_2O \qquad \log K = 11.23$$

$$\log [UO_2^{2+}]/[U^{4+}] = -\log K + 4pH + 2pe$$

where $pe = E_h/0.05916$

The ratio $[UO_2^{2+}]/[U^{4+}]$ is a function of one variable in the sections mentioned above.

By combining the preceeding equation with other equilibria in the system under study, one may obtain a complete representation of all concentrations.

$$U^{4+} + 4 OH^- \rightleftharpoons U(OH)_4(s) \qquad \log K = 56$$
$$UO_2^{2+} + 3CO_3^{2-} \rightleftharpoons UO_2(CO_3)_3^{4-} \qquad \log K = 23.8$$
$$U^{4+} + 5H_2O \rightleftharpoons U(OH)_5^- + 5H^+ \qquad \log K = -22$$
$$U^{4+} + 5CO_3^{2-} \rightleftharpoons U(CO_3)_5^{6-} \qquad \log K = 38.2$$

If we select $UO_2(s)$ as the reference state, we obtain

$$\log [UO_2(CO_3)_3^{4-}]/\{UO_2(s)\} = 10.7 - 3p[CO_3^{2-}] + 2pe$$
$$\log [U(CO_3)_5^{6-}]/\{UO_2(s)\} = 26.4 - 4pH - 5p[CO_3^{2-}]$$
$$\log [U(OH)_5^-]/\{UO_2(s)\} = -24 + pH$$

One may easily find equilibrium data in the literature for other complexing agents (F^-, SO_4^{2-}, HPO_4^{2-}) which occur in ground water. For the uranium system, one finds by using the method outlined above that

$$\log [UF_3^+]/\{UO_2(s)\} = 16.94 - 4pH - 3p[F^-]$$
$$\log [USO_4^{2+}]/\{UO_2(s)\} = 2.9 - 4pH - p[SO_4^{2-}]$$
$$\log [(U(HPO_4)_3^{2-}]/\{UO_2(s)\} = 28.8 - 4pH - 3p[HPO_4^{2-}]$$

In a typical granitic ground water (Table II), one finds
$$p[F^-] \simeq 4$$
$$p[SO_4^{2-}] \simeq 3.8$$
$$p[HPO_4^{2-}] \simeq 6$$

Table II. Analytical data for some Swedish granitic ground waters. The data are presented as "probability" intervals.

pH	7.2 - 8.5		HCO_3^-	60 - 400	ppm
Ca^{2+}	25 - 50	ppm	CO_2	0 - 25	"
Mg^{2+}	5 - 20	"	Cl^-	5 - 50	"
Na^+	10 - 100	"	SO_4^{2-}	1 - 15	"
Fe-tot	1 - 20	"	PO_4^{3-}	0.01 - 0.1	"
Fe^{2+}	0.5 - 15	"	F^-	0.5 - 2	"
Mn^{2+}	0.1 - 0.5	"	HS^-	< 0.1 - 1	"

The only species which have to be taken into account are $U(OH)_5^-$, $U(CO_3)_5^{6-}$ and $UO_2(CO_3)_3^{4-}$; their concentration as a function of pe or E_h are shown in Figure 4.

We must point out once more that a correct estimate of the maximum concentration of soluble species and the composition of the

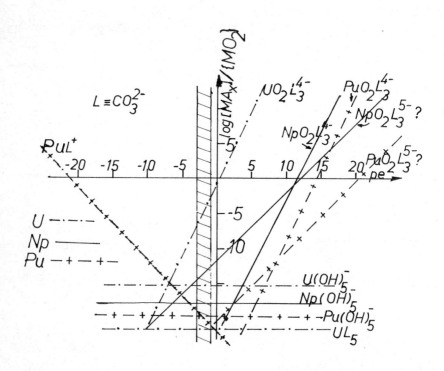

FIGURE 4. LOGARITMIC CONCENTRATION - PE DIAGRAM ILLUSTRATING
THE PREDOMINANT URANIUM, NEPTUNIUM AND PLUTONIUM SPECIES IN
GROUND WATERS WITH PH=8.5 AND PCO_3^{2-}=3.85. THE VERTICAL BAND
DENOTES THE RANGE OF PE-VALUES IN TYPICAL REDUCING GROUND WATERS

thermodynamically stable phases is only obtained if <u>all</u> relevant equilibria are taken into account.

4.   EXAMPLES OF THE USE OF THERMODYNAMIC DATA

All actinide(IV) oxide hydrates are extremely insoluble, a fact that may limit the transport of <u>soluble</u> actinide species in ground waters. Thus, it is of prime importance to determine

- the range of stability of the various oxidation states of the actinides,

- the solubility product of the oxide hydrates and their variation with crystallinity,

- the possible occurence of soluble actinide(IV) species.

The dependency of the solubility on the crystallinity of the solid oxide hydrate is to a large extent unknown and care must be teken in any quantitative interpretation of diagram 4. However, the qualitative features, with the possible exception of the behaviour of the actinide(V) species, should be correct.

4.1 Actinide species in granitic ground waters

All data presented in this section refer to a granitic ground water with pCO_3 = 3.85 and pH = 8.5. The crystalline oxide $MO_2(s)$ has been used as a reference in all cases.

Figure 4 indicates that the predominating species in reducing ground water -300 mV < E_h < -100 mV (the vertical band in the Figure) are $UO_2(CO_3)_3^{4-}$, $NpO_2(CO_3)_3^{5-}$ (?) and $Pu(OH)_5^{-}$.

The relative importance of $M(OH)_5^{-}$ and $M(CO_3)_5^{6-}$ deserves a comment. The concentrations of the $M(OH)_5^{-}$ species are larger than those of the corresponding $M(CO_3)_5^{6-}$ species under the conditions studied. This conclusions is based on the published equilibrium constants for the reaction

$$MO_2(s) + OH^- + 2H_2O \rightleftarrows M(OH)_5^{-}$$

However, the $M(OH)_5^{-}$ species are extremely prone to oxidation which may lead to an erroneous solubility and an equilibrium constant that is too high. It is highly desirable that some of these systems are reinvestigated by using coulometric technique, where it is more easy to exclude oxidants.

The total concentrations of dissolved species is strongly dependent on the solubility product of $MO_2(s)$ as mentioned before.

The stability constant for the $MO_2(CO_3)_3^{5-}$ complexes are estimated values and should be determined experimentally. However, there is no doubt that this oxidation state is of prime importance for migration of neptunium, but hardly for the other actinides.

4.2 Solubility limitations of the concentration of actinide species in solution

The case of actinide(IV) species was considered in previous paragraph. In this section, we will use thermodynamic data to interpret the fairly strong sorption of actinide(III) species on calcite containing minerals. The experimentally observed decrease in concentration of dissolved actinide(III) species [11] may be due to a precipitation of sparingly soluble $M_2(CO_3)_3(s)$.

The maximum concentration of M may be estimated from the

following equilibrium data

$$Ca^{2+} + CO_3^{2-} \rightleftharpoons CaCO_3(s) \qquad\qquad \log K = 8.2$$

$$2M^{3+} + 3CO_3^{2-} \rightleftharpoons M_2(CO_3)_3(s) \qquad \log K = \approx32$$

$$M^{3+} + CO_3^{2-} \rightleftharpoons MCO_3^{+} \qquad\qquad \log K = \approx6.5$$

$$CaCO_3(s) + 2MCO_3^{+} \rightleftharpoons M_2(CO_3)_3 + Ca^{2+} \quad \log K = \approx10.8$$

$$\frac{[Ca^{2+}]}{[MCO_3^{+}]^2} = 10^{10.8} \; ; \; \text{i.e.} \; [MCO_3^{+}] = \left(\frac{[Ca^{2+}]}{10^{10.8}}\right)^{1/2}$$

In solutions saturated with calcite of $p[CO_3^{2-}] = 3.85$, we have $p[Ca^{2+}] = 4.35$ and $p[MCO_3^{+}] \simeq 7.6$. Many migration experiments are made with Am^{3+} concentrations in the range 10^{-8} M and precipitation can in these cases not be ruled out as a "sorption" mechanism.

In this context, it may be pointed out that the size of the Ca^{2+} ion is very near those of the tervalent lanthanoids and actinoids, hence, Ca^{2+} in $CaCO_3(s)$ may be isomorphically replaced by these ions. It would be of value to investigate the "ion-exchange" equilibrium

$$CaCO_3(s) + xM^{3+} + xA^{-} \rightleftharpoons Ca_{(1-x)}M_xCO_3A_x(s) + xCa^{2+}$$

e.g. by using a regular solution model for the activity of the "solid solution", $Ca_{(1-x)}M_xCO_3A_x(s)$. Regular solution models could perhaps also be used for modelling the co-precipitation of actinide(IV) oxide hydrates.

4.3 Dissolution of uranium dioxide under ground water conditions

A number of studies have demonstrated the usefulness of thermodynamic data, both when leaching irradiated UO_2 fuel [11] and in various geochemical contexts [12,13]. It may be of interest to compare the E_h - pH diagrams in these studies with those based on our own results.

The main differences are the overall value of the stability constants of $UO_2(CO_3)_3^{4-}$ and the large predominance area of $(UO_2)_3(CO_3)_6^{6-}$. The main consequence is that the uranium(VI) predominance area is extended towards lower E_h-values, a fact which might effect the dissolution and migration of $UO_2(s)$. Re-precipitation of U(IV) will also require somewhat lower E_h-values than previously thought.

REFERENCES

(1) Allard, B. et al. : " Expected Species of Uranium, Neptunium and Plutonium in Neutral Aqueous Solutions", J. Inorg. Nucl. Chem. 42, 1015-1027 (1980).

(2) Skytte Jensen, B. : "The Geochemistry of Radionuclides with Long Half-Lives, Their Expected Migration Behaviour", p. 33, Risø-R-430, November 1980.

(3) Ciavatta, L. et al. : "The First Acidification Step of the Tris(carbonato)dioxouranate(VI) Ion, $UO_2(CO_3)_3^{4-}$ ", Inorg. Chem. 20, 463-467 (1981).

(4) Ferri, D. et al. : "Dioxouranium(VI) Carbonate Complexes in Neutral and Alkaline Solution", Acta Chem. Scand. A35, 165-168 (1981).

(5) Ciavatta, L. et al. : "Studies on Metal Carbonate Equilibria. 3. The Lanthanum(III) Carbonate Complexes in Aqueous Perchlorate Media", Acta Chem. Scand. A35, In press (1981).

(6) Baes, C.F. et al. : "The Hydrolysis of Cations", Wiley, Toronto, 1976.

(7) Sillén, L.G. et al. : "Stability Constants of Metal Ion Complexes", The Chem. Soc., London, 1971.

(8) Ciavatta, L. et al. : "Dioxouranium(VI) Carbonate Complexes in Acid Solution", J. Inorg. Nucl. Chem. 41 1175-1182 (1979).

(9) Langmuir, D. : "Uranium Solution - Mineral Equilibria at Low Temperatures with Applications to Sedimentary Ore Deposits", Geochim. et Cosmochim. Acta 42 547-569 (1978).

(10) Rai, D. et al. : "Solubility of Plutonium Compounds and Their Behaviour in Soil", Batelle Northwest, PNL-SA-7892.

(11) Johnson, L.H. et al. : "Mechanism of Leaching and Dissolution of UO_2 Fuel", Unpublished results, Whiteshell Nuclear Research Establishment, Atomic Energy of Canada.

(12) Hoestetler, P.B. et al. : "Transportation and Precipitation of Uranium and Vanadium at Low Temperatures, with Special Reference to Sandstone-type Uranium Deposits", Econ. Geol. 57, 137-167 (1962).

(13) Grandstaff, D.E. : "A Kinetic Study of the Dissolution of Uranite", Econ. Geol. 71, 1493-1506 (1976).

DISCUSSION

A.T. JAKUBICK, Canada

The main trouble with thermodynamics, as we all know, is that it does not consider the time dependence of solubility. However time is an important factor when considering solubilities as examplified by ageing of Pu solutions. It seems to me that specifying the time for which the thermodynamic calculations are made would be enough to make them more applicable.

D. RAI, United States

We have seen the solubility of Pu(IV) hydrous oxide converge towards the Pu(IV) oxide value in approximately 4 years.

I.R. GRENTHE, Sweden

I agree with your general remarks. However, my main concern with phase transformations of hydrous oxides is that *precipitation* always occurs via a more soluble phase. If the repository, or far field, conditions are such that the solubility product of this phase is not exceeded then no precipitate will form, and thus there will be no solubility limitation on the transport. Solubility products of crystalline hydrous oxides are of very little use for estimating precipitation.

Once a precipitate has formed one does not have such a problem since the rate of phase transformation is rapid in comparison with confinement times and most water flowrates.

RADIONUCLIDE SOURCE TERMS FOR IRRADIATED UO$_2$ FUEL[*]

B.W. Goodwin, L.H. Johnson and D.M. Wuschke

Atomic Energy of Canada Limited
Whiteshell Nuclear Research Establishment
Pinawa, Manitoba, Canada   ROE 1L0

ABSTRACT

The Canadian concept for nuclear fuel waste disposal involves the isolation of immobilized wastes in a deep underground
vault located in crystalline rock in the Precambian Shield. The most
likely mechanism for the release of this material to the environment
is through groundwater transport, and therefore the first step in an
environmental analysis involves an estimation of radionuclide source
terms.

The source terms used in our current environmental analysis
studies are based on estimates of the container lifetime and the assumption that the UO$_2$ fuel matrix dissolves congruently to the thermodynamic solubility limit of uranium in a deep underground environment.
An additional modification is required to account for the "instantaneous" release of several volatile radionuclides, such as ^{129}I.
Preliminary experimental evidence tends to support this approach.

---

[*]Issued as AECL-7352

1.   INTRODUCTION

The Canadian concept of nuclear fuel waste disposal involves the containment of radioactive wastes in corrosion-resistant metallic containers, and the emplacement of these containers in a deep vault located in a crystalline pluton in the Canadian Precambrian Shield. Subsequently the vault will be backfilled with a suitable packing material, such as a highly impervious clay, and permanently sealed [1].

An environmental assessment of this concept is currently underway, and will address a variety of potential release processes, including those that could occur during the pre-closure or operations phase and those that could occur during the post-closure phase. This report deals with the latter phase, and specifically relates to a groundwater release mechanism, in which groundwater percolating through the geosphere sequentially penetrates the vault backfill, corrodes the metal container and dissolves the radioactive wastes. It is also assumed that the radioactive wastes consist of irradiated UO_2 fuel bundles from a CANDU[1] reactor.

For this groundwater-release scenario, the first step in our post-closure assessment is to provide an estimate of the radionuclide concentrations in the groundwater immediately surrounding the waste containers. The source-term model described here takes into account estimates for:

a)   the rate of failure of the metallic containers,

b)   the mechanism by which radionuclides are released from the UO_2 fuel matrix, and

c)   the radionuclide inventory of the irradiated fuel.

Other steps in the assessment deal with the subsequent transport of radionuclides in the near-field region (the vault) and in the far-field region (the geosphere and biosphere).

For an estimation of a near-field source term, we are particularly interested in the sub-models used to describe the container lifetime and the dissolution of the UO_2 fuel matrix. The third aspect of the source-term model, the radionuclide contents of the fuel, will not be discussed in detail since it may be readily deduced from a knowledge of the irradiation history of the fuel [2].

2.   THE CONTAINER SUB-MODEL

At present, the characteristics of the metallic containers are not fully defined, and there is a gap in the experimental data for the long-term corrosion resistance of most candidate metals. As a consequence, we have developed a container sub-model that is based on engineering judgment.

The basic assumption of our container sub-model is that the containers fail according to a truncated normal distribution, with a mean failure time, t_m, and a standard deviation, σ, equal to $t_m/2$. The distribution function is truncated at time t=0 so that all containers are assumed to be initially intact. We have also assumed that a failed container plays no further role in the containment of the radioactive wastes.

With these assumptions, the rate of failure of containers, $r_c(t)$, expressed as a fraction of the total number of containers, is

_____
(1)   CANada Deuterium Uranium

given by

$$r_c(t) = 0 \qquad t \leq 0$$

$$r_c(t) = N \exp\left[-\frac{(t-t_m)^2}{2\sigma^2}\right] \qquad t > 0 \qquad (1)$$

where N is a normalization constant that takes into account trunca-
tion at t=0. The fraction of failed containers at any time t, $f_c(t)$,
is given by the integral of eq. (1):

$$f_c(t) = \int_0^t r_c(t') \, dt' \qquad (2)$$

with $f_c(t) = 0$ for $t \leq 0$ and $f_c(t) \approx 1$ for $t \geq t_m + 4\sigma$.

In our integrated source-term model, we also require an
expression for the fraction of containers contributing to the release
of radionuclides as a function of time. The contributing fraction,
$F_c(t)$, is given by the fraction of failed containers less the fraction that have already been depleted, or:

$$F_c(t) = f_c(t) - f_c(t-t_D) \qquad (3)$$

where it is assumed that it takes t_D years to dissolve the contents
of a container (see Section 3.3). Figure 1 shows a plot of $F_c(t)$
for the case where $t_D \gg \sigma$.

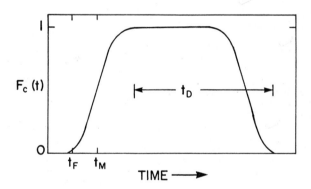

Figure 1: Fraction of containers contributing to the release of
radionuclides, $F_c(t)$, vs. time for the case $t_D \gg \sigma$. t_D
is the time required to completely dissolve the contents of
a container and σ is the standard deviation of the container-failure function. Times t_M and t_F are the times of
mean failure and first failure, respectively.

Equations (1) and (3) are both required in our current
post-closure environmental assessment studies, as described further
in Section 4. These studies take into account the uncertainty and
variation in the data [3], and we have assumed that values of t_m
have an equal probability of occurring in the range 1000 to 5000
years [4]. For a typical case study, using $t_m = 2000$ and $\sigma = 1000$ years,
approximately 2% of $r_c(t)$ is truncated at t=0.

3.   THE DISSOLUTION OF THE UO_2 FUEL MATRIX

3.1   General Discussion

The container sub-model can be used to estimate the time before groundwater reaches the UO_2 fuel matrix. The next requirement of the source-term model is to describe the processes by which radionuclides are released from the irradiated fuel to the groundwater.

Our dissolution sub-model is based partly on experimental data and partly on extrapolation of this data to the conditions expected in the vault. We shall start by describing the expected vault environment and its effect on calculated uranium solubilities. This leads to a description of fuel dissolution, based on a congruent dissolution mechanism, which appears to be applicable for most radionuclides in the fuel. Further consideration of the experimental data suggests a modification for some radionuclides that require special treatment.

3.2   Uranium Solubility in the Vault Environment

The proposed Canadian disposal vault will be located at a depth of approximately one kilometre in a crystalline rock, such as a granitic pluton. At this depth, we can expect that the groundwater entering the vault will have a pH buffered between 5-10 and a reducing electrochemical potential (Eh) buffered near the magnetite-hematite redox equilibrium [5] (magnetite and hematite are common granitic materials). Addditives to the vault backfill might be used to reinforce these expected conditions. The vault will also be designed to limit the maximum temperature to $\sim 100°C$.

In this environment, and considering a variety of ground-water compositions, the calculated stable uranium solid is UO_2 and the corresponding uranium solubility ranges from about 10^{-11} to 10^{-8} molal, with a very small temperature dependence [6,7]. Typical solubility curves, shown in Figure 2, have been calculated using an equilibrium thermodynamic approach, which does not consider non-equilibrium effects, such as kinetics, radiolysis or the formation of colloids. In spite of these limitations, equilibrium solubility data should provide a reasonably accurate value for uranium concentrations when other factors are taken into account. Two of these factors are the extremly low water velocity expected for deep groundwaters, which increases the time available for equilibration, and the limited supply of water and the reducing Eh in the vault, which may limit the effects of radiolytic oxidation. Recent experimental data [8] under conditions approaching the expected vault environment also support this conclusion.

3.3   The Long-Term Behaviour of the Fuel Matrix

Given the low solubility of UO_2, and because irradiated CANDU fuel is $\sim 99\%$ by weight UO_2, we can expect that the fuel matrix will also have a very low solubility. Furthermore, since UO_2 is likely the stable thermodynamic solid in a deep underground environment, it might be expected that the original fuel matrix will persist for a long time, and also that the release of radionuclides trapped in the fuel matrix will be limited by the slow disintegration of the matrix as the uranium oxide dissolves.

This dissolution mechanism presupposes that there are no major matrix-destruction processes at work. One such process could be a geothermal convection loop, in which uranium would be dissolved at the hot fuel-water interface and precipitated at a cooler water-rock interface, thereby releasing its radionuclide inventory at a faster rate. However, this process is unlikely in view of the low dependence of uranium solubility on temperature.

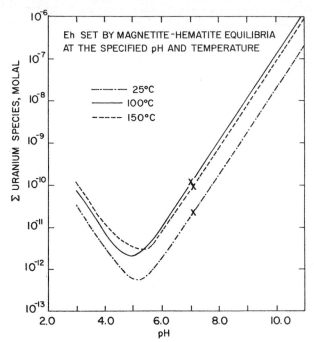

Figure 2: Maximum uranium solubilities in "granite" groundwater. The
abscissa represents the Eh-pH equilibrium boundary of the
magnetite-hematite reaction (only pH values are shown).
The crosses show the expected vault Eh-pH conditions at the
three temperatures. Data taken from reference [6].

We can also find geologic evidence for the long-term stability of uranium minerals. Perhaps the most relevant evidence comes
from observations of the natural reactor discovered at Oklo [9],
where it has been deduced that irradiated uraninite grains have
remained intact for over a billion years [10]. Although the geologic
history of the Oklo ores is not completely understood, they must have
been exposed to a considerable supply of groundwater over this long
period of time, and it appears that a congruent dissolution mechanism
must have been operative for the most part.

There is also some experimental evidence that the congruent
dissolution mechanism applies to spent CANDU fuel. Since little
information is available for dissolution of irradiated fuel under
reducing conditions, we must use data obtained under oxidizing conditions. At 25°C in air-saturated groundwater, dissolution of CANDU
fuel appears to approach a congruent limit after the first few years.
This is illustrated in Figure 3, which shows the fractional release
rate for several important radionuclides. After about one year of
dissolution, the fractional release rates of U, ^{99}Tc, ^{137}Cs and ^{90}Sr
are all within one order of magnitude of one another, even though the
initial rate differences were greater than three orders of magnitude.
Other experimental data [11] show similar patterns of behaviour.
These observations suggest that the long-term dissolution of irradiated CANDU fuel is congruent, but that some modification is required to describe the short-term dissolution behaviour.

3.4   The Modified Sub-Model for UO_2 Fuel Dissolution

The initial incongruent leaching behaviour of irradiated
fuel is, in fact, not surprising, since the radionuclides produced

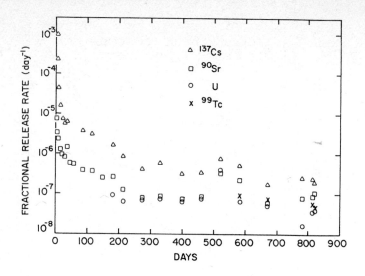

Figure 3: Fractional rate of release of some radionuclides vs. time
for irradiated CANDU fuel in contact with air-saturated
water at 25°C, and containing about 5 x 10⁻³ molal total
carbonate. The rates have been normalized relative to the
original radionuclide contents of the irradiated fuel.

are not homogeneously distributed. Some elements, such as cesium,
are relatively volatile at high temperatures [12], and tend to migrate during irradiation to gaps within the fuel and to gaps between
the fuel and its sheath. We have, therefore, divided the radionuclides in CANDU fuel into two groups: the "uniform" group, which
we regard as being uniformly distributed throughout the fuel matrix,
and the "instant" group, whose members are predicted to be somewhat
enriched at certain sites in the matrix. The radionuclides in each
group are listed in Table I.

Table I also gives the fraction of several important "instant" radionuclides estimated to be present in the fuel-sheath gap
at the time of container failure. These estimates are dependent in a
complex manner on parameters such as the fuel irradiation temperature, irradiation history, enrichment, density, etc., as well as the
reactor type [4,13]. The values given in Table I are averages for
the expected fuel inventory in the Canadian waste-disposal vault.

These considerations lead to a modification of the congruent dissolution sub-model, which allows that, at the time of
container failure, the instant radionuclides in the fuel-sheath gap
begin to dissolve in the surrounding groundwater, and are completely
dissolved within one year from failure. Experimental data on the
leaching of ¹³⁷Cs from fuel show that ∿ 80% of the calculated fuel-
sheath inventory was released during the first year of dissolution,
supporting the short-term assumptions of the modified dissolution
sub-model.

The long-term behaviour of the fuel is assumed to be governed by a congruent dissolution mechanism, as discussed in Section
3.3. If the dissolution of the fuel matrix begins when the container
is breached, and if there is an equal and steady flow of groundwater
past all containers, the time, t_D, required to dissolve the uranium
oxide inventory of any container is given by

TABLE I

THE UNIFORM AND INSTANT GROUPS OF RADIONUCLIDES IN CANDU FUEL

Uniform Group	
Actinides ⎫ Lanthanides ⎭	Probably exist as solid solutions in UO_2
^{90}Sr* ^{95}Sr ^{99}Tc ^{106}Ru ^{107}Pd	Matrix-bound particulates (as oxides, alloys, etc.)

Instant Group (and estimated fraction in the fuel-sheath gap)	
Inert gases	(0.7%)
^{79}Se	
^{90}Sr*	(0.02%)
^{125}Sb	
^{129}I	(0.7%)
^{134}Cs, ^{135}Cs, ^{137}Cs	(0.7%)

* most of the ^{90}Sr is matrix bound; less than
1% is believed to be redistributed due to the
volatility of its parent, ^{90}Rb.

$$t_D = \frac{\text{mass of } UO_2 \text{ per container}}{\text{transport rate of uranium}} \qquad (4)$$

In calculating the uranium transport rate, we take into account both
advection and diffusion, with uranium at its solubility limit. For
diffusional transport, we also use the approximation that the concentration gradient of uranium is constant within the vault, with a
value given by the uranium solubility at the fuel-backfill interface
and a value of zero at the vault-geosphere interface [4].

4.   THE INTEGRATED SOURCE-TERM MODEL

The container sub-model and the modified dissolution sub-
model have been combined in our assessment to yield a source-term
model for radionuclide concentrations in the groundwater in the
vault. Figure 4 illustrates the predicted concentration of an instant nuclide near a single, isolated container. Initial concentrations are assumed to be zero, and the spike occurs over the year
following failure of the container. Subsequent lower concentrations
are determined by the solubility of uranium and the ratio of the
radionuclide to uranium masses in the irradiated fuel (taking into
account the leached out fuel-sheath inventory). A concentration
profile for the uniform group of radionuclides would be similar to
the lower concentration line extrapolated backwards to t_F.

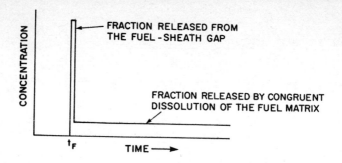

Figure 4: Concentration vs. time profile for an instant radionuclide
in the groundwater near a single isolated fuel container.
The container is assumed to fail at time t_F, which is
determined by the container-failure function.

Source terms for the entire vault also require the parameter for the time required to dissolve the UO_2 in a container. The
rate of discharge of uranium, $R_u(t)$, from all containers is then
given by:

$$R_u(t) = F_c(t) \frac{M_u}{t_D} \qquad (5)$$

where M_u is the total amount of uranium emplaced in the vault. The
rate of release of a uniformly distributed radionuclide, $R_n(t)$, is
given by:

$$R_n(t) = F_c(t) \frac{M_n(t)}{t_D} \qquad (6)$$

where $M_n(t)$ is the total inventory of radionuclide "n", taking into
account radioactive decay. Finally, the rate of release of an instant radionuclide, $R_i(t)$, is given by:

$$R_i(t) = r_c(t) M_i(t)p_i + F_c(t) \frac{M_i(t)}{t_D} (1-p_i) \qquad (7)$$

where p_i is the fraction of the instant radionuclide "i" in the fuel-
sheath gap, and $M_i(t)$ is its total inventory, taking into account
radioactive decay. Note that the first term on the right-hand side
of eq. (7) involves the rate of failure of the containers (from eq.
(1)), and the second term involves the fraction of containers contributing to the congruent release (from eq. (3)).

Equations (5-7) have been used in our current post-closure
environmental assessment of the proposed Canadian waste disposal
vault [4]. As noted previously, this analysis takes into account
uncertainty and variation in the data, and typical analyses use
several thousands of cases in which t_m is varied between 1000 and
5000 years, and uranium solubility is varied between 10^{-11} and 10^{-8}
molal. The assessment also takes into account variations in the
parameters used to calculate material transport in the near and far
fields, and produces, as its result, plots showing the frequency
distributions of selected variables [4]. Some results of this preliminary assessment relating to the source-term model are reproduced
in Figures 5 and 6.

Figure 5 shows the calculated release of ^{129}I for three
specific cases in which all the parameters were fixed, except those

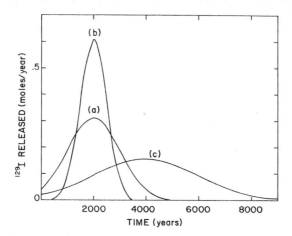

Figure 5: Calculated source term for the release of ^{127}I from the vault. The three curves were computed using:

a) t_M = 2000 and σ = ½t_M

b) t_M = 2000 and σ = ¼t_M

c) t_M = 4000 and σ = ½t_M

The instant fraction of ^{129}I is much greater than the uniform fraction in all cases.

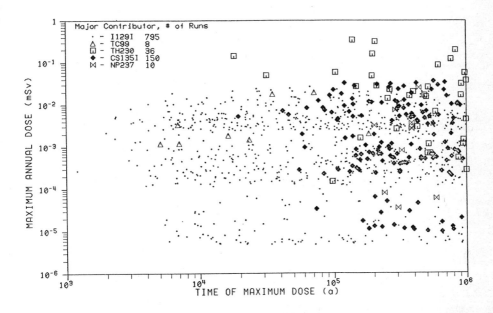

Figure 6: Calculated distribution of the estimated time of occurrence of maximum dose for 1000 test cases. Each case is identified with a symbol showing the major contributor to the dose. Calculated results reflect the source-term model described here, and models for the geosphere and biosphere [4].

pertaining to the container sub-model. In all three cases, the ^{129}I release is dominated by the release of its instant inventory. A comparison of curves (a) and (b) shows that, for a given container lifetime, the maximum rate of release of ^{129}I is significantly affected by the spread of container failures (σ). A comparison of curves (a) and (c) shows that t_m also affects the maximum rate of release, although in this case we can attribute the effect to the dependence of σ on t_m (in curve (a) $\sigma = t_m/2 = 1000$ years, and in curve (c) $\sigma = t_m/2 = 2000$ years). Similar observations can be made for the other instant radionuclides.

The release rates for the uniform radionuclides and the instant radionuclides dispersed throughout the fuel matrix are affected very slightly by the parameters of the container sub-model. The effect occurs only over a short period of time, and is not observable for times greater than $\sim t_m$. These rates of release are, however, significantly affected by the solubility of uranium, and results not given here show that the rates of release are proportional to the values of the solubility chosen for UO_2.

Figure 6 shows the calculated maximum annual dose versus the time of occurrence derived from 1000 cases in a typical analysis. The major contributor to the dose in each case is identified. Two features of this figure are:

1) the most common major contributor to dose is the instant radionuclide ^{129}I, which is assumed to move with the groundwater without retardation by chemical sorption processes (another instant radionuclide, ^{135}Cs, is significantly retarded in most cases); and

2) radionuclides from both groups that are released by congruent dissolution of the fuel tend to become important only at longer times.

It is difficult to deduce specific conclusions from these last results that relate directly to the source-term model, since the assessment includes other major components (such as the geosphere-transport model), and a sensitivity analysis is not yet complete. We can make some qualitative statements, however, taking into consideration the characteristics of the source-term model.

1) The sub-model describing the container suggests that the container is an important barrier in limiting the overall release of the instant group of radionuclides, since the rate of container failure controls the duration of their release and, therefore, effectively dilutes some important species, such as ^{129}I.

2) The container does not appear to be an important barrier in limiting the release of the uniform group of radionuclides, since its effect will be mostly to delay the initiation of release of these radionuclides. An increased delay is important only when the half-life of a radioactive species is much less than the mean container failure time.

3) The fuel-dissolution sub-model suggests that the congruent dissolution mechanism for UO_2 fuel is an important factor in limiting the release of the uniform radionuclides and the fraction of instant radionuclides bound in the fuel matrix.

5.   CONCLUSIONS

The groundwater source-term model used in our current post-closure environmental assessment has been described, along with some validation of the sub-models involved.

The sub-models employed are simple representations of complex processes, and can be expected to evolve as our knowledge improves. For example, experimental studies are now in progress to help define more accurately a container-failure model. Other experimental studies are directed towards the dissolution of irradiated fuel in a reducing environment, and towards a more accurate estimation of the fractional release of the instant radionuclides. Of particular interest is the experimental determination of the fraction of ^{129}I in the fuel-sheath gaps, and its subsequent leaching behaviour.

Preliminary results from our analyses also show that both components of the source-term model have substantial impacts on limiting the calculated dose to man. The container appears to be most important in prolonging the release of the instant radionuclides, in particular ^{129}I, and the fuel-dissolution mechanism appears to be most important in limiting the rate of release of the radionuclides trapped within the fuel.

ACKNOWLEDGEMENTS

The authors of this report are grateful for the suggestions and constructive criticism provided by a number of people at WNRE, including T. Andres, K.W. Dormuth and R.B. Lyon.

REFERENCES

[1] J. Boulton, Ed., "Management of Radioactive Fuel Wastes: The Canadian Disposal Program", Atomic Energy of Canada Limited Report, AECL-6314 (1978).

[2] L.J. Clegg and J.R. Coady, "Radioactive Decay Properties of CANDU Fuel. Volume 1: The Natural Uranium Fuel Cycle", Atomic Energy of Canada Limited Report, AECL-4436/1 (1977).

[3] K.W. Dormuth and R.D. Quick, "Accounting for Parameter Variability in Risk Assessment for a Canadian Nuclear Waste Disposal Vault", Atomic Energy of Canada Limited Report, AECL-6999 (1980).

[4] D.M. Wuschke, K.K. Mehta, K.W. Dormuth, T.E. Andres, G.R. Sherman, E.L.J. Rosinger, B.W. Goodwin, J.A.K. Reid and R.B. Lyon, "Environmental and Safety Assessment Studies for Nuclear Fuel Waste Management. Volume 3: Post-Closure Assessment", unpublished work, Atomic Energy of Canada Limited Technical Record, TR-127-3 (in preparation).

[5] M. Sato, "Oxidation of Sulfide Ore Bodies. 1. Geochemical Environments in Terms of Eh and pH", Econ. Geol. 55, 928 (1960).

[6] B.W. Goodwin, "Maximum Total Uranium Solubility Under Conditions Expected in a Nuclear Waste Vault", unpublished work, Atomic Energy of Canada Limited Technical Record, TR-29 (1980).

[7] J. Paquette and R.J. Lemire, "A Description of the Chemistry of Aqueous Solution of Uranium and Plutonium to 200°C Using Potential-pH Diagrams", Nucl. Sci. Eng. (in press). Also issued as AECL-7037 (1981).

[8] P.R. Tremaine, J.D. Chen, G.J. Wallace and W.A. Boivin, "The Solubility of Uranium (IV) Oxide in Alkaline Aqueous Solutions to 300°C", J. Solution Chem. (in press). Also issued as AECL-7001 (1981).

[9] International Atomic Energy Agency, Proceedings of a Symposium
     on the OKLO Phenomenon, IAEA, Vienna, 1975 (STI/PUB/405) and
     Proceedings of a Meeting of the Technical Committee on Natural
     Fission Reactors, IAEA, Vienna, 1975 (STI/PUB/475).

[10] G.A. Cowan, "Migration Paths for OKLO Reactor Products and
     Applications to the Problem of Geological Storage of Nuclear
     Wastes", IAEA-TC-119/26, p. 693 (1978).

[11] L.H. Johnson, D.W. Shoesmith, G.E. Lunansky, M.G. Bailey and
     P.R. Tremaine, "Mechanisms of Leaching and Dissolution of UO_2
     Fuel", presented at the Proceedings of the Second Annual Work-
     shop on Comparative Leaching Behaviour of Radioactive Waste
     Forms, Argonne National Laboratory, Sept. 3-4, 1980. Also
     issued as AECL-6992 (1980).

[12] J.R. Findley, "The Birth, Abundance and Movement of Fission
     Products Through Fuel", J. Brit. Nucl. Soc. 12, 415 (1973).

[13] W.B. Lewis, J.R. MacEwan, W.H. Stevens and R.G. Hart, "Fission-
     Gas Behaviour in UO_2 Fuel", Atomic Energy of Canada Limited
     Report, AECL-2019 (1964).

DISCUSSION

F. GERA, Italy

In the periodic table you have shown, Neptunium is indicated as not mobile. In a number of analyses neptunium is considered to be rather mobile in the geosphere. Could you comment ?

B.M. GOODWIN, Canada

The information persented in the periodic table derives from observation of the uraninite area at Oklo. Presumably Neptunium is immobile because it forms a solid solution with uraninite, and is released to the geosphere as it is exposed during the dissolution of the uraninite. Consequently, the apparent "immobility" of neptunium is due to the low solubility and relative immobility of uranium. As an isolated aqueous species, Neptunium may be mobile ; but it appears that as a contaminant in uraninite neptunium is relatively immobile.

I.R. GRENTHE, Sweden

Available thermodynamic data on Np(V) indicate that this element would not be immobile under most groundwater conditions. How will this affect your model calculations ?

B.M. GOODWIN, Canada

This question is partly a continuation of the previous one. Our source term model suggests that neptunium is initially immobile because it is one of the uniformly distributed radionuclides in UO_2. Once Neptunium enters the aqueous phase, its mobility in our calculations is described by the distribution coefficient approach, and I have not described here these aspects of our total system model. The range of values we use for the distribution coefficient of Neptunium are given in reference (4), and should reflect its apparent mobility in the aqueous phase.

H.C. BURKHOLDER, United States

The results from your SYVAC analysis seem to suggest that I and Cs are the most important nuclides from the overall system performance point of view. However, your analysis appears to have assumed a "time cut-off" of 1,000,000 years. What nuclides would the SYVAC analyses show to be the most important if a "time cut-off" was not assumed ?

B.W. GOODWIN, Canada

SYVAC analyses were cut-off at 10^6 years because our preliminary effort has been directed towards developing the analytical tools. We have not yet produced results beyond a million years, but I suspect that the long-lived actinides among others would become more important in the overall point of view. However it is still too early to say whether or not these other radionuclides will eclipse the importance of Cs and I.

DATA REQUIREMENTS FOR INTEGRATED NEAR FIELD MODELS

R. E. Wilems
INTERA Environmental Consultants, Inc.
Houston, Texas, USA

C. R. Faust
GeoTrans, Inc.
Herndon, Virginia, USA

A. Brecher
Arthur D. Little, Inc.
Cambridge, Massachusetts, USA

F. J. Pearson, Jr.
INTERA Environmental Consultants, Inc.
Houston, Texas, USA

The coupled nature of the various processes in the near field require that integrated models be employed to assess long term performance of the waste package and repository. The nature of the integrated near field models being compiled under the SCEPTER program are discussed. The interfaces between these near field models and far field models are described. Finally, near field data requirements are outlined in sufficient detail to indicate overall programmatic guidance for data gathering activities.

1. INTRODUCTION

Under contract with the U. S. Department of Energy through the Office of Nuclear Waste Isolation, Battelle Memorial Institute, INTERA Environmental Consultants, Inc. is conducting the SCEPTER program - Systematic Comprehensive Evaluation of Performance and Total Effectiveness of Repositories. INTERA is being assisted in this program by GeoTrans, Inc. and Arthur D. Little, Inc. The objective of this program is "...to integrate, develop and apply modeling technology to evaluate the effectiveness of nuclear waste isolation systems in preventing adverse radiological effects to present and future human beings and their environments." To facilitate the accomplishment of this objective, the sphere-of-interest for repository performance assessment has been arbitrarily divided into near field and far field. The near field is defined as that region in which waste-induced or excavation-induced effects are significant enough that coupled models may be required. The focus of this paper is on these near field models, their

o   Nature
o   Interface with Far Field Models
o   Data Requirements.

2. NATURE OF NEAR FIELD MODELS

The nature of any model depends on the properties of the components being considered, the environment of the components, the processes of interest, and the role or application of the model.

For modeling convenience, the near-field models within the SCEPTER program have been structured along the lines of two subsystems: waste package and repository subsystems. The waste package subsystem consists of the radioactive waste in some form and the assemblage of engineered barriers which will be emplaced within the host rock. For purposes of post-closure performance assessment, the repository subsystem is defined as all man-made underground structures which permit access to the underground facilities (boreholes, shafts, drifts, etc.), the materials used to fill these openings and the natural barriers (host rock formation) in the vicinity of the openings. As indicated by the definition of the near field presented above, the repository models consider the natural barriers "in the vicinity of the repository openings" throughout the region in which waste-induced or repository-induced effects are of such magnitude that coupled models may be required.

2.1 Waste Package Model

The focus of the waste package model role is to predict the performance of a single waste package in a repository environment; that is, the degradation of each barrier and the history of leaching and release of radionuclides from the package. These predictions are to be used in two basic types of applications:

o   To define a radionuclide release source term for repository design and site assessments;

o   To support waste package design and development.

The latter of these two applications may include pre-test design and test data interpretation, identification and assessment of data bases, integrated characterization of waste package designs, setting research and development priorities, and absolute and relative performance assessments.

The processes of concern in these predictions include:

o   Radiation transport and effects
o   Heat generation and transport
o   Internal and external mechanical stresses and their effects
o   Corrosion of metallic components
o   Leaching of nuclides from waste form as influenced by the solubilities of nuclides and their transport.

Within the SCEPTER waste package model these processes are simulated within radiation, thermal, mechanical, corrosion and leach models. The allocation of the various processes within the SCEPTER waste package model structure is presented in Figure 1. Each of the process models may be coded in two level of complexity. The more complex description takes into account details of the physiochemical interactions within each process to allow in-depth analysis and understanding of the process evaluation. The simpler description of the process concentrates on representing the effects of the process through empirical equations that are dependent upon various waste package environmental parameters. The more complex descriptions are particularly useful in generating a detailed understanding of component performance and in the delineation of research and development priorities. The simpler descriptions are more appropriate to assess the coupled effect of the various processes working together to define waste package performance and will so be used in the integrated waste package subsystem model.

The SCEPTER waste package subsystem model structure serves as a framework within which the five simpler process models reside as subroutines. The subsystem structure illustrated in Figure 1 permits the coupling of the five degradation-process models through a sequential operation of each process model at an appropriate time step. The time step is presently supplied by a logarithmically increasing time-interval function that can be readily adapted to specific applications. The coupling is effected through input-output streams between each process model and various data blocks within a data bank. This waste package model structure allows the integrated consideration of all waste package components, including the effects of one component's performance on another.

2.2   Repository Subsystem Model

The function of the repository subsystem model is (1) to predict the ground water flow through the repository region (engineered features and surrounding host formation) considering waste-induced and excavation-induced effects, as well as the effect of natural or human-caused perturbations; (2) to integrate the predicted radionuclide release from an array of waste packages (which could vary from considering a few packages and their specific relative locations to considering the entire emplacement horizon as a single, homogeneous source block); and (3) to predict the transport of the radionuclides through the repository region to define the time-dependent nuclide flux to the geosphere (which models the ground water flow and transport through that region of the environment not significantly affected by waste or excavaction induced effects), or to the biosphere. Analogous to the waste package, these repository model predictions are to be used in two basic types of applications: to define boundary conditions for other models and to support repository design and development.

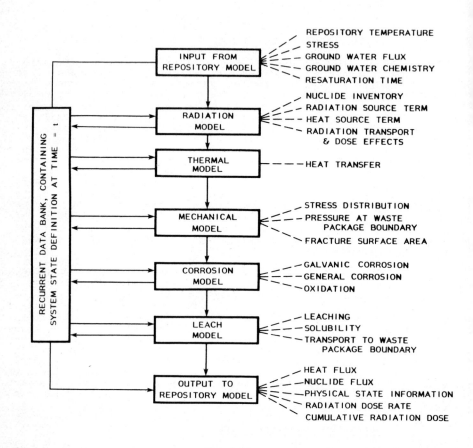

FIGURE 1.   WASTE PACKAGE SUBSYSTEM INTEGRATION

The repository model may be used to define boundary conditions for either the waste package model or the site model. The boundary conditions that may be defined for the waste package model include ground water flux and chemistry, external mechanical stresses, and, in some cases, boundary temperatures. The boundary conditions defined for the site model include the flux for each species of radionuclide that reaches the repository-geosphere boundary.

The use of the repository model to support repository design and development parallels the support to the waste package design and development outlined in Section 2.1. However, the demands on the repository model require more flexibility because the scope of the tests to be conducted to accomplish repository development are much broader. The laboratory and field tests may include assessments of fluid migration, fluid flow perturbed by thermal or thermo-mechanical effects, similarly perturbed nuclide transport, excavation technology development, borehole or shaft sealing, drift backfilling, etc. The physical scale of interest in many cases will require field testing rather than laboratory testing. In field tests, the uncertainty in characterizing the system becomes much more important than in the waste package tests.

The processes that the repository model must consider include heat transfer, ground water flow, rock mechanics, species transport and chemical effects on species transport. The data transfer relationship among these various processes within the repository model and the transfer of data between these repository processes and the waste package and geosphere model are illustrated in Figure 2.

Whereas in the waste package model each process was modeled in a separate code and these codes were all closely coupled through the waste package subsystem model structure, the situation is much more complex in defining the structural relationships within the repository subsystem model. The repository subsystem model must handle a wide variety of problems from analyzing the very closely coupled processes in the vicinity of the waste package to analyzing the flow and transport considerations close to the repository/geosphere interface where the processes are much more loosely coupled. This spectrum of problems in which various levels of process coupling are appropriate gives rise to a number of different process codes that incorporate different levels of process model detail and different coupling configurations. The present SCEPTER configuration includes the following codes:

- o 3-D Thermal-Hydrologic
- o 3-D Thermal-Hydrologic-Transport
- o Network Flow-Transport
- o 2-D Linear Thermal-Mechanical
- o 2-D General Thermal-Mechanical
- o 3-D Linear Thermal-Mechanical
- o 3-D General Thermal-Mechanical
- o 2-D Thermal-Hydrologic-Mechanical-Transport
- o 2-D Dual Porosity Flow-Transport

In addition, the SCEPTER configuration includes a steady-state flow model to provide boundary conditions to the waste package model and a complex geochemical model.

The operational configuration of the repository system model differs from the waste package model in that presently it is

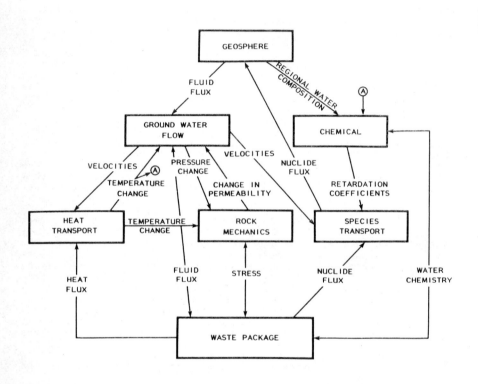

FIGURE 2. REPOSITORY SUBSYSTEM INFORMATION FLOW

not intended to couple any two process codes; rather, any given code is expected to interface within data bases or equivalent simple relationships that represent the effects of the other processes. For example, the thermal-hydrologic-transport code might interface with a data base generated by a thermal-mechanical code to account for repository and formation thermal-mechanical effects, and with a data base generated by the geochemical model to account for chemical retardation of nuclide transport.

2.3 Handling of Uncertainties

The SCEPTER approach to uncertainty analysis recognizes two fundamental types of uncertainty: (1) uncertainty in the adequacy of performance assessment models; and (2) data uncertainties. Within the SCEPTER program, every attempt will be made to validate the performance assessment models; however, once the models are validated, the assessments will proceed as if the models are true. On the other hand, attempts will be made to quantify data uncertainties, of which there are several types:

o   Uncertainties related to adequacy of data to characterize the system

o   Uncertainties introduced by data interpolation to fill in areas where data are sparse

o   Data measurement errors.

Although the latter listed uncertainties are better understood and easier to quantify, they are also expected to be of much less magnitude, and consequently, of less importance than the first type of data uncertainty.

The long time frames (at least thousands of years) over which waste isolation is required renders predictive analytical models a cornerstone of performance demonstration. Therefore, the importance of model validation cannot be overstressed. Five basic techniques are anticipated to be used for model validation:

o   Comparison of model predictions with laboratory and field test data -- a technique particularly useful for models based on scientific principals that do not change over time.

o   Comparison of model predictions with natural analogs -- a method of validating predictions over longer lengths of time.

o   Comparisons with known asymptotic or limiting behavior -- to lend credence to a model when data are not available to assess full adequacy.

o   Use of peer review -- a technique that at least adds credence that nothing important has been overlooked.

o   Comparison of predictions from two different models -- a technique that is particularly useful if the two models use different constitutive relationships to represent a physical process.

Once the models are validated, uncertainty analysis focuses on quantification of uncertainties in input data and the processing of these uncertainties to define uncertainties in the model predictions.

Depending on the amount and type of data available, any one of three techniques might be used to quantify uncertainties in input data:

- o Identification of reasonable upper and lower bounds
- o Estimation of a mean and variance
- o Development of a probability density function.

A basic premise within the SCEPTER approach to uncertainty analysis is that all of the input data uncertainty is represented by the bounding interval, mean and variance, or probability density function (PDF), and that the interval, mean and variance, or PDF itself is correct.

After reviewing a number of techniques to translate input variable uncertainties into a definition of output variable uncertainty, two techniques have been selected for application:

- o An ordered sampling of input variable uncertainties for processing through a deterministic model or set of models taking into account appropriate correlations to produce sets of output variables (e.g., Modified Monte Carlo or Latin Hypercube Sampling).

- o Use of perturbation analysis techniques, or Taylor series expansion approximations, to convert a deterministic description into a stochastic model through which to define output uncertainties.

The selection of which of these two techniques to use for a specific analysis will depend on the type of model being used and the data available. The latter of the two techniques is only practical when used within a single code; therefore it will be most useful when applying individual process models representings the complex physics and chemistry of a process. The ordered-sampling technique will be more useful within system or subsystem models consisting of integrated simpler codes.

3.   <u>INTERFACE BETWEEN NEAR FIELD AND FAR FIELD MODELS.</u>

As indicated earlier, the distinction between near field and far field has been arbitrarily established such that the near field is that region in which waste-induced or excavation-induced effects are significant enough that coupled models may be required. This approach has allowed the specification of three subsystems:

- o Waste Package Subsystems
- o Repository Subsystem
- o Site Subsystem

The first two of these constitute the near field and have been discussed earlier. The site subsystem consists of the geosphere, the biosphere and descriptions of externally-imposed natural or human-caused events that may affect repository performance (e.g., faulting or future drilling).

The interface between the near field and far field models is between the repository model and the site model. The site models define for the repository model:

- o Ground water potentials at the repository/geosphere boundary versus time o   Internal property or geometry parameter changes versus time

These latter parameter changes are most often associated with externally-imposed events. For example, the parameters may be associated with changes in formation hydrologic properties caused by glacial overburden or with changes to describe a human intrusion event, such as holes drilled through or near the repository.

On the other hand, the repository model defines the time-dependent flux of each species of radionuclide at the repository/geosphere boundary. This radionuclide flux is sufficient for the site model to compute radionuclide concentrations and then track these concentrations through the geosphere and biosphere. Defining the flux is sufficient to compute the concentrations, because by definition, the repository/geosphere boundary is located where waste and excavation effects on the ground water flow are insignificant.

4.   DATA REQUIREMENTS

The data requirements to be discussed in this paper are those which need to be specified by a user of the integrated system of near field models. Specific data requirements will depend on the detail of the user's conceptualization of the system and the specific codes the user employs. Because of length considerations, this discussion will be limited to general types of data required with sufficient detail to indicate overall programmatic guidance to data base gathering activities.

The input data requirements can be categorized as boundary conditions, initial conditions, and properties. The term "property" refers to a quantity whose value is assumed to be either a constant or a known function of time, space or system state (e.g., temperature), and which can be used in solving an equation of state. In general, whether any given input data set is a function of time, space or system state depends on the detail of the user's description of the system.

A large standard data base is not being compiled under the SCEPTER program. However, each application of the SCEPTER model technology may draw upon the data bases from previous applications.

4.1   Waste Package Model Data Requirements

The waste package model data requirements consist of a description of the waste package, boundary conditions from the repository model and various property data.

The basic data required to describe the waste package include the number of barriers, the barrier materials, and the height and inner and outer radii of each barrier. Note that the waste form is considered to be a barrier, as are any void spaces or coatings on other barriers.

The waste package property data can be divided into waste property data and material property data for each barrier. The required waste property data include: waste type, nuclide inventory, thermal power vs. time, waste age at emplacement, total activity, initial waste mass, and waste loading ratio. The barrier property data can be divided into radiation (transport and damage), thermal, mechanical, corrosion and leaching properties. Not every barrier exhibits each type of property, and thus, some barriers do not have all five property files; for example, most

- 55 -

backfill materials and waste forms being considered by the NWTS program do not exhibit corrosion. A comprehensive, but not necessarily exhaustive, list of barrier property data is:

o   Radiation Properties - attenuation coefficients, radiation damage initiation doses and response coefficients, e.g., gamma dose to activate radiolysis

o   Thermal Properties - density, conductivity, specific heat or heat capacity, surface emittance

o   Mechanical Properties - stress and strength coefficients, fracture toughness coefficient, compaction constants, number of initial flows or fractures

o   Corrosion Properties - empirical parameters for stress corrosion cracking, general corrosion rate, oxidation corrosion rate law constants, galvanic corrosion rate

o   Leaching Properties - leach rate coefficients, saturation concentrations, activation energies, diffusion and sorption coefficients, backfill porosity and density.

The waste package boundary conditions specified by the repository model include vertical and radial stresses, fluid density and flux, initial fluid composition, time to re-wet repository, repository thermal properties, initial repository temperature and package emplacement geometry.

4.2   Repository Model Data Requirements

The repository model data requirements consist of a description of the repository configuration, boundary conditions from both the waste package and the site models, and property data for the various natural and engineered materials within the repository.

The data required to describe the repository depends on the scale of the problem being analyzed. If flow or transport in the vicinity of a waste package or an excavation (e.g., shaft) are to be analyzed, details of host formation, fractured regions, package emplacement or excavation geometries, and perhaps details of spatial dependence of property data would be required. On the other hand, if flow and transport within the entire repository region were of interest (e.g., to define nuclide flux for site evaluation), then total repository geometries, averaged emplacement densities, and averaged property data would be used. In either case, such data as types of materials and geometries of boundaries will describe the system being analyzed.

The repository material property data can be divided into thermal, mechanical hydrologic and solute transport data requirements. In addition, since a detailed geochemical model is available within the set of repository models to provide input to the transport models, a chemical model data base is required when the geochemical model is to be used. Since radiation effects do not seem to be important in determining the performance of repository components outside the waste package for presently envisioned repository designs, there is no requirement for a radiation transport or damage data base in the repository model. A comprehensive, but not necessarily exhaustive, list of repository material property data is:

o   Thermal Properties - initial temperature gradient, density, conductivity and heat capacity for repository materials and adjacent strata

o   Mechanical Properties - initial stresses, strains and body forces, surface tractions and displacements at boundaries, stress and strain coefficients, thermal expansion coefficient, cohesion and angle of friction, compressive, tensile and shear strength, creep parameters, and fracture response coefficients

o   Hydrologic Properties - initial fluid pressure, permeability, porosity, dispersivity, compressibility, fluid viscosity and density (for material matrix or fractures)

o   Solute Transport Properties - identity of chemical species, nuclide chain identities and half-lives, fluid pressure or Darcy velocity.

In addition, if the detailed geochemical model is to be exercised, then the following data will be required:

o   Geochemical Data - Identity of initial chemical species, compositions and masses, descriptions of solution reactions and reaction equilibrium constants, solid phase descriptions and associated formation equilibrium constants, non-equilibrium relative reaction rates, solution volume and initial oxidation potential.

The boundary conditions specified by the waste package model include waste package height and radius, heat flux and nuclide mass flux through the waste package boundary, and hydraulic conductivity of the backfill. The boundary conditions from the site model (presented in Section 3) include ground water potentials at the repository/geosphere boundary and changes in process or geometric parameters because of externally-imposed events.

DISCUSSION

F. GERA, Italy

How do you handle in your models the changes that might take place in the future ?

R.E. WILEMS, United States

By changes in the future I presume you mean changes that are not associated directly with waste-induced processus (e.g. thermal expansion of the rock) or excavation-induced processes (e.g. subsidence of the rock) as these are handled directly by the models presented. Other future changes, such as climatic changes and their effects on the hydrology or human intrusion into the site via boreholes are handled through the Site Model. One of the major components of the Site Model is the far field State Model which simulates the probability and time of occurrence and the effects of such future processes and events. Data from executing the far field State Model will be input into the Near Field Models.

R.H. KOSTER, Federal Republic of Germany

You mentioned a long list of relevant data for integrated models. Your presentation concerned corrosion models for waste containers including galvanic, general corrosion, oxidation. I missed selective corrosion effects like pitting, stress corrosion and hydrogen brittleing which are of great importance.

R.E. WILEMS, United States

Parameters for stress corrosion were mentioned in the list. We have considered and, in fact, drawn up performance and design specifications for other selective corrosion processes. However, the initial review of the performance specifications generated several comments that if such corrosion processes are found to be dominant the package would either be scrapped or the problem engineered out of the system. If for a specific package concept, such corrosion processes are deemed important and performance data are available, the models can be introduced.

R.G. BACA, United States

What is your plan for verifying and validating the numerical models (in particular the hydrologic and transport models) ?

R.E. WILEMS, United States

Within the SCEPTER Program we devine verification as "assuring that the model is properly coded, that is, the code meets the design specification" and we define validation as "assuring that the model properly represents the processes or effects that we intended to model, that is, the performance specifications are appropriate and the code meets these specifications. Verification will be accomplished by a careful process of review outlined in our Quality Assurance Program through which the code is compared with the design specifications. There are five techniques outlined in the paper to accomplish validation. These range from comparisons with natural occurrences and comparisons with laboratory and field experiments to peer review. Some models, particularly the system level models will probably be validated only through peer review. On the other hand, we are in the process of a validation exercise on the hydrologic and transport models which involves a comparison with data on a natural uranium ore-body migration in the Carrigzo Sandstone Formation in West Texas. Of course, a model may be subjected to more than one validation technique.

R.G. BACA, United States

What is the status of documentation on these models ?

R.E. WILEMS, United States

Since, we are integrating models developed by others and some developed by us, the status of documentation varies from model to model. However, our goal is to have the entire package presented here today plus the codes in the Site Model fully documented by July 1982.

A.T. JAKUBICK, Canada

You mentioned that you do not intend to couple chemical modelling with hydrological modelling. The results of the chemical modelling will, obviously, have to be considered in the transport model somehow. Could you give us an idea how will the chemical data enter the transport model ? Are you considering a one parameter or a more parameter type of consideration ?

R.E. WILEMS, United States

Through an analysis of the results of the chemical model we will specify a retardation factor for each species, that may be space and time dependent. These retardation factors will be used by the transport model.

C. McCOMBIE, Switzerland

A very extensive or exhaustive list of data requirements was shown. What work is currently being undertaken to identify the most important or most urgent areas requiring work ? In particular have you yourselves done work to justify the reduced importance attached, according your earlier answer to Mr. Koster, to problems of corrosion ?

R.E. WILEMS, United States

There are a number of sensitivity studies being performed throughout the US Nuclear Waste Terminal Storage Program (including some within our program) and by programs in other nations. Our modeling efforts will continue to be responsive to the findings of these studies in identifying the most important parameters. The most important parameters requiring additional research are those to which the system response is sensitive and for which large uncertainties exist.

We have not done any specific work to justify reducing the importance of certain selective corrosion processes other than soliciting an initial peer review. Our models are so structured that we can reverse our decision readily, if later it seems appropriate to do so.

THE UNDERGROUND RESEARCH LABORATORY -
A COMPONENT OF THE CANADIAN NUCLEAR FUEL
WASTE MANAGEMENT PROGRAM*

Gary R. Simmons
Applied Geoscience Branch
Atomic Energy of Canada Limited
Whiteshell Nuclear Research Establishment
Pinawa, Manitoba, Canada R0E 1L0

ABSTRACT

An underground research laboratory (URL) has been proposed for geotechnical research in the Canadian Nuclear Fuel Waste Management program. A series of experiments is being developed for the URL that will satisfy the in situ testing needs of the program. Several of the proposed experiments are discussed in this paper.

* Issued as AECL-7349

1. INTRODUCTION

A major objective of the Canadian nuclear program is the safe management of radioactive wastes, ensuring that there will be no significant adverse effects on man or the environment at any time. The goals of the Nuclear Fuel Waste Management research and development program are to assess the concept that the disposal of highly radioactive wastes in deep, stable geologic formations will achieve this objective and, if the concept is proven, to establish the capability for safely disposing of these wastes.

The geotechnical portion of this research and development program includes activities related to hydrogeology, geophysics, geology, rock properties and geomechanics. The various segments of the program are directed towards evaluation of plutonic bodies, defining the undisturbed hydrogeology and rock-mass parameters and predicting long-term changes in these characteristics due to construction, loading and closure of an underground vault. The development of techniques for predicting the subsurface geologic environment and modelling the geochemistry, hydrogeology and thermal rock mechanics of a rock mass are key activities.

The mechanical, thermal, hydrogeologic and geochemical properties of deep, stable, geologic formations that are representative of potential disposal-vault sites must be understood to develop mathematical models. However, geologic formations suitable for a waste-disposal vault are not generally of interest to other organizations (e.g., mining or petroleum companies) because valuable minerals are not present in commercial quantities. Thus, to develop the necessary understanding of the properties and mechanisms that control the properties of large, intact rock masses, a suitably located research laboratory is required.

To meet the needs of the Canadian program, an underground research laboratory (URL) is being planned to provide a representative geologic environment in which to pursue the research program. This laboratory should be constructed in a previously undisturbed crystalline rock pluton, at a location where there is substantial surface exposure of the rock, and be sufficiently deep to avoid surface influences. Atomic Energy of Canada Limited (AECL) is presently exploring a site that meets these requirements. It is on the Lac du Bonnet batholith, near AECL's Whiteshell Nuclear Research Establishment, in eastern Manitoba. Figure 1 shows the proposed location. Construction of the URL, shown conceptually in Figure 2, is scheduled to begin in 1983. Experiments will begin during construction and continue until the land lease terminates in the year 2000.

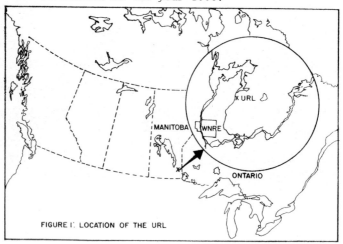

FIGURE 1. LOCATION OF THE URL

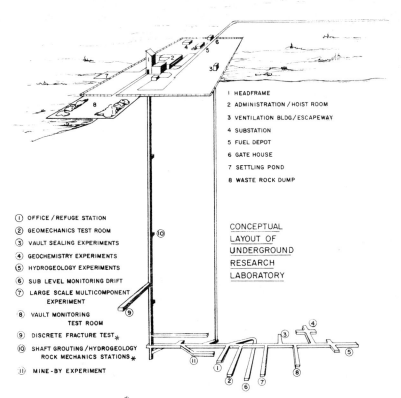

1 HEADFRAME
2 ADMINISTRATION / HOIST ROOM
3 VENTILATION BLDG / ESCAPEWAY
4 SUBSTATION
5 FUEL DEPOT
6 GATE HOUSE
7 SETTLING POND
8 WASTE ROCK DUMP

① OFFICE / REFUGE STATION
② GEOMECHANICS TEST ROOM
③ VAULT SEALING EXPERIMENTS
④ GEOCHEMISTRY EXPERIMENTS
⑤ HYDROGEOLOGY EXPERIMENTS
⑥ SUB LEVEL MONITORING DRIFT
⑦ LARGE SCALE MULTICOMPONENT
   EXPERIMENT
⑧ VAULT MONITORING
   TEST ROOM
⑨ DISCRETE FRACTURE TEST *
⑩ SHAFT GROUTING / HYDROGEOLOGY
   ROCK MECHANICS STATIONS *
⑪ MINE - BY EXPERIMENT

CONCEPTUAL
LAYOUT OF
UNDERGROUND
RESEARCH
LABORATORY

*LOCATION DEPENDS ON GEOLOGY

FIGURE 2: CONCEPTUAL DESIGN OF THE URL

2.   OBJECTIVES OF THE URL

A vast amount of geoscience information is required for the
Nuclear Fuel Waste Management program. Much of this information is
not presently available. Although the waste-disposal research programs now in place are intended to answer the questions, there is a
practical limit to how much can be learned through a borehole, on the
laboratory bench, or by computer simulation. Ultimately, it is essential to go underground to assess the accuracy of the analytical techniques and the mathematical models.

The URL will contribute experience and data in the areas of
rock mechanics, hydrogeology, geochemistry, vault sealing, geology,
geophysics and instrumentation development. Only a portion of the URL
program, however, falls within the "near field" and, therefore, is relevant to this workshop. This paper describes concepts for several
near-field experiments being considered to study the mechanical, thermal, hydrogeologic and geochemical response of a rock mass.

The exact procedure to be followed in Canada for licensing a
demonstration waste-disposal vault has not been established, but environmental and safety assessment studies will be a necessary part of
the process. Predictive methodologies are currently being developed
and have not yet been applied to an underground facility. The URL
will offer the first real test for these methodologies.

3.   EXPERIMENTAL PROGRAM

The experimental program being developed for the URL will be described in a Program Document [1], presently in preparation. The program includes some of the experiments suggested in a conceptual design study recently completed by Golder and Associates [2] and others proposed by program participants. In the following sections, several of the experiments are described.

3.1   Assessment of rock-mass response to excavation

A mine-by experiment, similar to the one done as part of the Spent Fuel Test at the Climax Mine [3], has been proposed to assess our rock-mass modelling capability. This experiment will also be designed to include an assessment of the excavation damage caused by blasting, which will continue the work begun by the Colorado School of Mines [4]. The objectives are to assess the capability of the mechanical rock-mass models to assess the response of a rock mass to excavation, and to assess the rock-mass damage caused by selected blast-round designs.

The physical arrangement for the proposed experiment is shown in Figure 3. In this concept, an upper and lower instrumenta-

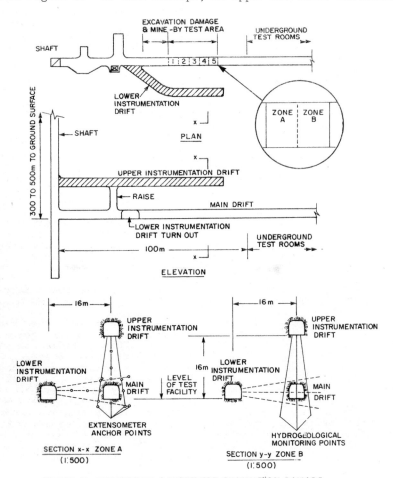

FIGURE 3: CONCEPTUAL LAYOUT FOR EXCAVATION DAMAGE
AND MINE-BY EXPERIMENTS

tion drift will be excavated before the main drift. Extensometers, stress meters and hydrogeologic monitoring holes will be installed from these drifts in arrays around the path of the main drift. The main drift will be excavated using a different blast design in each of five test zones and the effect on the rock mass assessed. Measurements of displacement, stress field and hydraulic conductivity will be made before, during and at intervals after completion of the main drift excavation. The changes in these parameters will be compared to the predictions of the mathematical models and will also provide the basis for assessing the effectiveness of various blast designs.

3.2   Heater experiments

Single- or multiple-heater tests (see Figure 4) are proposed

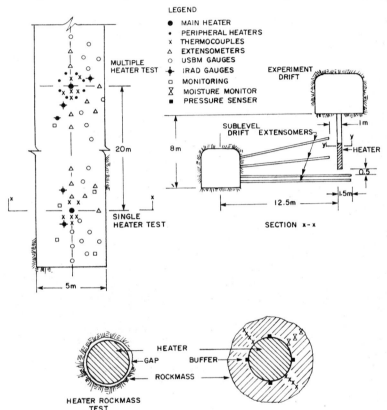

FIGURE 4: CONCEPTUAL LAYOUT OF HEATER TESTS

as a means of determining the in situ thermal properties of the rock mass and assessing the accuracy of the thermal-mechanical models. Since similar experiments have been done in the Stripa Mine in Sweden [5] and in the Near Surface Test Facility (NSTF) [6] at Hanford, with varying degrees of success, these experiments will be used to confirm that our rock-mass response can be predicted and to determine the thermal properties necessary to model later experiments.

For the heater tests, the heater(s) will be installed in the floor of a drift and surrounded by instruments to monitor temperature distributions, displacements, and stress-field variations that occur during the tests. In addition, monitoring holes will be provided for geophysical and fluid conductivity measurements to be made at various intervals during the test program.

When the buffer material and the immobilized waste-canister dimensions for the Canadian program have been specified, a heater-buffer test will be initiated. This experiment may be similar to the buffer-mass test presently being installed in the Stripa Mine [7]. It would provide data for assessing the mathematical models of the heater-buffer-rock mass system and data on buffer/rock mass/groundwater interactions.

3.3 Block experiment

The objective of the block experiment will be to provide thermal, mechanical and coupled thermal-mechanical response data for a large block of rock. Information on hydraulic conductivity and mass transport along a discrete fracture may also be obtained if the rock being tested contains a suitable fracture.

Block experiments have been conducted at the Colorado School of Mines [8] and are to be run at the Near-Surface Test Facility [6]. A variation of this test was proposed in the conceptual design study [2]. Figure 5 shows the conceptual layout of the modified block test

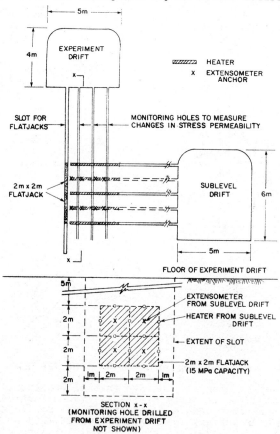

FIGURE 5: CONCEPTUAL LAYOUT OF MODIFIED BLOCK TEST

proposed by Golder and Associates. In this experiment a single slot, at least six metres long, is drilled to a depth below the sublevel instrumentation drift. Four 2-metre square, high-capacity (15 MPa), flat jacks are grouted into the slot at the elevation of the sublevel drift. Twelve heaters and four multipoint extensometers are installed horizontally from the sublevel drift into the volume of rock under test. Vertical monitoring holes for stress and permeability measurements are drilled vertically into the test rock.

Extensive scoping calculations are still necessary to ensure that the modified block concept will provide the desired data. Either this or the more standard block test [6,8] will be included in the URL program.

3.4 Hydrogeology and geochemistry experiments

Several hydrogeology and geochemistry experiments are being proposed in the URL program [1]. One concept, which includes a thermal component, is shown in Figure 6. The object of the experiment is

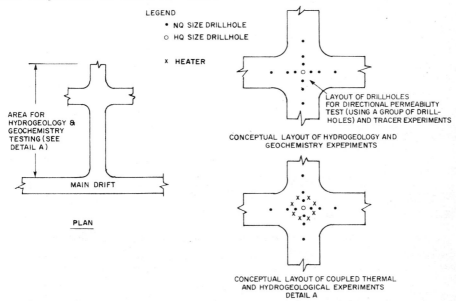

LEGEND
- NQ SIZE DRILLHOLE
o HQ SIZE DRILLHOLE

x HEATER

AREA FOR
HYDROGEOLOGY &
GEOCHEMISTRY
TESTING (SEE
DETAIL A)

MAIN DRIFT

PLAN

LAYOUT OF DRILLHOLES
FOR DIRECTIONAL PERMEABILITY
TEST (USING A GROUP OF DRILL-
HOLES) AND TRACER EXPERIMENTS

CONCEPTUAL LAYOUT OF HYDROGEOLOGY AND
GEOCHEMISTRY EXPEPIMENTS

CONCEPTUAL LAYOUT OF COUPLED THERMAL
AND HYDROGEOLOGICAL EXPERIMENTS
DETAIL A

FIGURE 6: CONCEPTUAL LAYOUT OF AN HYDROGEOLOGY AND GEOCHEMISTRY EXPERIMENT

to determine the directional hydraulic conductivity and mass-transport properties of the rock mass. A drift would be excavated to provide space for a cross-shaped array of 100-metre deep boreholes. Water or chemical tracers would be injected into various sections of the central hole while monitoring selected intervals of the surrounding array of holes.

The hydraulic conductivity tests would be run without, and then with, the heaters activated to assess the effect of a thermal field. The power level of the heaters would be regulated to vary the rock-mass temperature and the thermal gradient, to provide a wide range of data. The tracer tests would follow, with tests at conditions ranging from ambient to maximum temperature.

The data from this experiment would provide the parameters necessary for the hydrogeologic and geochemical modelling of later URL experiments.

3.5 Large-scale multicomponent experiment

When the program of individual parameter experiments in the URL is essentially complete, and when the containers, buffer and backfill materials have been specified for the Canadian Disposal Program, a large-scale multicomponent experiment will be designed. A concept for such an experiment is shown in Figure 7. This experiment could

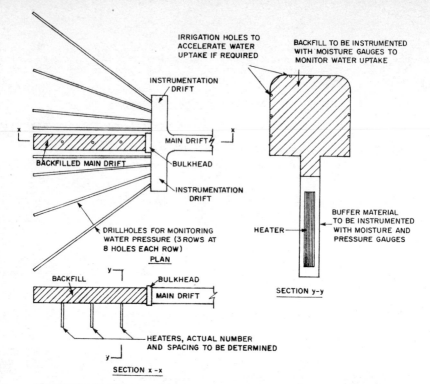

FIGURE 7. CONCEPT OF A LARGE - SCALE MULTI-COMPONENT EXPERIMENT

simulate, at full scale, a section of a disposal-vault room, with heaters designed to simulate, geometrically and thermally, the proposed waste containers.

Tests could be run to determine the thermal and mechanical response of the experiment room to the heaters and buffer material installed in boreholes in the floor. The room could then be backfilled, with the material proposed for disposal-vault backfilling, and instrumented to monitor the temperature response and the rewetting transient in the room.

Such an experiment would provide a thorough test of the modelling and assessment capability developed through the next 10 to 12 years in the waste-disposal program.

4.   CONCLUSION

A program of experiments is being developed for the underground research laboratory that is proposed as a major element in the Canadian Nuclear Fuel Waste Management program. Some of the proposed experiments have been outlined in this paper. Details of the URL and its experiments will evolve over the next two to five years. Each experiment will be designed to enhance the existing state of the art at the time it is to be conducted, by taking full advantage of experience developing both in Canada and abroad.

5.   REFERENCES

[1]   Simmons, G.R., and Soonawala, N.M., Editors, "Underground Research Laboratory: Underground Experimental Program", Atomic Energy of Canada Limited Technical Record, TR-153, in preparation.

[2] H.Q. Golder and Associates Ltd., "A Conceptual Design Study for an Underground Research Laboratory", H.Q. Golder and Associates Ltd. Report 791-1187, 1980 December (to be issued as Atomic Energy of Canada Limited Technical Record, TR-160).

[3] a) Ramspott, L.D., et al, "Technical Concept for a Test of Geologic Storage of Spent Reactor Fuel in Climax Granite, Nevada Test Site", Lawrence Livermore Laboratory Report UCRL-52796, 1979.

b) Carlson, R.C., et al, "Spent Fuel Test - Climax: Technical Measurements Interim Report, FY 1980", Lawrence Livermore Laboratories Report UCRL-53064, 1980.

[4] Hustrulid, W., et al, "Mining Technology Development for Hard Rock Excavation", Proceedings of the Workshop on Thermomechanical-Hydrochemical Modeling for a Hard Rock Waste Repository, Lawrence Berkeley Laboratories Report LBL-11204, 1980.

[5] Schrauf, T., et al, "Instrument Evaluation, Calibration, and Installation for Heater Experiments at Stripa", Lawrence Berkeley Laboratories Report LBL-8313, 1979.

[6] Basalt Waste Isolation Program, "Near-Surface Test Facility Test Program (Phase 1 & Phase 2)", Report No. BWI02TP0101 Rev 1, 1980 March.

[7] Carlsson, H., "The Stripa Project, A Multilateral Project in the Management of Radioactive Waste Storage", Subsurface Space, Rock Store 80, Stockholm, Sweden, Pergamon Press, 1980.

[8] Voegele, M., et al, "Site Characterization of Joint Permeability Using the Heated Block Test", Proceedings of the 22nd U.S. Symposium on Rock Mechanics, Massachusetts Institute of Technology, 1981 June.

DISCUSSION

C. McCOMBIE, Switzerland

The Canadian URL is unique case of a facility being introduced into a properly representative, undisturbed geology. What are the plans for monitoring the transient hydrologic effects caused by introduction of the mined repository ?

G.R. SIMMONS, Canada

A hydrogeologic monitoring program is planned for the URL to establish the undisturbed hydrogeologic environment and to monitor the perturbation that occurs during and after URL construction. The monitoring system will utilize four geologic boreholes drilled during the 1981 field season to depths between 400 m and 1100 m. As well, at least six monitoring boreholes will be drilled and instrumented during each of the 1982 and 1983 field seasons.

The monitoring systems will provide base-line information on the undisturbed hydrogeology prior to the beginning of construction in 1984 and the transient beyond that time.

A. MULLER, United States

In the hydrogeologic/geochemical experiments which you described :

1) which inert and reactive tracers do you intend to use, and

2) how representative do you think your geochemical results will be of actual *in situ* conditions ? I am thinking primarily of the effects of pressure release on noble gas content, the effect of oxygen injection on redox reactions and the influence of the drilling fluids (if wet drilling is to be used) on fracture fill minerology and chemistry ?

G.R. SIMMONS, Canada

We intend to use both inert and reactive tracers in the URL experiments. The exact materials have not been decided. Tritium has been suggested as a inert tracer and radionuclides will be used as reactive tracers. The radionuclides will be selected when specific experiments are designed and these experiments will be licenced by the Atomic Energy Control Board in Canada.

I cannot address your second question. The person to answer this question is Dr. P. Pearson of the Applied Geoscience Branch at WNRE. I suggest you address your question to him.

H. NIINI, Finland

Do you include in the activity programmes for this test site any possibility that this same site be used as a real repository for final disposal of nuclear wastes ?

G.R. SIMMONS, Canada

The URL site is not being considered as a site for a disposal vault. The demonstration disposal vault and commercial disposal vaults proposed in the Canadian Nuclear Fuel Waste Management Program will be sited in the Canadian Shield in the Province of Ontario.

AECL has a 21 year lease on the URL site and at the end of the lease, in the year 2000, we will seal the laboratory, remove surface facilities and return the land to the Province of Manitoba.

F. GERA, Italy

You mentioned that the lease on the land where the URL is located only lasts until the year 2000. Since this seems to give you barely the required time and provides no flexibility in case of unforeseen changes of program I wonder if it would not be wise to negotiate a longer lease.

G.R. SIMMONS, Canada

The termination of our lease in the year 2000 may limit our flexibility at the URL but the Canadian Nuclear Fuel Waste Management Program contains plans for a demonstration disposal vault. This demonstration vault will be committed in the late 1980's or early 1990's and will become the focus for geotechnical R & D. When the demonstration vault is available work at the URL will decrease ; with this plan we maintain flexibility.

CONSIDERATIONS OF NEAR-FIELD PHENOMENA IN THE INTERNATIONAL ATOMIC ENERGY AGENCY'S UNDERGROUND DISPOSAL PROGRAMME

V. Tsyplenkov and K.T. Thomas
International Atomic Energy Agency
Vienna, Austria

ABSTRACT

The IAEA Underground Disposal Programme has so far addressed near-field phenomena only generally in the context of the more important general system considerations. The need for more technical reviews and discussions on subjects of specialised importance in this area is highlighted.

1.   Introduction

The International Atomic Energy Agency has an ongoing programme in Underground Disposal of Radioactive Wastes which was stepped up after 1977 because of the importance of the subject and the overwhelmingly positive interest shown by Member States. During a ten year period beginning in 1978, a number of technical and safety series reports, guides and codes are planned for publication in the important subjects connected with the various options of underground disposal. These reports cover basic aspects of guidance and technical information on topics which are primarily of interest to national authorities and programme managers. Other supporting technical reports are also being considered to cover in more detail topics or sub-areas of specific importance for the various experts involved. There are many areas the adequacy of research and development work of which have to be evaluated and assessed by scientists in national and international meetings.
Consideration of 'near field' phenomena, being of increasing importance, is relevant in this context. Whereas from previous studies and the literature there is considerable information on the subject, there has been no specific systematic attention paid to discuss it internationally. The exceptions are, the NEA meeting on 'In-Situ' Heating Experiments in Geological Formations, Sweden 1978, and the session on thermal aspects at the IAEA/NEA symposium on Underground Disposal of Radioactive Wastes held in Otaniemi 1979 (1). The present NEA workshop is therefore timely and welcome.

2.   Background Consideration

What are near-field phenomena? In assessing the long-term integrity of and for designing an engineered geological repository, in particular a repository containing conditioned high-level and alpha-bearing wastes, an understanding as to the potential causes of release and transport of radionuclides to the biosphere is necessary. Some of these causes or processes and phenomena occur in the 'near field' and some in the 'far field'. There may not be any precise boundary which can be demarcated between regions affected by these phenomena. A discussion on an adequate definition of what are 'near-field' phenonena at this workshop would be welcomed. The physical and chemical effects of heat and radiation from the waste are predominantly limited to the near-field, due to their impacts on repository design and construction and the waste package. The transport phenomena in the entire rock mass caused by heat effects, natural causes and from human actions after the repository is closed and which usually appear in the geosphere and biosphere outside of the repository are due to long term effects. Hence both near-field and far field performance are important while considering the integrity of a repository disposal.

In view of the complexities of processes which may occur in a
geological repository, assessment of its performance in the near
field must take into account the properties of the host rock and the
effect of mechanical stresses, heat flow, chemical interactions and
radiation in the host rock and the multiple barriers. Development
of heat transfer, thermomechanical, and chemical models have made
much progress both in salt formations, crystalline rocks and clay.
Chemical interactions at elevated temperature in the presence of
radiation and water, have been and continue to be studied in
relation to transport, corrosion, sorption and alteration of
chemical phases and properties. Thermal models based on physical
laws provide an accurate portrayal of heat flow and changes in
temperature. The thermomechanical effects in rock masses, and an
understanding of the functional dependence of strain on stress are
fields of study which have to be continued for a complete
understanding of the various parameters. Coupling of thermal and
thermomechanical models to models of water flow in the near-field
and of groundwater flow in the far field needs much more work and
experience before such models are reliable for predicting the
effects.

3.   Reflections of near-field aspects in the IAEA programme
In the current stage of the underground disposal programme
near-field phenomena has been addressed only generally in the
documents completed so far (2-6), where they have to be seen in the
context of the more important general system considerations. With
more details available on definition of repository designs, site
selection, safety assessment etc., the need and importance of more
information in this area grow. The following effects have been
identified in connection with considerations on release scenario for
waste repositories, as relevant to 'near-field' waste and repository
effects.

Thermal effects
. Differential elastic response
. Non-elastic response
. Fluid pressure, density, viscosity changes
. Fluid migration

Chemical effects
. Corrosion
. Waste package-rock interactions
. Gas generation
. Geochemical alterations

Mechanical effects
. Canister movement
. Local fracturing

Radiological effects
. Material property changes
. Radiolysis
. Decay product gas generation
. Nuclear criticality

The importance of thermal and radiological effects on host rock, backfilling, sealing materials, waste package and the solidified waste is well recognised as also the aspects of corrosion, waste-rock interactions, diffusion, dispersion and sorption in the near-field. Rock mass movement and deformation, canister displacements, fracture mechanics and propagation, strain stress relationships, groundwater circulation and behaviour in complex fractures and geochemistry at great depths are some other of the many parameters needing due consideration. The importance of "in-situ" experiments and the need for reliable field data gathering for an understanding and modelling of both short and long term effects is to be specially stressed in this context.

The report on Site Investigations for Repositories for Solid Radioactive Wastes in Deep Geological Formations (4) covers in some detail the need for information on waste-water-rock interactions and the effects of heat, radiolysis and waste chemistry. In view of its relevance, the appropriate portions from the report are reproduced in the Annex.

Future plans of the IAEA foresee the review of the state-of-the-art in various areas and publish them as Technical documents as a supporting basis for proceeding later to formulating codes and guides on the design and construction and on waste acceptance criteria for repositories. Some of the most important areas are:

- Effects of heat and radiation on repository preformance (models, parameters and results of heat dissipation calculations; role of radiation effects)

- Evaluation and assessment of technology and performance of barrier materials

- Groundwater chemistry and chemistry in the near-field

- Studies in geology and geohydrology of the different rock types

Better understanding in qualitative and quantitative terms is necessary in all these areas and in particular as those related to site specific influence, selection of barrier materials, and in the design of repositories. This is related to the methodology to perform in-situ investigations and design of laboratory experiments.

The Agency, being greatly interested in the subject of near-field phenomena in geological repositories for radioactive waste, would look forward to the recommendations for future investigations and studies emerging as a result of the workshop.

REFERENCES

1. "Undergrond Disposal of Radioactive Wastes",. Proceedings of an IAEA/NEA Symposium at Otaniemi, Finland, July 1979, Vol 11, STI/PUB/528 (1980).

2. IAEA "Safety Assessment for the Underground Disposal of Radioactive Wastes", Safety Series Report No. (under publication).

3. IAEA "Underground Disposal of Radioactive Wastes", Basic Guidance - Safety Series Report No. 54 (1981).

4. IAEA "Site Investigations for Repositories for Solid Radioactive Waste In Deep Continental Formations" Technical Report Series No. (under publication).

5. IAEA "Safety Analysis for Radioactive Waste Repositories in Continental Geological Formations" (Draft), Vienna 1980.

6. IAEA "Site Selection Factors for Repositories of Solid High Level and Alpha-Bearing Wastes in Geologic Formations". Technical Reports Series No. 177.

Excerpts from the report on "Site Investigations for Repositories
in Deep Geological Formations"

"WASTE-WATER-ROCK INTERACTIONS
The range of chemical and physico-chemical interactions
between the immobilised waste form, the canister and the
back-fill material on the one hand and the repository
environment (specifically the rock and water) on the other,
will need to be evaluated for all potential repository sites.
Thus the site investigation programme should include the
acquisition of relevant data from both field and laboratory
tests and measurements. The effects which should be
considered include:
- canister corrosion resulting from rock-water-canister
interactions,
- leaching of nuclides from waste as the result of
rock-water-waste interactions followed by subsequent migration
of nuclides through the man-made and natural barriers,
- chemical and mineralogical transformations in the wastes
and rocks due to rock-water-waste interactions, especially at
elevated temperatures and pressures, and
- deterioration of the rock-water-waste due to high radiation
fields.

The geochemical and physico-chemical characteristics to be
studied in this context are those of the host rock and the
water it contains, and include:

- chemical, radiochemical and mineralogical composition of
rocks,
- sorption capacities of minerals and rocks,
- chemical and radiochemical composition of groundwater,
- electrochemical properties of groundwater, such as
reducing-oxidation potential, pH and conductivity, and
- effects of radiation and decay heat on the rock and on the
groundwater.

These characteristics can be measured by a variety of
methods. Most measurements are made in the laboratory but
some need to be studied in the field. For sites at which
reliable data cannot be acquired during preliminary
investigation stages, supporting studies may be needed from
other sites to ensure the acquisition of data from similar
geological conditions.

Effects of Decay Heat, Radiolysis and Waste Chemistry
Significant studies should be directed towards determining the
effects of decay heat, of radiolysis and of waste form
chemistry under repository pressures. Solid rock
geochemistry, should be examined as well as chemistry of both
mobile and non-mobile groundwater, including water of
hydration and fluid inclusions. In salt the occurrence,
content and mobility of fluid inclusions will be particularly
relevant.

In addition to mechanical effects, heat from the decay of
radionuclides in the waste may cause complex chemical changes
in the groundwater. Some of these changes may affect natural
barriers and therefore radionuclide transport rates. A
temperature gradient in salt may cause decrepitation and/or
brine migration towards the heat source due to differential
thermal solutioning. Heat may alter geochemical properties by
causing mineralogic changes such as dehydration, which may
alter the sorptive characteristics of the rock. Also, the
solubilities of many radionuclides in water increase with
increasing temperature, thus permitting greater concentrations
of radionuclides to enter into solution. Increases in
temperature of groundwater will also decrease the viscosity
and thus increase the flow rate of water.

Although it is generally of small significance radiation could
adversely affect the repository environment in several ways.
Two major possible effects are (a) the production of reactive
chemical species including gases, and (b) the storage of
energy. The possible reactive materials produced may include
species such as peroxides, oxygen, and nitric acid; and in
brine, chlorates and bromates. These oxidised species could
accelerate corrosion rates on waste canisters, or they may
react with minerals in the backfill or the rock and thereby
degrade their sorptive characteristics. The production of
stored energy in the rock may also cause mineralogic
transformations, which could result in volume changes and in

decreased radionuclide sorption capabilities. Volume changes
or excessive gas production may cause the rock to fracture.
The combined effect of both these possibilities is likely to
be offset, however, by the major increase in the number of
sorption sites on the surfaces of the newly-created fissures.

In addition to radiation and heat, certain chemical substances
that might be in the waste package may inhibit sorption
processes. Such materials include: (a) chelating or
complexing compounds which can form stable solution complexes
with actinides; (b) some metals, which upon dissolution into
groundwater may be preferentially adsorbed by the geologic
medium; and (c) some waste forms which could give rise to
acidic or alkaline groundwater, causing degradation of the
rock or backfill.

These effects of heat, radiation, and waste chemistry will be
highly dependent upon the geological environment. It is
necessary to assess their significance in relation to the
geology of each proposed repository site."

Session 2

GRANITE AND CRYSTALLINE ROCKS

Chairman - Président

A. BARBREAU

(France)

Séance 2

ROCHES GRANITIQUES ET CRISTALLINES

NEAR-FIELD PROCESSES AND RELEASE MODELS FOR GRANITIC ROCKS

N.A. Chapman, I.G. McKinley and D. Savage
Environmental Protection Unit, Institute of Geological Sciences,
Harwell Laboratory, Oxfordshire, U.K.

ABSTRACT

The system of engineered barriers contained in the near-field is the only part of the multibarrier system over which direct control can be exercised, in terms of its effect on containment of the waste. A combination of experiment and modelling of release processes which treats the principal mechanisms involved in the breakdown of the waste form and in subsequent migration, can be used to help define the behaviour of these barriers during the thermal and post-thermal periods.

A review of the results of such studies is presented, principally of the hydrothermal dissolution behaviour of borosilicate glass in a granitic environment, the effects of waste-rock-water interaction, and of the sorption mechanisms for specific nuclides in contact with altered granite fracture surfaces. These results are combined in a model of release from the near-field, and migration through the far-field.

INTRODUCTION

The near-field of a high-level waste repository in granite has been variously described as the immediate locality of a waste container, or that part of the repository which experiences a significant thermal transient as a result of the decay heat of the waste. Here we use the latter definition, which implies that the whole repository volume is at some time in the near-field, since host-rock temperatures approach a uniform maximum in a 'cubic' repository (see Beale et al, 1980) a few decades after waste emplacement.

This paper is a brief report of the results obtained so far by the IGS Environmental Protection Unit on the geochemical assessment of processes leading to the release and subsequent migration of radionuclides from the near-field of a granite repository. This work combines laboratory studies of hydrothermal waste glass-granite-water interactions, and sorption mechanisms of specific nuclides on altered granite surfaces, with modelling of glass dissolution mechanisms in very low flow environments where saturation processes dominate waste breakdown and nuclide release. The aim of the work is to progressively refine the release source term in order to facilitate more realistic risk assessment and comparison of crystalline rock environments. In addition it is hoped eventually to define more rational temperature limits for the near-field than those currently used in design studies. If the principal release mechanisms can be confidently linked to release rates; for example rate control by groundwater flow volume; then it should be possible to define the potential performance as a function of time for the engineered barriers of the near-field, and indeed to define which of these barriers are really necessary to ensure adequate containment.

This latter aspect is important, since many potential roles have been defined for components of the near-field, in particular the backfill. For example it would be possible to design a backfill which acted as a groundwater conditioner, a flow barrier, a sorbing medium for released nuclides, and a 'speciator' to control the movement of specific nuclides by buffering redox conditions on a very localised scale. Modelling of release processes and their effect on a risk assessment may however indicate that some of these functions are unnecessary, or that some need only operate within certain time limits or for short periods, thus aiding in design optimisation.

Experimental background to the study

Since the near-field will experience elevated temperatures under high hydrostatic or lithostatic pressures, experiments have been performed on waste-rock-water interaction which attempt to replicate these hydrothermal conditions (Savage and Chapman, 1980, 1981). Data on the dissolution behaviour of reference UK borosilicate glasses (e.g. M22/209) as a function of temperature (over the range 100-350°C) and time have been derived under closed system conditions. In all experiments to date non-active borosilicate glasses have been used, doped with fission product analogues and uranium, and analyses have been made of solution chemistry and solid run-products. The techniques and detailed results of this work, which includes granite-water, glass-water, and granite-glass-water experiments, will be reported by Savage and Chapman (1981).

In the present programme it is intended to produce the simplest possible realistic source term model, which takes account of only the principal controlling factors, whilst ignoring the potential roles of the engineered barriers. These roles can then be tailored to cope with any specific problems which are seen to arise as a result of applying the basic model. For this reason, no experiments have yet been performed on the geochemical interactions of potential backfill pore-waters with waste glass, although this will clearly be necessary at a later stage.

In addition to laboratory simulation of release processes, a parallel research project has studied the mechanisms of sorption of specific nuclides onto degraded granite surfaces. Since it is anticipated that many of the fissures which take part in groundwater flow through granite may have surfaces dominated by low-grade alteration products produced by weathering, late-stage hydrothermal events, or by the hydrothermal epoch of the repository itself, the results of this study have been combined with data on release to construct a mobilisation and migration model. The experimental study of sorption processes on weathered granite is described in detail by McKinley and West (1981).

Implications of the experimental results

Before considering the release model itself, several important conclusions drawn from the experimental studies and pertinent to the near-field are outlined.

1. Low temperature granite-water interactions (including those involving glass) tend to induce mildly alkaline conditions. Groundwater pH values increase markedly with increasing reaction temperature and duration, and theoretical neutral pH decreases to accentuate this effect. Groundwater chemistry is dominated by feldspar dissolution equilibria, and its evolution can be predicted using thermodynamic data. The alkaline conditions are favourable for preservation of stainless steel canister materials, at least during the first millenium after emplacement. In addition Cl^- and $SO_4^=$ are present in very low concentrations in reacted solutions, also minimising corrosion potential. The implication of these data is that no geochemical buffer material is required to condition groundwater chemistry and prolong canister-overpack life; the granite is a sufficient buffer in itself.

2. Hydrothermal degradation of granite-glass and glass systems results in the formation of montmorillonite at temperatures up to at least 150°C. This would suggest that bentonite is a geochemically stable backfill and should prove suitable even at quite elevated temperatures (above its one-atmosphere breakdown temperature, owing to confinement at high pressure in a saturated environment). Further data are however required on its Na-Ca exchange behaviour in warm pore waters.

3. The very low flow rates anticipated in the near-field indicate that groundwater-glass ratios will remain low, and that consequently saturation effects would dominate the leachate solution chemistry. Silica saturation is seen as the controlling mechanism of glass dissolution (Chapman et al, 1981) and, although it has been pointed out (Johnson and Wikjord, 1981) that glasses should have no equilibrium solubility value, experimental solution data indicate that silica values approximate to those for saturation in the amorphous silica-water system at the relevant temperature (Savage and Chapman, 1981; Paul, 1977). Whilst the mechanism and rate of glass matrix breakdown may thus be controlled, the release of radionuclides into solution may take place congruently or incongruently, and is unlikely to be controlled by saturation of individual elements.

4. The domination of breakdown processes by the low fluid-solid ratios is reflected in very low and protracted release episodes compared with those derived from 'dynamic' flow-through or Soxhlet type leach tests. The latter methods are intended only as comparative rapid-sorting techniques and data derived from them are wholly inappropriate for use in realistic risk assessments.

5. The rate of release of nuclides is controlled by diffusion processes and sorption or uptake in the altered layer developed on the corroding glass. Leach rates of individual elements are seen to diminish with time at constant temperature. The generally amorphous alteration layer becomes more crystalline at higher temperatures, consequently reducing the amount of colloidal silica available as a sorbing and transport medium. The crystalline reaction products are principally sodium and neodymium borosilicates, albite, smectites, acmite and riebeckite, with a phillipsite analogue and exotic phases such as barium molybdate. Certain waste elements enter these phases as stable mineralogically bound forms. For example Cs and Zr can comprise up to 20 wt% of the smectites, which also concentrate and incorporate up to 1 wt% uranium. $BaMoO_4$ is particularly interesting since in fully active waste glasses radium would be expected to follow the behaviour of barium. It is interesting to speculate on the possibility that solid solutions with radium molybdate, $RaMoO_4$, might be formed, hence

isolating one of the critical radionuclides in this unusual mineral (Fig.1). However, this phase has only yet been found in experiments at high temperatures, and its stability field is not known. The 'secondary waste form' comprising all these minerals is thus of some potential in controlling subsequent releases.

6. Each element is released at a different rate in the early stages of glass dissolution. Bulk glass dissolution rates which imply congruent release processes are thus misleading. Three independent mechanisms may exist; the real behaviour probably being a combination of incongruent and congruent dissolution, and varying with time. These mechanisms are shown for the model case of Cs release in Fig. 2, where it can be seen that each leads to a different rate of Cs release.

Fig. 1 Single crystal of BaMoO$_4$ formed by hydrothermal reaction of
borosilicate glass with water. Under similar conditions this would
be expected to form a solid-solution with iso-structural RaMoO$_4$

Release and migration modelling

A preliminary model of release in a low groundwater availability environment (Chapman et al, 1981) has been further developed to examine the effect of time and temperature dependency in the early life of a repository, and the possible migration behaviour of Cs released from the glass. The basis of this revised model is the use of the experimental results described above in conjunction with a simple estimation of groundwater fluxes anticipated in a chosen granite disposal environment. In common with the preliminary study, this revised model considers the behaviour of a cubic metre of unencapsulated glass surrounded by granite, and might thus be considered conservative in allocating no role to the engineered barriers. The maximum temperature of hydrothermal interactions as a function of time has been calculated from an empirical relationship derived by D. Hodgkinson (pers. comm.) which relates to realistic sized waste blocks emplaced in a cubic array in a granite repository.

The breakdown behaviour of the glass matrix is taken to depend on the solubility of silica in the limited volumes of water available over the range of temperatures predicted by the thermal decay relationship. The rate of release of Cs from the glass has been modelled using the three mechanisms mentioned in point (6) above and illustrated in Fig. 2. Caesium has been used in this model because its geochemical behaviour is simple compared with the albeit more radiologically significant actinides, and because adequate data not only on its release behaviour but also on its sorption behaviour are available. Whilst it is generally taken to be of minor importance in safety assessments it is used in this model to demonstrate potential release mechanisms, which might be applicable to other elements, and their effects on migration. In fact, although the bulk of

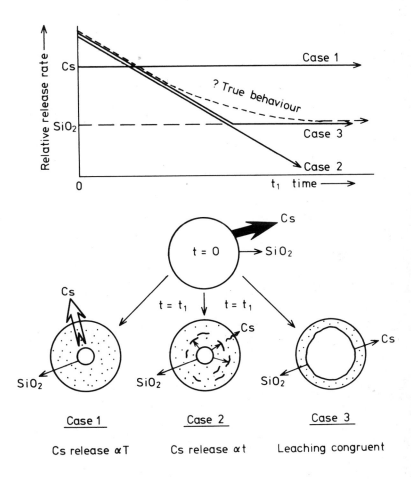

Fig. 2 Potential borosilicate glass leach mechanisms for caesium.
At time (t) = 0 (initial leaching) Cs leaches incongruently and very much
faster than the silica matrix dissolves. At time = t_1 three potential
mechanisms exist:- CASE 1 where incongruent leaching persists as a
function of temperature only. There is no hindrance of Cs release by the
leached layer. CASE 2 where Cs release decreases with time at a steady
rate. Cs release is hindered by the development of the leached layer,
and in the longer term by the formation of alteration products which sorb
or incorporate it into their structures. CASE 3 where Cs release
decreases until dissolution is congruent. Initial preferential Cs
leaching decreases to a steady state when the rate of penetration of the
depleted layer into the matrix equals the rate of matrix dissolution.

the radiocaesium decays within the first 1000 years or so, ^{135}Cs comprises 18% of the original Cs inventory and has a half-life of 2.3×10^6 years and may consequently be of significance.

The release rates estimated from the three mechanisms were used as input to a simple analogue model of groundwater migration described by McKinley and West (1981). This compartmentalises the migration path through the far-field into small 'boxes' in which equilibrium between rock surface and solution is taken to occur. Caesium retardation is calculated from an empirical, concentration dependent isotherm, and the same flow parameters as were used in the leach model are applied to the migration assessment. Hydraulic dispersion can be modelled by increasing the cross-sectional area of consecutive boxes. The results of this combined study will be reported by Chapman, McKinley, Savage and West (1981, in prep.), and the principal conclusions are outlined below and in Fig. 3.

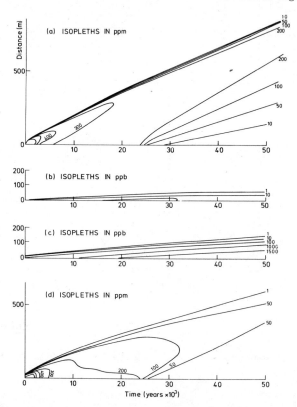

Fig. 3   Modelled Cs migration plumes (concentration in solution as function of distance from source) for the various assumed leach mechanisms shown in Fig. 2. (a) CASE 1; (b) CASE 2; (c) CASE 3; (d) CASE 1 with addition of hydrodynamic dispersion

Implications of modelled release and migration behaviour

1.   There are quite clear limitations to the use of existing data on release mechanisms. The migration plumes for ^{135}Cs are shown in Fig. 3. Depending on the type of leaching mechanism assumed, a single sorption isotherm has predicted Cs releases which, at a nominal distance of 1000 m from source, vary in concentration by up to five orders of magnitude.

2.   The times to breakthrough at 1 km for the three leach models, although varying between 65,000 and 330,000 years, are less significantly different than the concentrations at the leading edge of the plumes. Taking this

together with point 1, ^{135}Cs could be regarded as being of significance, or not, in a risk analysis of this type of disposal.

3. Cs is a relatively well-sorbed element, yet the times to breakthrough arising from the model are nowhere greater than a few hundred thousand years, for what is probably a quite reasonable pathlength in crystalline rock. There is thus evidence from this type of work that the period of complete isolation of wastes from the biosphere in a fractured rock environment is likely to be measured in tens or hundreds of millenia, with subsequent releases taking place at extremely low levels over very protracted time spans. The importance of far-field flow behaviour cannot be overstated, and models such as this cannot be validated until some truly site-specific studies have been performed using thorough hydrogeological characterisation of the rock volume.

4. Calculated releases to the near-field are very protracted, and waste form lifetimes of millions of years or longer seem quite possible. Even without taking account of hydraulic dispersion during subsequent migration the output peak profiles of individual nuclides of long half-life will be very extended and attenuated, and calculated doses will tend to be additive at any given time, rather than occurring at more or less discreet time intervals. The influence of dispersion will be specific to each element owing to the complexity of the sorption isotherms which will eventually be derived.

Discussion

The most appropriate approach to assessing the behaviour or the performance requirements of the waste form, or any of the near-field engineered barriers, is to do so in the context of the performance of the complete barrier system. It is clearly pointless to design and develop any one of these components as though it were to function in isolation. Disposal options, including the non-geological ones, must be compared on a rational basis as total systems, not as a series of isolated barriers. The near-field assessment must thus be seen simply in terms of the source term for release modelling. This approach has been used in the experimental and modelling studies reviewed here, to examine the interaction of the far-field rock barrier with releases of various types arising from an unencapsulated block of borosilicate glass. The results should be useful in subsequent design of near-field engineered barriers.

For example, the several potential roles of the backfill can now be seen to be dominated by the aquitard function, whereas corrosion conditioning is apparently unnecessary in the environment considered here. Reduction or prevention of flow such that diffusion is the only means of water movement through the backfill has a massive effect on numerical values of release. Beyond this, particularly when trying to assess the value of a nuclide scavenging role for the same backfill, there is some difficulty associated with a lack of data on the release behaviour of the important nuclides such as ^{226}Ra, ^{237}Np and ^{99}Tc. This information must be gathered by experiments on fully active or doped glasses in closed systems, and low fluid-solid ratio environments.

There are some attractions in the potential fixing and retardation functions of a crystalline, high-temperature, secondary waste form. Whilst a 'thousand year canister' may be attractive, this indicates that it may not be necessary. Our own model takes no account of any engineered barriers at any time, yet produces very protracted releases. There is evidence from safety assessments that, if release processes rather than events are considered, there is little value in trying to assure short to medium term containment within the engineered barriers.

From the geochemical viewpoint a maximum temperature for any part of the near-field of between 100 - 200°C would appear to be tolerable, and it is considered that the engineered barriers could be designed to cope with this with little or no modification to existing concepts.

In conclusion it is emphasised that there is a need to treat the near-field as part of a complete disposal system, and to design experimental assessments accordingly. The release modelling presented here can thus only be of real value when linked to similarly realistic assessments of far-field migration, and to do the latter comprehensive site-specific hydrogeological and geochemical data are required. In this respect there seems to be a long way to go, as such investigations are only just starting, and only in a limited number of crystalline rock environments, in a limited number of countries.

Acknowledgements

This work was funded by the Department of the Environment and the Commission of the European Communities as part of their research programmes into the disposal of radioactive wastes, and is published by permission of the Director, IGS (NERC).

References

[1] Beale, H., Engelmann, H.J., Souquet, G., Mayence, M. and Hamstra, J. Conceptual design of repository facilities. In: Radioactive Waste Management and Disposal. Simon, R. and Orlowski, S. (Eds), Harwood, CEC, pps. 488-510. 1980.

[2] Chapman, N.A., McKinley, I.G. and Savage, D. The effect of groundwater availability on the release source term in a low hydraulic conductivity environment. Proceedings of the NEA workshop: Radionuclide release scenarios for geologic repositories, NEA/OECD, Paris, 91-103. 1981.

[3] Chapman, N.A., McKinley, I.G., Savage, D. and West, J.M. Mechanisms of dissolution of radioactive waste storage glasses and caesium migration from a granite repository (submitted to Scientific Basis for Nuclear Waste Management, 4, Boston 1981).

[4] Johnson, L.H. and Wikjord, A.G. The rate of mobilisation of radionuclides from nuclear fuel and reprocessed wastes. AECL-TR-79 pps. 91-104. 1981.

[5] McKinley, I.G. and West, J.M. Radionuclide sorption/desorption processes occurring during groundwater transport:- Progress Report October 1979 - December 1980. Rept. Inst. Geol. Sci. ENPU 81-6, 37 pp. 1981.

[6] Paul, A. Chemical durability of glasses; a thermodynamic approach. J. Materials Science, 12, 2246-2268, 1977.

[7] Savage, D. and Chapman, N.A. Geochemical factors controlling the nuclide release source term in granite: dissolution of the waste form. Rep. Inst. Geol. Sci., ENPU 80-12, 27 pp. 1980.

[8] Savage, D. and Chapman, N.A. Hydrothermal behaviour of simulated waste glass, and waste-rock interactions under repository conditions. Chemical Geology (in press). 1981.

DISCUSSION

R. STORCK, Federal Republic of Germany

Did you consider any other radionuclides for drawing your conclusions for temperature limits and canister life-times ?

N.A. CHAPMAN, United Kingdom

Our model uses ^{135}Cs as a means of demonstrating the relative significance of the various mechanisms of waste form breakdown. Our conclusions on temperatures combined with canister lifetimes are not drawn from this approach, rather from an assessment of overall waste form behaviour, for example the development of a secondary waste form.

H.C. BURKHOLDER, United States

Your presentation stressed the importance of considering overall system performance in developing a basis for deciding what engineered components should be present in the disposal system and what performance levels they should have. However, the statement of your conclusions seemed to suggest that long-lived containers were beneficial because they gave short-lived fission products, like ^{90}Sr, time to decay and because they prevented water from contacting the glass waste form until the temperature at the waste form surface was lower. Does not your own work, the work of your colleagues, and the work of many others in the literature suggest that *overall* system performance is insensitive to container lifetime when the measure of performance is maximum radiation dose to future individuals ?

N.A. CHAPMAN, United Kingdom

We merely suggest that a container that survives the initial thermal period may be useful, in that it reduces the early fast waste dissolution. In the relatively benign granite environment, this should be straightforward to provide. It is certainly not essential, however, and I agree that long-lived containers are generally not justified by safety assessments.

In our conclusions we state that there may be attractions in a high temperature secondary waste from produced by hydrothermal interactions. This would allow earlier waste disposal, but would only apply in cases where early canister failure occurred. The time of canister failure thus again becomes irrelevant if this approach is adopted. You will note of course that in both this, and our previous model, we assume no engineered barriers whatsoever are present, and still compute very slow releases and protracted waste form lifetimes. Consideration of total system performance is indeed paramount rather than fixations with arbitrary component performance criteria.

MINERALS AND PRECIPITATES IN FRACTURES AND THEIR EFFECTS ON THE RETENTION OF RADIONUCLIDES IN CRYSTALLINE ROCKS

B. Allard[a], S.A. Larson[b], Y. Albinsson[a], E.L. Tullborg[b],
M. Karlsson[c], K. Andersson[a], and B. Torstenfelt[a].

[a] Department of Nuclear Chemistry, Chalmers University of Technology
[b] Swedish Geological Survey
[c] Department of Geology, Chalmers University of Technology and
University of Gothenburg,
Gothenburg (Sweden)

ABSTRACT

Fracture filling materials in a plagioclase-quartz rich granitic bedrock were identified, and most abundant were calcite and prehnite. The cation exchange capacities were determined for some fracture minerals as well as for some major components of granite. Distribution coefficients were measured for Cs, Sr and Am between geologic materials (fracture minerals, rock forming minerals), fresh metal hydroxide precipitates (Cu(II), Pb(II), Fe(III)) and groundwater. A significantly reduced sorption was obtained for Sr (up to a factor of five) and Cs (up to two orders of magnitude) on calcite/prehnite in comparison with granite. The sorption of Cs and Sr on Cu(II)- and Pb(II)-hydroxides was very poor. The sorption of Am was very little affected by the mineral composition.

1. INTRODUCTION

In the present Swedish concepts for disposal of high-level radioactive waste or spent nuclear fuel a storage in a deep underground repository in crystalline rock is envisualized [1]. A number of barriers such as the use of waste forms of low solubility, a non-corrosive canister and a back-fill that prevents free water flow around the waste canisters will delay the release and transport of radionuclides from the repository. The major barrier preventing the radionuclides from reaching the biosphere would be the host-rock itself.

Material released from the repository can be transported by the groundwater flow, either in true solution or as particulates. Although dispersion and diffusion into microfissures may turn out to be the dominating retention mechanism [2], the chemical interaction between the radionuclides and the water-exposed rock surfaces would be of a great importance, and retention factors (nuclide velocity/water velocity) of several orders of magnitude will be expected for e.g. the lanthanides and actinides in their lower oxidation states [3].

The most important pathways would be open water carrying fractures. Fracture minerals in crystalline rocks are usually quite different from the bulk of the rock. For igneous rocks like granite the major rock forming minerals would be quartz, feldspars (orthoclase, plagioclase, etc.), micas (biotite, muscovite) and amphiboles (hornblende), and in basic rocks like basalt also pyroxene and olivine. In the young fractures essentially three categories of minerals will be found:
1. Weathering and alteration products of e.g. micas, feldspars and amphiboles (clay minerals, etc.).
2. Precipitates and crystallization products from aqueous solutions, not necessarily under hydrothermal conditions.
3. Metamorphic products.
Some of the precipitation products may be formed at low temperatures and be of quite recent origin. The formation of such fracture precipitates would be expected to be fairly sensitive to changes in groundwater composition and temperature.

In order to allow a quantitative description of the radionuclide retardation it is essential that chemical properties of the pathways in the bedrock/groundwater system are characterized in detail. Therefore the composition and properties, including possible formation conditions and age, of the fracture filling material should be studied, as well as the local effects of changes in temperature and possibly groundwater composition due to the presence of an underground repository. Also any potential effects of the dissolution and precipitation of metals from the repository (e.g. lead, iron, possibly copper, etc.) on the radionuclide retention in the near field should be considered.

In a current project the fracture filling materials in granitic bedrock at various depths (0-600m) are characterized, the corresponding groundwater analysed and the chemical interaction between pertinent radionuclides and the fracture filling materials studied. Some data from this project are discussed in this paper.

2. FRACTURE MINERALS IN GRANITE

2.1 The Finnsjön area

As a part of the geochemical-hydrological studies of potential waste repository sites a drilling program is in progress in the Finnsjön area. The bedrock in this area is a quartz-plagioclase rich granite, fairly poor in dark minerals, with the average composition given in Table I. Groundwater samples have been taken from various depth down to about 500m in totally seven different drilling holes. The concentration of Na and Cl are higher than for average granitic groundwaters (c.f. Table II) and is increasing with depth in some of the drilling holes, indicating the presence of old saline water. Also the concentration of Ca and Mg are far above the levels in most groundwaters. All the groundwaters are saturated or some times over-saturated with respect to $CaCO_3(s)$.

Table I Mineralogic composition of Finnsjön granite [4]

Mineral	Concentration,%
Quartz	27-39
Plagioclase	26-35
Microcline	15-23
Hornblende	8-11
Biotite	4-6
Chlorite	3-7
Epidote	1-4
Others[a]	1

[a] Calcite, apatite, prehnite, titanite

Table II Groundwater composition at Finnsjön [5]

	Concentration, mg/l					
	a	b	c	d	e	f
Ca^{2+}	22-60	107-1790	0.7	25-50	188	18
Mg^{2+}	4-9.5	16-110	0.1	5-20	599	4.3
Na^+	13-124	224-1460	0	10-100	4980	65
K^+	1.4-3.1	1.8-10.0	0.9	1-5	179	3.9
Fe(tot)	2.9-21	0.6-9.2	0.08	0.5-20		
Fe(II)/Fe(tot)	0.8-1.0	0.25-1.0				
Mn	0.05-0.31	0.06-0.74	0.04			
Cl^-	13-124	380-5500	3	5-100	8950	70
SO_4^{2-}	1-46	35-325	5-7	3-40	1250	9.6
HCO_3^-	322-395	39-295	0	60-400	140	123
PO_4^{3-}	0.03-0.26	0.03-0.17	0.01	0.01-0.5		
F^-	1.4-3.0	0.7-2.3	0.2	0.01-5	0.66	
SiO_2(tot)	6.0-18	7.7-14	1	5-60		12
Org.C	6.2-11.0	1.2-6.2	7-20			
Tot.	530-750	1300-9200	30		16335	306
pH	7.1-8.8	7.7-8.4	5	7.2-8.5	8	8.2

a Fi1, Fi2, Fi4, Fi7:123
b F7:**301**, Fi5
c Precipitation
d Probable concentration interval in granitic non-saline groundwater [6]
e Baltic Sea water
f Synthetic groundwater used in the sorption studies

2.2 Fracture minerals

Fracture zones were located in the drill cores, and the fracture minerals, both in open and healed fractures, were identified by chemical analysis and X-ray diffractometry [4] (c.f. Table III). In healed fractures the dominating materials were quartz and calcite, where calcite appeared to be youngest.

In open fractures the dominating minerals were calcite and prehnite, and in some fractures also chlorite, epidote and zeolites like laumontite and stilbite were identified. Occasionally also dolomite, gypsum and pyrite were observed, as well as clay minerals (kaolinite, montmorillonite, illite, etc.) Besides the clay minerals only calcite-dolomite, gypsum and possibly pyrite would be expected to origin from low-temperature aqueous solution.

Measured m_{Sr}/m_{Ca} - ratios and $\delta^{18}O$ - $\delta^{13}C$ - data, both for the calcite from the fractures and from the groundwaters, indicate that most of the calcite was either formed at temperatures above $100°C$ or it was crystallized from a water with a different composition than today. Some of the samples, however, could have been formed under present conditions. It seems evident that several different generations exist. The variation in water composition, including m_{Sr}/m_{Ca} - ratios, and between the different drilling holes as well as the variation with depth within one single hole and studies of fluid inclusions indicate a very slow or negligible exchange of water. (A detailed discussion of m_{Sr}/m_{Ca} - ratios as well as isotopic data is given elsewhere [5].)

Table III Fracture minerals in granite [4]

Mineral	Mineral class	Occurrence[a]
Calcite	Carbonate	A
Dolomite	"	S, Me
Gypsum	Sulfate	P
Pyrite	Sulfide	A
Epidote	Soro silicate	Me, H, S
Prehnite	"	Me, H
Chlorite	Phyllo silicate	A
Kaolinite	"	W, H
Montmorillonite	"	S, H
Illite	"	W, H
Quartz	Tecto silicate	A
Laumontite	"	S, H
Stilbite	"	S, H
Analcime	"	S, H

A = occurs in all geologic environments
S = in sedimentary rocks
Me= in metamorphic rocks
W = as weathering products
H = as hydrothermal products

3.   SORPTION OF Cs, Sr AND Am ON FRACTURE MINERALS AND METAL HYDROXIDES

Pure minerals, not taken from fractures but representing the various products identified in granite fractures as well as major rock-forming minerals were collected. The purity was checked by X-ray diffractometry only. For comparison two minerals taken from fractures in the Finnsjön granite were studied (calcite and prehnite [4]) and the unweathered granite itself. Also three metal hydroxides (Cu(II), Pb(II), Fe(III)) were included in the study.

3.1 Distribution measurements

The minerals were crushed and sieved. The size fraction 0.045-0.090 mm was washed twice in synthetic groundwater (Table II) and equilibrated with this water for about one week. Active stock solutions were added (^{85}Sr, ^{134}Cs or ^{137}Cs, ^{241}Am) and pH adjusted back to the equilibrium value. The solid/liquid ratio was 20g/1 and the total Sr-, Cs- and Am-concentrations $1x10^{-8}M$, $6x10^{-9}M$ and $5x10^{-9}M$, respectively. The activity remaining in solution and the corresponding pH was measured as a function of time. (The procedure for distribution measurements is described in detail elsewhere [7]). The metal hydroxides were prepared from $Cu(NO_3)_2$, $Pb(NO_3)_2$ and $FeCl_3$, respectively, which were dissolved in water. The hydroxides were precipitated with NaOH at pH 8-8.5, washed thoroughly and dried at $105^{\circ}C$ before washing and equilibration with the synthetic groundwater. The sorption (in % sorbed on the solid) is given in Table IV.

3.2 Cation exchange capacities

Apparent cation exchange capacities (CEC) were determined for the various minerals by an isotopic dilution batch technique using ^{22}Na, and the total uptake of Na as a function of pH (5-9) was measured [8].

4.   RESULTS AND DISCUSSIONS

The sorption of both Cs and Sr would be expected to be highly dependent on the CEC of the sorbent (at a constant pH and nuclide concentration), which is evident from Figure 1. For Am there is no apparent correlation between sorption and CEC, and no such correlation would be expected for a highly hydrolyzed cation [3].

Both the CEC and distribution coefficient K_d are significantly higher (at least by a factor of two) for the natural rock than can be calculated from data for the pure minerals, assuming contributions from the individual components in proportion to the stoichiometric composition (c.f. Table V). However, it is likely that the mineral faces in the natural rock are altered, and there are probably alteration products, (e.g. clay minerals with a high CEC) in the grain boundaries in the rock not easily determined in a conventional mineral analysis of a heterogenous rock. The presence of 0.5% of a high-capacity mineral like montmorillonite would be enough to account for the high observed CEC of the rock. Less than 1% of such high-capacity minerals in the rock would also account for the observed distribution coefficients for Cs, but not quite for Sr.

The CEC, and thus the distribution coefficients, are lower for pure minerals representing the most abundant fracture minerals than for the granite which would give a lower retardation of non-hydrolyzed cationic radionuclides in a typical granite fracture than expected considering the bulk composition. However, it is evident that the pure minerals are not representative of the fairly unpure mixed products that precipitate in a fracture. Also the measured distribution coefficients for the most abundant fracture minerals calcite/prehnite using materials taken from fractures, are considerably lower than for the granite itself or even for the calculated granite values, Table V. For Sr the calcite/prehnite data are a factor <5 lower than the values for granite, while for Cs the difference may be as high as 1-1.9 orders of magnitude.

The sorption of Am is very little affected by the fracture mineralogy, and is almost quantitative in the pH-range studied (8-8.5). The presence of minor amounts of apatite, as well as clay minerals, may account for the difference between measured and calculated values for granite [3].

Table IV Sorption of Cs, Sr and Am

(pH 8-8.5, temperature $22\pm2^{\circ}C$, contact time 1 week, solid/liquid ratio 20g/1)

Mineral		Sorption, %		
		Cs	Sr	Am
Calcite	(Ca)	3	2	98
Calcite[a]		11	8	>99
Dolomite	(D)	9	5	99
Gypsum	(Gy)	10	5	>99
Pyrite	(Py)	6	6	97
Epidote	(E)	28	16	96
Prehnite	(Pr)	20	17	98
Prehnite[b]		49	24	99
Chlorite	(Ch)	72	50	99
Kaolinite	(K)	54	37	>99
Montmorillonite	(Mo)	52	53	2
Illite	(I)	>99	87	99
Laumontite	(L)	70	37	98
Stilbite	(S)	95	79	>99
Quartz	(Q)	2	3	94
Plagioclase	(Pl)	18	8	99
Microcline	(Mi)	78	19	99
Hornblende	(H)	79	27	>99
Biotite	(B)	91	31	>99
Apatite	(A)	6	13	>99
Granite	(Gr)	88[c]	26[c]	>99[d]
Cu(II)-hydroxide	(Cu)	0	1	47
Pb(II)-hydroxide	(Pb)	0	0	81
Fe(III)-hydroxide	(Fe)	17	96	99

a From Finnsjön, Fi6:309m
b From Finnsjön, Fi7:519m, contaminated with 10% calcite
c From Finnsjön, unweathered; composition according to Table II.
d From Climax Stock [3].

Table V Measured and calculated CEC and distribution coefficients (K_d) for Finnsjön granite

	CEC, meq/kg	K_d, dm^3/kg		
		Cs	Sr	Am
Measured granite	12	390–490	16–20	>15000[e]
Calculated granite[a]	2.4–8.5 6.7–13[b]	52–270	7–10	7900–10000[e]
Calcite/Prehnite[c]	<2	6–48	4–15	>20000
Calcite/Prehnite[d]	<1	1–15	1–10	3000–14000

a Considering the possible composition variation, Table I
b Assuming 0.5% high-capacity components (CEC=850 meq/kg)
c From fractures
d Pure minerals
e For Climax Stock granite [3]

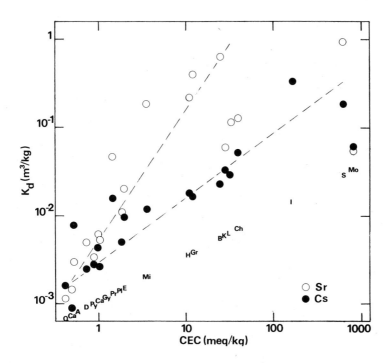

Figure 1. Distribution coefficients (K_d) vs CEC.

For a predominantly ion exchange process determining the sorption of radionuclides at trace concentrations the presence of cations origin from e.g. canisters (Fe, Pb, Cu, etc.) in the water would possibly affect the quantitative uptake. Such cations would also precipitate as hydroxides (Cu(II) and Fe(III)) or carbonates-hydroxycarbonates (Pb(II)) at concentration levels above 10^{-6}–10^{-7}M. As can be seen in Table IV the sorption of Sr and Cs on fresh hydroxide precipitates of Cu and Pb is negligible. Both these hydroxides, however, were highly colloidal.

5.   CONCLUSIONS

In granitic bedrock the mineralogy of water carrying fractures, where the predominant radionuclide transport would be expected, is entirely different from the host-rock itself. Most abundant fracture minerals are calcite and prehnite in the Finnsjön granite. Some of the fracture calcite could be of recent origin and would probably precipitate from groundwater of present composition. The CEC of the most common fracture minerals calcite and prehnite is lower than for the rock itself although some minor constituents (clays, zeolites) would have a higher CEC. A reduced sorption of Sr (by a factor of less than five) and of Cs (by up to two orders of magnitude) can be feasible. Am, which would be highly hydrolyzed, is very little affected of the mineralogic composition of the fracture. Fresh precipitate of Fe-hydroxide gives a very high sorption of Sr and Am. Fresh Cu- and Pb-hydroxides form gelatinous colloids with poor sorptive capacity which can not solely be attributed to poor phase separation (centrifugation, 4000rpm, 1h).

It should be stated that these conclusions are based solely on static batch-type sorption determinations and short contact times. The possibility of radionuclide transport by particulates in fractures, e.g. by clay colloids or by metal hydroxide colloids, etc. should be considered as well as the sorption properties of aged metal hydroxide precipitates. Continued studies of sorptive properties of fracture minerals as well as metal hydroxide precipitates are in progress.

6.   ACKNOWLEDGEMENTS

This project is financed by the National Council for Radioactive Waste (PRAV) and the Natural Science Research Council (NFR).

7.   REFERENCES

1. Handling of Spent Nuclear Fuel and Final Storage of Vitrified High-level Reprocessing Waste, Kärnbränslesäkerhet, Stockholm (1977), and

   Handling of Nuclear Fuel and Final Storage of Unreprocessed Spent Fuel, Kärnbränslesäkerhet, Stockholm (1978).

2. Neretnieks, I: "Diffusion in the Rock Matrix: An Important Factor in Radio-nuclide Retardation?", J. Geophys. Res. 85, 4379-96 (1980).

3. Allard, B., Beall, G.W. and Krajewski, T: "The Sorption of Actinides in Igneous Rocks", Nucl. Techn. 49, 474-480 (1980).

4. Larson, S.Å., Tullborg, E.L. and Lindblom, S: "Sprickmineralogiska under-sökningar", PRAV 4-20, Stockholm (1981).

5. Hultberg, B., Larson, S.Å. and Tullborg, E.L.: "Grundvatten i kristallin berggrund", SGU Dnr 41.41.-81-H206-U, Uppsala (1981).

6. Jacks, G.: "Groundwater Chemistry at Depth in Granites and Geisses", KBS Technical Report 88, Stockholm (1978).

7. Allard, B., Andersson, K. and Torstenfelt, B.: "Technique for Batch-wize Studies of Radionuclide Sorption on Geologic Media", KBS Technical Report, in prep.

8. Francis, C. and Grigal, D.F.: "A Rapid and Simple Procedure using Sr-85 for Determining Cation Exchange Capacities of Soils and Clays", Soil Sci. 112, 17-21 (1971).

DISCUSSION

B. SKYTTE-JENSEN, Denmark

Danish work has demonstrated that CaCo₃ in tuff is an efficient sorber for Cs, Sr and Eu.

B.M. ALLARD, Sweden

For Am the observed sorption in calcite was very high. However, both for Cs and Sr the sorption on crystalline calcite was poor (c.f. Table IV). It is not unlikely that $CaCo_3$ in the form of amorphous limestone or chalk would exhibit other sorptive properties in accordance with your observations. Although calcite has been iden- tified as one component in the $CaCo_3$-precipitate (and not aragonite), the degree of crystallinity is not known so far.

I.R. GRENTHE, Sweden

Is there any amorphous silica in the fractures ? How will this affect the sorption ?

B.M. ALLARD, Sweden

The most common fracture minerals are calcite and prehnite in the samples studied. However, there are always at least 5-10 % of other constituents present, such as clay minerals and zeolites as well as amorphous silica. All these additional components would enhance the sorption.

R.H. KOSTER, Federal Republic of Germany

Your 5th conclusion was, that the sorption of Cs, Sr on Cu, Pb- hydroxides is poor. This is in complete agreement with the normal waste management experiences. The decontamination factors for Cs and Sr for precipitation processes of HLW and MLW concentrates are poor. This means, sorption of Cs, Sr on commonly used hydroxides is poor.

B.M. ALLARD, Sweden

The cooloidal nature of the freshly precipitated Cu and Pb hydroxides should be pointed out, and possibly the sorptive properties of aged hydroxides could be different. Further studies of radionuclide sorption on metal hydroxides as well as on geologic media in the pres- ence of metals in solution are in progress within the present project.

NEAR-FIELD THERMAL TRANSIENT AND
THERMOMECHANICAL STRESS ANALYSIS OF
A DISPOSAL VAULT IN CRYSTALLINE HARD ROCK

K.K. Tsui, A. Tsai and C.F. Lee
Geotechnical Engineering Department
Ontario Hydro, Toronto, Ontario, Canada

ABSTRACT

The Canadian Nuclear Fuel Waste Management Program currently focuses on the development of a disposal vault in crystalline hard rock at a reference depth of 1 km below the surface in a suitable pluton in the Canadian Shield. As part of Ontario Hydro's technical assistance to the Atomic Energy of Canada Limited in this program, studies are being carried out to determine the effects of radiogenic heat on the near-field behaviour of a disposal vault.

This paper presents the study results obtained to date. Temperature and stress fields were computed and cross-checked by several finite element codes. A comparison between vertical and horizontal borehole emplacement concepts is made. The effects of material non-linearity (temperature dependence) and three-dimensionality on the thermomechanical response are evaluated. Case histories of thermal spalling or fracturing in rock were summarized and discussed to illustrate the possible mechanisms and processes involved in thermal fracturing. An assessment of the thermomechanical stability of the rock mass around a disposal vault under a state of high horizontal in-situ stress is also presented.

1. INTRODUCTION

The Canadian Nuclear Fuel Waste Management Program currently focuses on the development of a disposal vault in crystalline hard rock at a reference depth of 1 km below the surface in a suitable pluton in the Canadian Shield. Under the auspices of the Atomic Energy of Canada Limited (AECL), a series of conceptual design studies for the disposal vault had been carried out by Acres Consulting Services Limited, in association with RE/SPEC Inc., and Dilworth, Secord, Meagher and Associates (Mahtab and Wiles, 1980, Burgess, 1980). This series of studies include the near-field or room-and-pillar thermal rock mechanics analyses for the immobilized waste (IW) and irradiated fuel (IF) disposal vaults. A number of variables, such as extraction ratio (ER) and waste container pitch (S), were considered in the studies. Geological and construction features, such as discrete and ubiquitous joints and blast fracture zones, were also taken into account in the analyses. In addition, different backfilling and cooling options were also examined.

As part of the technical assistance to AECL, the Geotechnical Engineering Department of Ontario Hydro has carried out an independent near-field thermal and thermomechanical analysis of the same conceptual design of the disposal vaults to study some aspects not covered by the aforementioned study. These aspects include the option of waste emplacement in horizontal boreholes, the effects of material non-linearity and three dimensionality on temperature and stress, and a comparison of different finite element computer codes. Typical results of Ontario Hydro's analyses obtained to date are presented and discussed in this paper.

In addition, case histories of thermal spalling or fracturing in rock are summarized and discussed in this paper to illustrate the possible mechanisms and processes involved in thermal fracturing. The thermomechanical stability of the disposal vault is also examined in view of the combined effects of thermal loading and high horizontal in-situ stresses.

2. REFERENCE DESIGNS OF THE DISPOSAL VAULT

The reference design of the conceptual repositories consists of a series of long horizontal rooms at a reference depth of 1 km from the surface. The tentative geometric layouts of the immobilized waste (IW) and irradiated fuel (IF) rooms and containers, as contemplated by Acres et al are illustrated in Figure 1. The IW containers would be placed in vertical boreholes drilled from the floor of the waste room, with the room to be backfilled immediately after waste emplacement. An alternative of emplacement considered in this paper is to place the IW containers in horizontal boreholes drilled from the sidewalls of the waste rooms.

In the case of IF, the containers would be placed in the room with at least 1 m of initial backfill material around the containers. The remainder of the IF room would be left open for the first 20 years after emplacement. Two subsequent backfilling and cooling options are under consideration: (a) no ventilation of the room after emplacement and initial backfilling, to be followed by complete backfilling at 20 years; (b) ventilation cooling starting at 20 years after emplacement, for a period of 10 years, prior to final backfilling.

The study results of Acres et al indicated that the optimal extraction ratio for both the IF and IW vaults is 25 percent, while the optimal canister pitches for IF and IW vaults are 2.5 m and 1.5 m respectively. The analysis described in this paper refers to these geometric layouts only.

3. METHODS OF ANALYSIS

For the analysis described in this paper, the rock mass is assumed to be a homogeneous, isotropic, elastic and intact granite. For the linear analysis, the material properties are assumed to be temperature independent as shown in Table I.

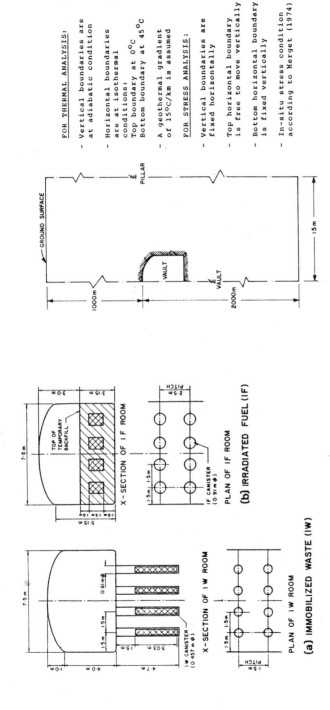

FOR THERMAL ANALYSIS:

- Vertical boundaries are at adiabatic condition

- Horizontal boundaries are at isothermal conditions:
  Top boundary at 0°C
  Bottom boundary at 45°C

- A geothermal gradient of 15°C/km is assumed

FOR STRESS ANALYSIS:

- Vertical boundaries are fixed horizontally

- Top horizontal boundary is free to move vertically

- Bottom horizontal boundary is fixed vertically

- In-situ stress condition according to Herget (1974)

FIG. 2 BOUNDARY AND INITIAL CONDITIONS FOR THERMAL AND STRESS ANALYSIS

(b) IRRADIATED FUEL (IF)

(a) IMMOBILIZED WASTE (IW)

FIG. 1 GEOMETRY OF IMMOBILIZED WASTE (IW) AND IRRADIATED FUEL (IF) ROOM AND CANISTER.

Table I

Temperature Independent Material Properties
Used in the Analyses

Properties	Unit	Rock (Granite)	Backfill
Thermal Conductivity	W/m°C	3	2
Specific Heat	J/Kg°C	800	800
Density	Kg/m^3	2800	2000
Young's Modulus	GPa	40	1.45
Poisson's Ratio	--	0.2	0.16
Linear Coefficient of Thermal Expansion	10^{-6}/°C	8	34

As mentioned previously, the reference design of the disposal vault calls for the construction of series of long, parallel tunnels or drifts as waste rooms at a depth of 1 km. This configuration enables the application of the unit cell concept to the analysis. Such a unit cell is illustrated in Figure 2, together with the boundary and initial conditions used in the thermal and stress analyses. The application of the unit cell concept and the adoption of boundary conditions as shown in Figure 2 introduce implicitly the following two assumptions: the disposal vault is large enough to be considered infinite in the horizontal extent and the emplacement of waste canisters is simulataneous and instantaneous. Direct contact between the waste canisters and rock mass in the case of IW, and between the fuel canisters and backfill in the case of IF are also assumed in the analysis.

For many of the Paleozoic and Precambrian rocks of Ontario, Canada, a state of high horizontal in-situ stresses exists. This has been confirmed by the results of in-situ stress measurements using overcoring and hydrofracturing techniques (Coates and Grant, 1966, Herget, 1974, Palmer and Lo, 1976, Haimson, 1978, Lee, 1981). For the stress analysis described in this paper, linear relationships between in-situ stresses and depth suggested by Herget (1974) were used. Thus, at the disposal vault horizon, the in-situ horizontal stress would be in the order of 48 MPa, with a corresponding vertical stress of about 28 MPa.

For both the thermal transient and stress analyses, the two vertical boundaries, one at mid-pillar and one at room centreline, of the model (Figure 2) are fixed in position due to symmetry requirements. Prior to the selection of the top and bottom horizontal boundaries, the effects of model size on temperature and stress were investigated by varying the positions of these boundaries in the analysis. It was found that the top boundary should be extended to the ground surface at 1 km above the disposal vault and the bottom boundary should be located at least to a distance of 2 km below the vault. By locating the horizontal boundaries as such, complicated boundary conditions could be avoided and the errors introduced in the temperature and stress field in the near field were insignificantly small. Such model size was also used by other investigators (Dames and Moore, 1978, Science Applications, Inc., 1978) in studying the near field phenomenon of nuclear waste repositories.

At the time of emplacement, both IF and IW are assumed to be 10 years out of the reactor core and having an initial thermal loading of 269 W per canister. Their rates of radiogenic heat decay with time are calculated according to the equations proposed by Prowse (1978) and are shown on Figure 3 with the power output being normalized by the values of 10 year old waste or fuel. The arrangements of canisters shown on Figure 1 corresponds to initial panel thermal loadings (PTL) of 24 W/m^2 and 14 W/m^2 for IW and IF vaults respectively.

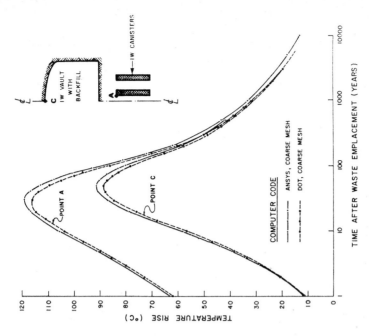

FIG. 4 COMPARISON OF TEMPERATURES FROM DIFFERENT COMPUTER CODES

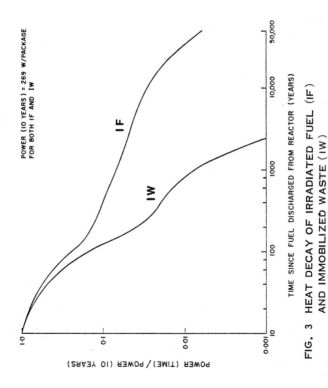

FIG. 3 HEAT DECAY OF IRRADIATED FUEL (IF) AND IMMOBILIZED WASTE (IW)

With the emplacement of waste or fuel canisters, the rock mass is subjected to an elevated temperature field. The temperatures obtained from the thermal transient analysis at a certain time after waste emplacement were used as thermal loads to calculate the thermal stresses. Then the thermal stresses were superimposed on the stresses due to excavation of the vault and the in-situ stresses to obtain the combined or resultant stresses. Thermal transient analysis was carried out up to 50,000 years after emplacement, while stress analysis was performed at frequent intervals up to 200 years after emplacement.

4. COMPARISON AMONG COMPUTER CODES

Thermal transient and thermomechanical stress analysis for the disposal vaults were carried out by employing several finite element codes. The code DOT (Determination of Temperature), developed at the University of California at Berkeley, was used to calculate the time-dependent temperature distributions. DOT was then interfaced with another finite element code, SAPIV (Structural Analysis Program) to determine the thermal stress and deformation. SAPIV was also utilized to compute the stress and deformation due to excavation of the vault. The results were then fed into a post processor XPOST, developed by Ontario Hydro, to obtain the combined stress and deformation distributions. The temperature and stress spatial distributions thus obtained were plotted by a CALCOMP plotter with the aid of the computer graphics package SURFACE II.

For cross-checking purposes, the finite element code ANSYS (Engineering ANalysis SYStem) was also used to calculate the temperature and thermal stress fields. In addition, an Ontario Hydro in-house finite element code EXAT (EXcavation And Thermal Stress Analysis) was also employed to determine the stress distributions in the rock mass around the disposal vault.

A comaprison of typical temperature time-histories obtained from DOT and ANSYS for an IW vault is shown in Figure 4. The maximum difference between the results for the entire analysis period of 50,000 years is about 2°C. This corresponds to only about 2 percent difference in temperature and is thus considered to be insignificant. Figure 5 shows the comparison of thermal stresses calculated by the above three different finite element codes. It can be seen that very close agreement among the codes is obtained, with a maximum difference of 3 MPa approximately or 3 percent between any two codes. Close agreement in the corresponding displacement is also observed.

Unlike ANSYS or DOT-SAPIV, EXAT requires the output temperatures from a separate code to calculate thermal stresse. Due to its proprietary nature, it is relatively difficult to use ANSYS to obtain the combined stresses and displacement. Therefore, the DOT-SAPIV package was chosen as the main computer code in performing the 2-D analyses described in this paper, while ANSYS was used in the study of 3-D effects.

5. COMPARISON OF HORIZONTAL AND VERTIAL BOREHOLE CONCEPTS

Temperature Comparison

The temperature time-histories for the vertical and horizontal borehole concepts are compared in Figure 6. The temperature at the rock/waste interface (Point A) for the horizontal borehole concept is considerably lower than the case of vertical borehole emplacement for the first 100 years. At the end of the first year, the difference in temperature exceeds 20°C. This difference diminishes with time, dropping off to less than 2°C at 100 years. The corresponding temperature at the crown is somewhat higher than the case of vertical borehole emplacement, the maximum difference being on the order of 8°C. However, the temperatures for the two concepts are almost identical after 200 years.

The above difference in temperature can be interpreted as follows. The waste canisters in vertical borehole emplacement are closer to the room centreline than those in horizontal borehole emplacement. Since the room

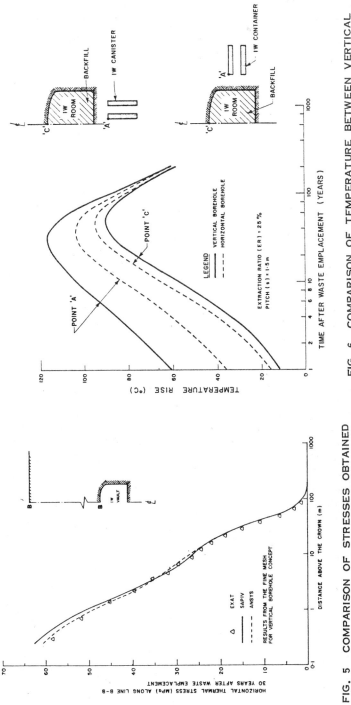

FIG. 6 COMPARISON OF TEMPERATURE BETWEEN VERTICAL AND HORIZONTAL BOREHOLE CONCEPTS

FIG. 5 COMPARISON OF STRESSES OBTAINED FROM DIFFERENT COMPUTER CODES

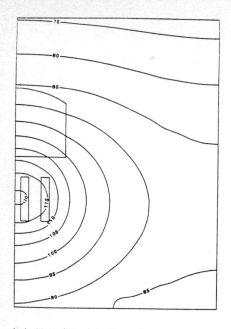

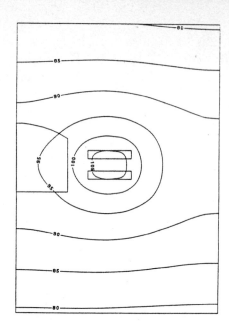

(a) Vertical Borehole Concept   (b) Horizontal Borehole Concept

FIG. 7   CONTOURS OF TEMPERATURE RISE (°C) AROUND AN IW ROOM
30 YEARS AFTER WASTE EMPLACEMENT

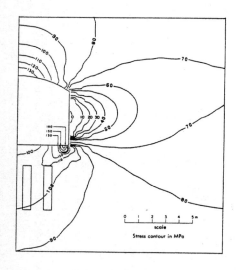

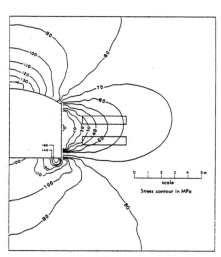

(a) Vertical Borehole Concept   (b) Horizontal Borehole Concept

FIG. 8   COMBINED HORIZONTAL STRESS AROUND AN IW ROOM
30 YEARS AFTER WASTE EMPLACEMENT

centreline is adiabatic due to symmetry, the rock/waste interface therefore registers higher temperatures for the case of vertical borehole emplacement. As for the crown, comparing the proximity of the waste canisters to the crown in both cases, it is obvious that the horizontal boreholes are located closer to the crown. They thus give rise to higher temperature at the crown.

Comparing the spatial variation of temperature at a particular time after emplacement, for example, 30 years for both cases (Figure 7), it is observed that the temperature gradients in the vicinity of the canisters for horizontal borehole emplacement is generally lower than those for vertical borehole emplacement. This is caused by the different thermal loading intensity in the two cases. In the vertical borehole emplacement, the canisters are arranged in a row of four across the room. While in the horizontal borehole emplacement, two canisters are placed in each side of the room, thus yielding a lower thermal loading intensity than the vertical borehole concept. As a result the temperature gradients in the vicinity of the canisters are lower for the case of horizontal borehole concept.

Stress Comparison

Figure 8 shows the comparison of stress distributions around the waste room 30 years after emplacement between the vertical and the horizontal borehole emplacement concepts. Except for some localized minor differences, the stress distributions for the two concepts are generally similar. One area where appreciable difference occurs is in the roof of the waste room. For early period of emplacement, horizontal borehole concept yields a higher horizontal stress in the roof, with a maximum difference of about 10 MPa at around 30 years (Figure 9). At 200 years, the stresses become practically the same for the two concepts.

It should be noted that the above comparison is restricted to the near field stress analysis which did not take into account the effects of drilling boreholes on stress. For the very near field or in the close vicinity of the canister boreholes, the stress distributions for the two concepts may be different due to the different orientations of the boreholes with regard to the in-situ stresses.

6. THERMOMECHANICAL STABILITY OF THE DISPOSAL VAULT

When a rock mass is subjected to a temperature rise, thermal stresses may develop as a result of one or any combination of the following three causes: (a) external constraints to thermal expansion, (b) differential thermal expansion due to temperature gradients, and (c) differential thermal expansion of neighbouring anisotropic grains or constituents of a rock. The thermal stresses due to the first and second causes can be considered as macroscopic stresses. The stress analyses presented in this paper only account for this type of macroscopic thermal stresses. The thermal stresses due to the third cause are microscopic in scale. The microcracks caused by these microscopic thermal stresses can significantly modify the properties of the intact rock. This may in turn affect the development of macroscopic thermal stresses (Johnson et al, 1978).

As mentioned previously, a state of high horizontal in-situ stress has been observed in many of the Paleozoic and Precambrian rocks of Ontario. Tunnels and caverns constructed in these rock types generally experience a concentration of horizontal compressive stresses in the roof and floor, as manifested by field observations. When the thermal stress in the rock mass around a heated underground opening is superimposed onto the in-situ and excavation stresses, the resulting stress may be high enough in some cases to cause compression failure of the crown and invert and may induce tension fractures in the sidewalls, along with some loosening and ravelling. This type of thermally induced rock failure has been oberved in Canada and elsewhere and is generally referred to as thermal spalling.

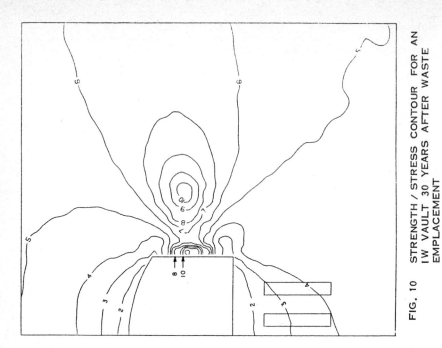

FIG. 10   STRENGTH / STRESS CONTOUR FOR AN
          IW VAULT 30 YEARS AFTER WASTE
          EMPLACEMENT

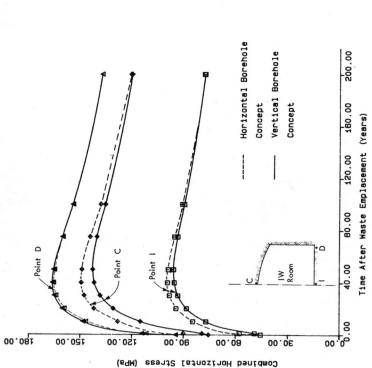

FIG. 9   STRESS TIME — HISTORY FOR SELECTED
         POINTS AROUND AN IW VAULT

Case Histories of Thermomechanical Failure

A review of case histories of thermal spalling in rock (Tsui, 1978) indicated that the number of well-documented case histories is limited. It was known, however, that various degrees of thermal spalling had occurred in some oil storage caverns in Sweden and Finland (Morfeldt, 1974; Johanson and Lahtinen, 1976), as well as in borehole heater experiments for nuclear waste disposal studies (Chan et al, 1980; Witherspoon et al, 1980). A case history of extensive thermal spalling involved a diesel-exhaust passage in an underground electric generation installation built in Precambrian granitic paragneiss at North Bay, Ontario (Gray, 1965). In this case, about 0.6 m of rock spalled from each wall and 1.8 m from the roof, changing the originally rectangular passage of 1.8 m wide and 2.1 m high to a rounded contour approaching an elliptical shape. A typical spall was a somewhat curved plate averaging about 95 mm in thickness and 0.3 m to 0.6 m in diameter. Similar mode of failure was also observed in a smaller test passage (0.8 m by 0.8 m), with the plate-like spalls being about one-third of the size and thickness of those seen in the diesel-exhaust passage. The principal mode of thermomechanical failure or thermal spalling in this case history appeared to be slab buckling and splitting parallel to the direction of maximum compression, as is commonly observed in uniaxial compression tests on brittle rock specimens and in local failure zones around some underground openings. Experience with the Stripa Mine heater experiments also indicated that possible thermal spalling mechanisms may involve dehydra·ion, degradation of rock strength due to large thermal deformation and time-dependent process in addition to stress-induced instability.

Factors Affecting Thermomechanical Stability

From the foregoing review of case histories and discussions on the mechanism of thermomechanical failure, and from the principle of heat transfer, the factors affecting the thermomechanical stability of heated rock caverns can be identified. It can be shown that the susceptibility of a rock cavern to thermomechanical failure depends on the thermal properties (diffusivity or a combination of conductivity, specific heat and density), thermoelastic properties (coefficient of linear thermal expansion, Young's modulus and Poisson's ratio), and mechanical properties and behaviour (strength parameters, stress-strain behaviour and discontinuities) of the host rock, in addition to the characteristics of thermal output of the heat source. Most of these properties and behaviour are temperature dependent and the effects of thermomechanical coupling may thus exist.

Since one of the governing factors in the development of thermal stress is external constraint, the stability of a rock cavern at elevated temperatures will therefore depend on the geometry of the cavern as demonstrated by the case histories. As discussed earlier, the intergranular thermal stresses, though microscopic in scale, may significantly modify the properties of the intact rock and the macroscopic thermal stresses. Therefore, the susceptibility of a rock cavern to thermomechanical failure will also depend on the mineralogical composition and texture of the host rock.

Evaluation of Thermomechanical Stability of the Disposal Vault

The distribution of combined horizontal stress around an IW vault at 30 years after waste emplacement has been shown in Figure 8. This corresponds to the time of occurrence of maximum temperature at the waste/rock interface. The stress distributions indicate that a state of horizontal compression prevails in the roof and floor, along with minor horizontal tension at midheight of the sidewalls. A minor vertical tension zone is also observed in the floor. The maximum tangential compressive stresses around the vault are on the order of 130 MPa in the crown and 160 MPa at the floor corners. Furthermore, the high stress at the corners occurs only in a very small localized zone. It is noted that approximately half of the horizontal compression in the roof and floor results from excavation in a high horizontal stress field, with the other half attributed to thermal loading. The uniaxial

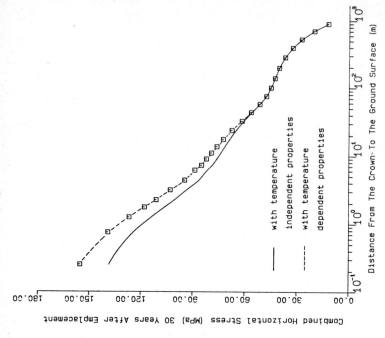

FIG. 12   EFFECTS OF TEMPERATURE DEPENDENT
MATERIAL PROPERTIES ON STRESS
DISTRIBUTION

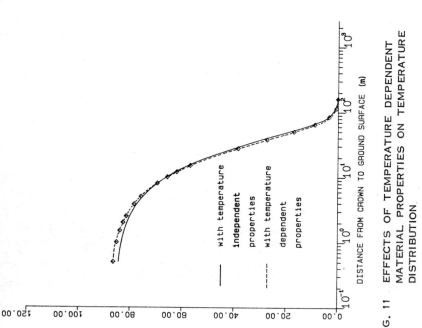

FIG. 11   EFFECTS OF TEMPERATURE DEPENDENT
MATERIAL PROPERTIES ON TEMPERATURE
DISTRIBUTION

compressive strength of granitic rock generally varies between 150 MPa and 200
MPa. On comparing this with the calculated tangential stresses, it is evident
that gross compressive failure of the rock mass around the vault is not
expected. The possibility, however, exists for some localized minor surficial
spalling and fracturing to occur due to the inherent variability of rock
properties. In addition, small zones of radial tensile stresses are
anticipated in the side walls and floor. This may result in some very
localized tensile fracture zones.

The above stability evaluation does not take into account the influence of
confining stresses. As such it can only apply to a thin layer of rock mass
around the vault opening where radial stresses are too small to be of any
significant influence. To include the influence of confining pressure, the
empirical strength criterion for granite and the procedures of stability
evaluation as proposed by Hoek and Brown (1980) are employed to further
examine the thermomechanical stability of the disposal vault. Based on a
typical uniaxial compressive strength of 170 MPa for granite, the contours of
strength/stress ratios at 30 years after waste emplacement are shown in Figure
10. Since the ratio is greater than unity anywhere in the rock mass around
the vault, no over-stressed zones or potential failure zones are expected.
The above evaluation was based on a linear elastic analysis. As such it did
not take into account progressive failure or post-peak stress-strain behaviour
of the rock mass. However, it provides a quick and inexpensive method of
evaluating the overall stability of a proposed excavation, especially in the
conceptual design stage.

7. EFFECTS OF TEMPERATURE DEPENDENCE

The results presented above are based on a 2-D linear analysis and the
assumption that the material properties are temperature independent. It has
been known, however, that most of the thermal and mechanical properties of
rock are temperature dependent, even in the temperature range expected in the
near field of the disposal vault. To study the effects of this temperature
dependence, the material properties as a function of temperature shown in
Table II were included in the thermal transient and stress analyses. The
values indicated in Table II are not site specific. They are largely
extracted from available literatures on granitic rocks and represent a general
trend in the variation of properties with temperature. Some preliminary
results are shown in Figures 11 and 12.

Table II

Temperature Dependence of
Properties for Granite

Temperature (°C)	Thermal Conductivity (W/m°C)	Specific Heat (J/kg°C)	Linear Coeff of Thermal Expansion (10^{-6}/°C)	Young's Modulus (GPa)	Poisson's Ratio --
15	3.00	800	8.0	40	0.20
25	2.86	814	9.1	39	0.20
50	2.74	827	10.2	38	0.19
100	2.51	855	12.4	34	0.17
150	2.31	881	14.6	30	0.13
200	2.14	910	16.8	25	0.10

Notes: 1. Density of granite is assumed to be independent of
          temperature.
       2. Properties of backfill are assumed to temperature independent.

Figure 11 compares the temperatures obtained from an analysis with
temperature dependent material properties to that with temperature independent
properties. It can be seen that close agreement exists between the two

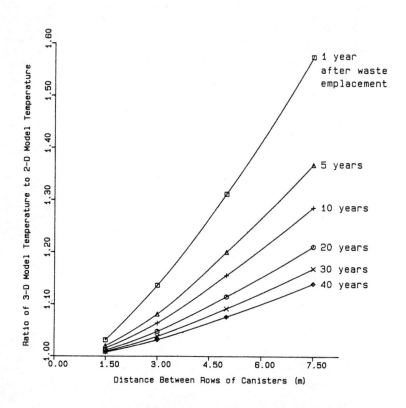

FIG. 13 3—D EFFECTS ON TEMPERATURE AS A
FUNCTION OF CANISTER SPACING AT
ROCK WASTE INTERFACE (IW VAULT)

analyses. The maximum difference is on the order of only 2°C at the crown of the waste room after 30 years of emplacement. Beyond a few metres from the room surface, no appreciable difference occurs. Thermal analysis assuming temperature independent material properties is thus believed to be sufficient for the conceptual design study of a disposal vault.

The corresponding comparison of stress as shown in Figure 12 indicates that the difference in stress is relatively more significant than that in temperature. With the temperature dependence of rock properties taken into account, the tangential compressive stresses around the waste room are found to be generally 10 percent to 20 percent higher than those with temperature independent properties. The diffrence in stress decreases with increasing distance from the room surface, becoming insignificant at about a few tens of meters from the room surface. It thus appears that temperature dependence of rock properties has a significant effect on the stress distribution in the near field of a disposal vault.

8. EFFECTS OF THREE DIMENSIONALITY

For the two-dimensional analysis described in this paper, each row of canisters along the length of the waste room is represented by a long slab as illustrated in Figure 1. In reality the thermomechanical response of the disposal vault is a 3-D problem. It has been shown that 2-D analysis could significantly underestimate the temperatures and thus the thermal stresses in the close vicinity of the waste canisters for the U.S. waste repositoties (Science Applications Inc., 1978; Davis, 1979). To investigate such effects for the Canadian disposal vault a preliminary 3-D thermal analysis has been carried out. The temperatures for a point at the rock/waste interface obtained from 3-D and 2-D analyses are compared in Figure 13. It is found that for pitches (longitudinal distance between adjacent rows of canisters, see Figure 1) smaller than 1.5 m, practically no difference is found between 3-D and 2-D analyses. With increasing pitch values, the difference between 3-D and 2-D models increases. This difference is also found to be decreasing with time after waste emplacement. For example, for a pitch value of 5 m and at one year after emplacement, the 3-D temperature at the rock/waste interface is about 30 percent higher than the corresponding 2-D temperature. At 40 years after emplacement, the 3-D temperature is only a few pecent higher. Since the reference conceptual design for IW vault calls for a pitch of 1.5 m, a 2-D analysis is believed to be adequate. For the IF vault which calls for a pitch distance of 2.5 m, the 2-D analysis is expected to underestimate the temperatures in the vicinity of the fuel canisters by less than 10 percent, according to the results shown in Figure 13.

9. SUMMARY AND CONCLUSIONS

(a) Several finite element computer codes: DOT, SAPIV, ANSYS and EXAT were used in the near field thermal transient and stress analyses. Comparisons of the results indicate that very close agreement in temperature, stress and displacement exists among the codes.

(b) Two emplacement concepts: vertical and horizontal boreholes for an IW vault were analysed and the resulting temperatures and stresses in the near field were compared. In the vicinity of the canisters, horizontal emplacement concept generally yields lower values in temperature and temperature gradients. The stresses obtained from the two concepts are comparable with no significant difference. After 200 years of emplacement, both the temperature and stresses in the near field become practically the same for the two concepts.

(c) An evaluation of the thermomechanical stability of the disposal vault indicates that no gross over-stressed or failure zones are expected. However, some potential minor surficial zones of thermal spalling and fracturing may exist.

(d) No significant difference in temperature is found between the thermal analyses with and without temperature dependence of thermal properties for rock. However, stress analysis with the temperature dependence of rock properties taken into account yields relatively significant higher compressive stresses around the waste room.

(e) Preliminary analysis indicated that in the vicinity of the canisters, temperatures from 3-D model is higher than 2-D results. The difference in temperature increases with the pitch distance but decreases with time. For pitch distance of 1.5 m, no significant difference is found between 3-D and 2-D results.

10. ACKNOWLEDGEMENT

The authors gratefully acknowledge Ontario Hydro and the Atomic Energy of Canada Limited for permission to publish this work. The work forms part of Ontario Hydro's technical assistance program for AECL. The program was co-ordinated by Dr. R.C. Oberth of Ontario Hydro's Nuclear Materials Management Department.

The authors wish to acknowledge Dr. T. Chan of the Lawrence Berkeley Laboratory for making the computer code DOT and SAPIV available to Ontario Hydro. Special thanks are due to Mr. N.L. Harris of the Geotechnical Engineering Department for performing the 3-D analysis described in the paper. Thanks are also extended to Messrs. M. Thomas, E. Lau and A. Lau for drafting the illustrations.

11. REFERENCES

1. Burgess, A.S. 1980, "Irradiated Fuel Vault: Room-and-pillar Thermal Rock Mechanics Analyses", AECL TR-55.

2. Chan, T., Littlestone, N., and Wan, O., 1980, "Thermomechanical Modelling and Data Analysis for Heating Experiments at Stripa, Sweden", Proc 21 st U.S. Symp on Rock Mech, Rolla, Missouri, pp 16 - 25.

3. Coates, D.F., and Grant, F., 1966, "Stress Measurement at Elliot Lake", Canadian Inst Min Metal Bul, Vol 59, pp 603-613.

4. Dames & Moore, 1978, "Technical Support for GEIS: Radioactive Waste Isolation in Geological Formations, Vol 20 - Thermomechanical Stress Analysis and Development of Thermal Loading Guidelines", U.S. Department of Commerce, NTIS Y/OWI/TM-36/20.

5. Davis, B.W., 1979, "Limitations to the Use of Two - Dimensional Thermal Modelling of a Nuclear Waste Repository", Lawrence Livermore Laboratory, UCID-18101.

6. Gray, W.M., 1965, "Surface Spalling by Thermal Stresses in Rocks", Proc Canadian Rock Mech Symp, Toronto, pp 85-106.

7. Haimson, B.C. 1978, "Hydrofracturing Stress Measurements, Hole UN-1, Darlington Generating Station", Ontario Hydro Design and Development Division Report No. 78250.

8. Herget, G., 1974, "Ground Stress Determination in Canada", Rock Mechanics, Vol 6, pp 53-64.

9. Hoek, E., and Brown, E.T., 1980, "Empirical Strength Criterion for Rock Masses", Journal of the Geotechnical Engineering Division, Proc ASCE, Vol 106, No. GT9, pp 1013-1035.

10. Johanson, S., and Lahtinen, R., 1976, "Oil Storage in Rock Caverns in Finland", Tunnelling 76, London, pp 41-58.

11. Johnson, B. Gangi, A.F., and Handlin, J., 1978, "Thermal Cracking of Rock Subjected to Slow, Uniform Temperature Changes", Proc 19th U.S. Symp Rock Mech, Stateline, Nevada.

12. Lee, C.F., 1981, "In-situ Stress Measurements in Southern Ontario", Proc 22nd U.S. Symp Rock Mech, Cambridge, Massachusetts, pp 435-442.

13. Mahtab, M.A., and Wiles, T., 1980, "Immobilized Waste Vault: Room-and-Pillar Thermal Rock Mechanics Analyses", AECL TR-54.

14. Morfeldt, C.O., 1974, "Storage of Oil and Gas in Unlined Caverns", SPE-European Spring Meeting of the Society of Petroleum Engineers of ASME, Amsterdam, the Netherlands, SPE4849.

15. Palmer, J.H.L., and Lo, K.Y., 1976, "In-situ Stress Measurements in Some Near-Surface Rock Formations - Thorold, Ontario" Can Geotechnical Journal, Vol 13, pp 1-7.

16. Prowse, D.R., 1978, "Initial Specifications for Design Study of Full Scale High Level Waste Repository", AECL Memorandum.

17. Science Applications Inc., 1978, "Technical Support for GEIS: Radioactive Waste Isolation is Geologic Formations, Vol 19 - Thermal Analyses", U.S. Department of Commerce, NTIS Y/OWI/TM-36/19.

18. Tsui, K.K. 1978, "A Study on the Thermomechanical Stability of Rock Caverns", Ontario Hydro Design and Development Division Report No. 78239.

19. Witherspoon, P.A., Cook, N.G.W., and Gale, J.E., 1980, "Progress with Field Investigations at Stripa", LBL-10559, SAC-27.

DISCUSSION

W.R. FISCHLE, Federal Republic of Germany

The fitures show a high stress on the corners of the drift.

Did you think about using other configurations ?

K.K. TSUI, Canada

The geometry of the room as shown in Figure 1 assumes a 90° angle for the corners. This is for the convenience of constructing the finite element model. In practice upon excavation of the room, these corners will probably be somewhat curved. Thus the high stress concentrations shown in the paper represent a somewhat of an overestimation of stress and can be considered on the conservative side.

K.R. SHULTZ, Canada

The diagrams in the paper imply a single room. What is the effect, particularly for horizontal emplacement, of a planar array of parallel rooms ?

K.K. TSUI, Canada

The diagram in the paper (Figure 2) shows the model used in the analysis. It is bounded by two vertical planes of symmetry, one at room centreline and one at pillar centreline. As such, it only shows one half of the room. But is implies that the disposal vault under consideration consists of a series of parallel drifts or rooms.

D.K. PARRISH, United States

Are the rooms modelled as having backfill in them ?

If so, it may explain why the horizontal stress components at the room wall are not zero (Figure 8).

If not, then I do not understand why these models are compared with Gray's (1965) study of rooms which are not backfilled.

K.K. TSUI, Canada

Yes, the IW rooms are modelled by assuming that the room is backfilled immediately after waste emplacement. The non-zero horizontal stress at parts of the surfaces of sidewalls is partly due to the averaging process in drawing the stress contours.

It is not the intention of the authors to compare directly the case history (Gray, 1965) with the disposal vault described in the paper. This case history is presented with the aim of illustrating one of the possible mechanisms of thermomechanical failure.

The maximum temperature of the exhaust gas in the case history described in our paper is 315°C. However, the temperature at which spalling occurs is believed to be about 93°C based on a back calculation by the author who originally presented the case history (Gray, 1965).

IMPORTANCE OF CREEP FAILURE OF HARD ROCK
IN THE NEAR FIELD OF A NUCLEAR WASTE REPOSITORY

James D. Blacic
Los Alamos National Laboratory
Los Alamos, New Mexico 87545, USA

ABSTRACT

Potential damage resulting from slow creep deformation intuitively seems unlikely for a high-level nuclear waste repository excavated in hard rock. However, recent experimental and modeling results indicate that the processes of time-dependent microcracking and water-induced stress corrosion can lead to significant reductions in strength and alteration of other key rock properties in the near-field region of a repository. We review the small data base supporting these conclusions and stress the need for an extensive laboratory program to obtain the new data that will be required for design of a repository.

1. INTRODUCTION

Design of a nuclear waste repository involves unique engineering and scientific challenges. In the long history of underground construction, assuring integrity of deep workings in a hot, wet environment for times so extensive that they approach the geologic has never been faced. The added societal constraint that man must be protected by an extremely high confidence in design success makes the challenge awesome.

A key aspect of the problem is that mine openings must be maintained for a minimum time of 100 years to allow retrieval of waste and monitoring of repository performance. The question then arises; what is the potential for creep failure of the host rock and how can allowance for rock response be incorporated in design?

Here we assess likelihood of brittle creep failure of the most prominently studied hard rock repository media: granite, basalt, and tuff. We show that, although data available are pitifully small, creep failure is likely in all hard rock media based on current preconceptual design parameters. This means that much expanded investigation of creep properties of hard rock media must begin so that early consideration can be given on how to control time-dependent rock deformation.

2. THE EVIDENCE

First, we establish a time frame for our considerations. Assume that the average low confining pressure failure strain (linear) of hard rock is approximately 0.005. The exact value is of small importance; a factor of 2 or 3 in either direction makes little difference to our conclusions. Current broad design constraints require controlled access to stored waste for a time period of 100 years. The strain rate of most immediate interest is then calculated simply as $0.005 \div 100$ years, or 1.6×10^{-12} per second.

Immediately our first problem arises: there is no constitutive relation for any crustal hard rock that allows extrapolation of mechanical response to 10^{-12} s^{-1} strain rates under the environmental conditions of a nuclear waste repository. The minimum magnitude of extrapolation in strain rate from the meager laboratory data available is five orders of magnitude! Clearly, it will not be easy to estimate potential rock failure at these extended times. However, we will show that there are sufficient published data to indicate the magnitude of a creep failure problem.

Next we consider mechanisms of creep deformation under temperature and pressure conditions expected near a waste repository. We recognize three fundamental mechanisms: (1) time-dependent tensile microcracking, (2) slow continuous or episodic frictional sliding on discontinuities, and (3) ductile flow of weak or thermodynamically unstable minerals. Water is an extremely important catalytic agent for all three mechanisms.

There is extensive evidence that the mechanism of compressive shear failure of hard rock in the brittle regime derives from cumulative tensile microcracking accompanied by dilatant volume strains [1] [2]. Most of this evidence originates from short term uniaxial or triaxial compression tests. However, Figure 1a [3] indicates that the same mechanism is responsible for time-dependent or creep failure of rock.

Figure 1a shows results of a uniaxial creep experiment on Westerly granite at room temperature. Axial (ε_z), radial (ε_r), and volume strains ($\Delta V/V_0$) are plotted along with acoustic emission energy (ΣE) as a function of time. Strain-time curves show the classic creep response. An initial primary creep region characterized by a decreasing rate of strain, followed by a long secondary or steady-state region in which strain rate is essentially constant, leading to a tertiary or accelerating creep phase in which strain rate increases rapidly to ultimate failure of the test sample. Note that these curves are mimicked by the microcrack acoustic emission curve. Also note that after an initial elastic compression (negative) the volumetric strain almost immediately becomes dilatant.

This evidence, together with the acoustic emissions, indicate that the mechanism of deformation must be time-dependent <u>tensile</u> microcracking, in as much as opening of tensile cracks is the only way sample volume can increase beyond its starting, uncompressed value. This observation is important because it means we can use much of the vast literature on slow tensile crack growth in glasses and ceramics to evaluate creep response of rock.

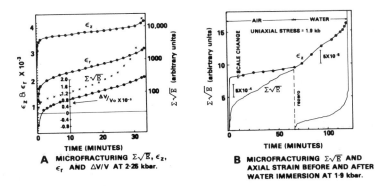

Figure 1. Creep of Westerly granite at room temperature [3].
A. Uniaxial creep at 2.25 kbar: axial strain (ϵ_z), radial strain (ϵ_r), volume strain ($\Delta V/V_0$), and acoustic emission energy (Σ E).
B. Effect of water stress corrosion. Uniaxial creep in air at 1.9 kbar followed by immersion in water after steady state axial strain is attained.

Figure 1b [3] illustrates the corrosive effect of water on rock. The first part of the creep curve in Figure 1b shows axial strain and acoustic emission into the steady-state creep regime for a sample compressed in air of normal humidity. When the sample is then immersed in water, strain rate increases and acoustic emission first drops, then increases dramatically to tertiary creep failure. Wu and Thomsen [3] find that at 25°C creep failure time of granite is reduced by 2 orders of magnitude by immersion in water. They also find that drying by heating their unconfined samples above 100°C increases failure times.

Stress corrosion action of water has been known and studied for some time for glass and ceramics [4] [5] [6]. For example, Figure 2 shows a 4 order of magnitude reduction in the static fatigue life (failure time) of aluminum oxide due to presence of air of normal room humidity compared to dry air -- a dramatic stress corrosion effect [7]. The mechanism of water stress corrosion [8] is essentially the same as that proposed to explain hydrolytic weakening of quartz in the plastic flow regime [9] [10]. Namely, local hydrolyzation of strong silicon-oxygen bonds and their replacement with weak, hydrogen-bonded bridges. This is a highly temperature-dependent and diffusion-limited process which explains why it is particularly active over extended times.

We see implications of these creep deformation mechanisms in the next two figures. Figure 3 shows an extrapolation of data summarized by Kranz [11] for creep of granite at room temperature. These results build on much earlier work of Scholz [12], Cruden [13], Rummel [14], and others. The extrapolations indicate at least a 30 to 50% reduction in compressive strength of granite over the 100-year time range critical to a repository. Note that the full weakening effects of water and temperature are not contained in these data. Figure 4 shows the problem from a slightly different point of view. Data shown in Figure 4 are from subcritical tensile crack growth tests of Halleck and others [15]. These tests measure tensile crack velocity vs mode I stress intensity factor at room temperature in granite. In the figure we recast the data in terms of failure time for 1-m blocks, i.e., the time for a crack to propagate 1 m, vs tensile strength relative to that observed in rapid direct pull or Brazilian tests. Solid parts of the curves represent data. Extrapolation to low crack rates is problematical. Strength must lie between a simple linear extrapolation of the data and a limiting value expected from similar

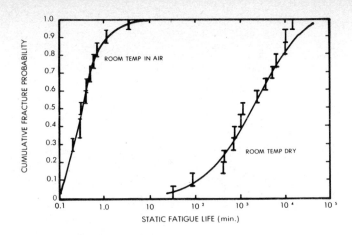

Figure 2. Static fatigue and stress corrosion of Al_2O_3 [7]. Static fatigue life as a function of probability of fracture for room temperature creep in dry air compared to air of normal room humidity.

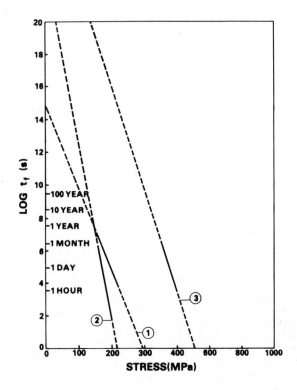

Figure 3. Room temperature creep failure stress of granite [11]. (1) Water saturated Westerly granite, 0.1 MPa confining pressure (data from Wawersik [19]). (2) Room humidity Barre granite, 0.1 MPa confining pressure. (3) Room humidity Barre granite, 53 MPa confining pressure.

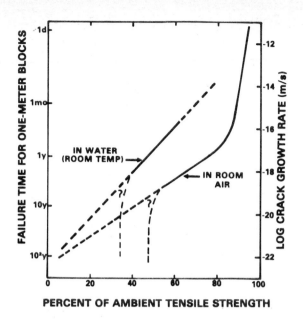

Figure 4. Subcritical crack growth of Berkeley granite at room temperature [15]. Failure time of one-meter blocks refers to the time required to extend a crack one meter based on the crack velocities measured in 3-point bend tests. Percent of ambient tensile strength is calculated from mode I stress intensities from the 3-point bend tests compared to the tensile strength measured in rapid direct pull or Brazilian tests.

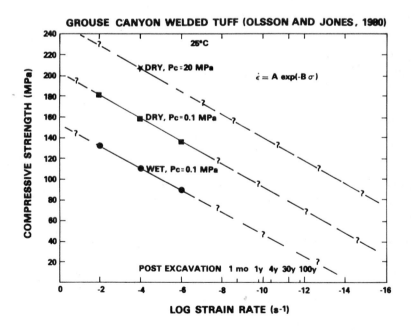

Figure 5. Compressive strength of welded tuff from the Nevada Test Site as a function of strain rate [17]. P_c is confining pressure; all tests at room temperature.

studies on glass and ceramics. Position of any limit for granite is unknown at present. In any case, these results indicate a very substantial reduction of tensile strength of granite within 100 years. Similarly, Waza and others [16] show the velocity of tensile cracks in water-saturated basalt to be 2 to 3 orders of magnitude greater than in room-dry samples.

Figure 5 shows preliminary data for compressive strength of Nevada welded tuffs as a function of log strain rate at room temperature [17]. Again note the marked weakening effect of water. We know that the simple exponential extrapolation shown is not correct, especially in view of the very limited experimental data. However, we can nevertheless use these data to estimate the impact of time-dependent rock weakening on a repository design.

Figure 6a shows results of a preconceptual thermomechanical model of a repository in welded tuff [18]. The model predicts thermoelastic stresses around a repository room at 800-m depth with a spent fuel waste canister embedded in the floor of the room. Heat load is 100 kw/acre and ground water boiling is not allowed. Calculated stresses are compared to a Coulomb-type shear failure criterion based on data similar to that shown in Figure 5. The results are represented

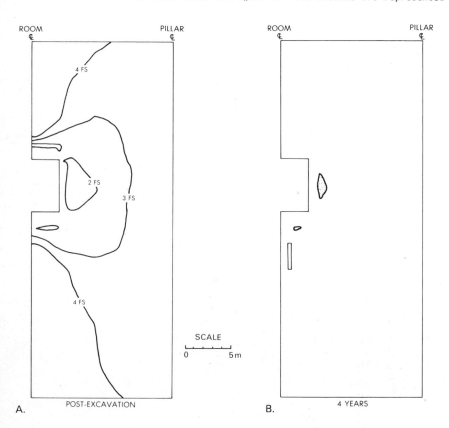

Figure 6. Estimates of creep failure around a high level nuclear waste repository in welded tuff. Failed regions are based on a comparison of extrapolated failure stresses from Figure 5 compared to thermoelastic stresses calculated for a preconceptual thermomechanical model of a repository in welded tuff [18]. Details are given in the text. A. Factors of safety (FS) immediately after excavation but before implacement of a simulated waste container in the floor of the model room. B. Interpreted creep failure zone (stippled) 4 years after implacement of waste. C. Creep failure zone after 30 years. D. Creep failure zone after 100 years.

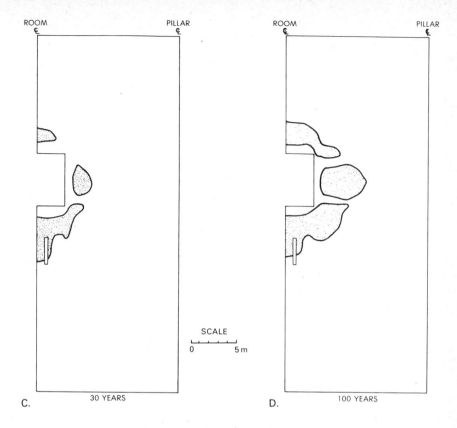

C. 30 YEARS

D. 100 YEARS

in Figure 6a as contours of factors of safety (FS) immediately after excavation but before implacement of the waste heat source. For example, a factor of safety of two indicates calculated stresses are half of those required for shear failure.

However, the failure criterion is based on laboratory tests at strain rates of 10^{-2} to 10^{-6} s^{-1}. If we compare stresses calculated at later times in the thermal history of this model and plot factors of safety, we can qualitatively evaluate possible time-dependent shear failure around the repository. For example, a factor of safety of two will outline a failure zone if the strength, based on extrapolations shown in Figure 5, has fallen by 50% at that time. Figures 6b through 6d show failure zones based on such extrapolations for the dry, 20 MPa confining pressure strength shown in Figure 5. Tuff strengths actually used in the model to calculate factors of safety are about a factor of four lower than those we use to estimate time-dependent failure, so if anything, we believe our approach is conservative. At 4 years (Figure 6b) we see a small failure zone in the rib of the model repository room. However, this failure zone grows dramatically at later times so that by 100 years the failed region completely encompasses the near field of the room and would likely prevent maintenance of access. Note that weakening effects of temperature and water are not included in this scenario.

Although these exercises are preliminary, we believe they illustrate the need to carefully evaluate effects of time on strength of waste repository host rocks.

3.   CONCLUSIONS

Contrary perhaps to one's intuition, creep deformation of hard rock <u>does</u> occur at nuclear waste repository conditions and <u>does</u> pose design problems. Other coupled phenomena that we do not discuss here make this conclusion even stronger.

For example, dilatant microcracking can cause a decrease in thermal conductivity leading to higher rock temperatures than those predicted by current models.

Clearly, we need an extensive laboratory investigation of the creep deformation of hard rock waste isolation media at repository conditions of temperature, stress, and water pore pressure. Primary deformation and secondary coupled effects on thermal conductivity, permeability, etc. require careful study. Constitutive deformation equations derived from laboratory experiments must be based on physical mechanisms identified in experiments and verified in natural deformations. Only in this way can we be confident in the long time extrapolations required in the ultimate design and performance models. By their very nature, these experiments are time consuming; therefore this work should be carried out immediately.

4. BIBLIOGRAPHIC REFERENCES

(1) Brace, W.F., B. Paulding, and C. Scholz : "Dilatancy in the Fracture of Crystalline Rocks," Pure Appl. Geophys., 11b, 807-839 (1966).

(2) Scholz, C.H. : "Mechanism of Creep in Brittle Rock," Jour. Geophys. Res. 73, 3295-3302 (1968).

(3) Wu, F.T. and L. Thomsen : "Microfracturing and Deformation of Westerly Granite Under Creep Conditions," Int. J. Rock Mech. Mining Sci. 12, 167-173 (1975).

(4) Widerhorn, S.M. and L.H. Bolz : "Stress Corrosion and Static Fatigue of Glass," J. Amer. Ceramic Soc. 53, 543-548 (1975).

(5) Wachtman, J.B. : "Highlights of Progress in the Science of Fracture of Ceramics and Glass," J. Amer. Ceramic Soc. 57, 509-519 (1974).

(6) Westwood, A.R.C. : "Control and Application of Environment-Sensitive Fracture Processes: Tewksbury Lecture," J. Mat. Sci. 9, 1871-1895 (1974).

(7) Chen, C.P. and W.J. Knapp : "Fatigue Fracture of an Alumina Ceramic at Several Temperatures," in Fracture Mechanics of Ceramics, ed. by R.C. Bradt, D.P.H. Hasselman, and F.F. Lange, Vol. 2, pp. 691-707 (1973).

(8) Hillig, W.B. and R.J. Charles : "Surfaces, Stress-Dependent Surface Reactions and Strength, in High Strength Materials," V. Zackay Ed., (John Wiley, New York, 1965) pp. 682-705.

(9) Griggs, D.T. and J.D. Blacic : "Quartz: Anomalous Weakness of Synthetic Crystals," Science 147, 292-295 (1965).

(10) Griggs, D.T. : "Hydrolytic Weakening of Quartz and Other Silicates," J. Roy. Astron. Soc. 14, 19-31 (1967).

(11) Kranz, R.L. : "The Effects of Confining Pressure and Stress Difference on Static Fatigue of Granite," Jour. Geophys. Res. 85, 1854-1866 (1980).

(12) Scholz, C.H. : "Mechanism of Creep in Brittle Rock," J. Geophys. Res. 73, 3295-3302 (1968).

(13) Cruden, D.M. : "A Theory of Brittle Creep in Rock Under Uniaxial Compression," J. Geophys. Res. 75, 3431-3442 (1970).

(14) Rummel, F. : "Studies of Time Dependent Deformation of Some Granite and Eclogite Rock Samples Under Uniaxial Constant Compressive Stress and Temperatures Up to 400°C," Z. Geofiz. 35, 17-42 (1969).

(15) Halleck, P.H. and A.J. Kumnick : "Subcritical Crack Growth in Berkeley Granite: Environmental and Orientation Effects," in preparation 1980.

(16) Waza, T., K. Kurita, and H. Mizutani : "The Effect of Water on the Sub-critical Crack Growth in Silicate Rocks," Tectonophysics 67, 25-34 (1980).

(17) Olsson, W.A. and Jones, A.K. : "Rock Mechanics Properties of Volcanic Tuffs from the Nevada Test Site," Sandia National Laboratories Report #SAND80-1453 (1980).

(18) Waldman, H. and Osnes, J.D. : "Near-Field Thermomechanical Analysis of a Spent Fuel Repository in Tuff," Sandia National Laboratory Report #SAND81-7151, (1981 in press).

(19) Wawersik, W.R. : "Time Dependent Rock Behavior in Uniaxial Compression," Proc. 14th Symp. Rock Mech., pp. 85-106, 1972.

DISCUSSION

A.T. JAKUBICK, Canada

Are you, in other words, suggesting that there will be a fractured zone around the repository after a certain time ? This is so, then what extent do you expect for the fractured zone ?

J.D. BLACIC, United States

What I am suggesting is that such a zone is possible, whether or not it develops will depend on the particular case under consideration. Factors such as thermal loading, *in situ* stress state, local rock strength, and others must all be considered on a case-by-case basis. The point is that time dependent effects should be included in the overall analysis.

D.K. PARRISH, United States

Linear extrapolations such as the ones shown in Figures 3 and 5 predict that the compressive strength of granite or tuff is zero at low (but finite) strain rates. How do you resolve this extrapolation with the fact that engineered structures and monuments composed of granite have existed for very long times under low stresses and low strain rates ? Your extrapolation appears to predict that these structures should have crumbled.

J.D. BLACIC, United States

It is likely that the exponential extrapolation (linear on a semilog plat) that I used would break down at very low stress. Since we do not know what this limiting stress might be I have not tried to take it into account. Nevertheless, I know of no engineered granite structure older than 5000 years. This would correspond to an average strain rate of about 3×10^{-14} s^{-1} for a failure strain of 0.005. Based on extrapolation of the data I showed for granite this would require a constant stress of about 100 MPa. I think stresses that high are unlikely in any structure constructed by man that might be as old as 5000 years.

T.J. CARMICHAEL, Canada

In terms of confirming the data in your paper, should there not be data of the in-situ Climax test, wherein a granite host rock surrounding a repository containing fuel bundles has been in place for more than 1 year and significant strains should have been measured in the repository walls if your data and assumptions are correct !

Would someone from Lawrence Lab. please comment on their data from the Climax test.

J.D. BLACIC, United States

In order to judge whether such effects should be expected at the Climax site within 1 year, one would need to know the creep properties of Climax granite under the conditions of temperature, stress, mixture content, etc., of the field test, as far as I know, such data does not exist. Some estimates might be made but I doubt that they would be conclusive.

H.S. RADHAKRISHNA, Canada

Time dependent effects, namely reduction in strength and deformation modulus, are dependent on the stress levels and stress ratio. Also some redistribution of stresses may take place due to the time dependent effects. Should not these be taken into consideration in estimating the critical factors of safety in the nearfield region ?

J.D. BLACIC, United States

Yes. Any detailed analysis of time dependent effects should take into account stress redistribution among many other factors. However, up until now, even the crudest estimates of time effects have not been done due to a lack of revelant data and sensitivity to the problem.

C. McCOMBIE, Switzerland

In view of the great scarcity of time dependent data which you mention, would it be worthwhile to study existing older hard rock structures and mined cavities or tunnels to attempt to set a lower bound on the expected reduction of strength with time ?

J.D. BLACIC, United States

Yes, it would be worthwhile to make such comparisons if the proper data could be found. Long term (years) measurements of displacement in a structure for which the stress field was also known would be required to make a meaningful lower bound estimate.

N.A. CHAPMAN, United Kingdom

Your argument on the significance of time dependent processes hinges on the present requirement for retrievability. Some concepts do not require the repository to remain open. What do you feel is the significance of the processes you describe on the physical and hydrologic properties of the rock in the case where immediate backfilling takes place ?

J.D. BLACIC, United States

Even if retrievability is not an issue I believe there are possible significant effects on release rate, transport, thermal field development, and even constructability in some cases as a result of time dependent deformation and its coupled effects. These would, of course, have to be treated on a case-by-case basis. My point is that the data needed to do this in a detailed fashion do not presently exist for hard rock media.

F. GERA, Italy

Would it be possible to estimate how far the fractures would extend ?

J.D. BLACIC, United States

I think it would be possible in specific cases, but I do not know how to give an estimate as a generalization.

F. GERA, Italy

You have mentioned a second effect of microcracking, that is the decrease of thermal conductivity. I do not think this would apply to a backfilled repositories since it would be water saturated and any microcracks would be full of water.

J.D. BLACIC, United States

I believe that thermal conductivity might be significantly altered due to deformation-induced porosity under saturated conditions, but this needs to be measured experimentally.

IMPORTANCE OF NEAR-FIELD ANALYSIS IN THE OVERALL SAFETY ASSESSMENT OF THE DISPOSAL OF SPENT FUEL IN THE PRECAMBRIAN BEDROCK

E.K. Peltonen and K.A. Heiskanen
Technical Research Centre of Finland
Nuclear Engineering Laboratory
Helsinki, Finland

ABSTRACT

This paper presents preliminary results of a study aiming at clarifying the role of the near-field phenomena in the overall safety of the final disposal of spent-fuel. The principle of the methodology employed was the following: By using certain acceptable limits for radiological impact and different scenarios for migration and pathways acceptable release rates of the radionuclides from a repository into the groundwater in the surrounding rock can be evaluated. At the same time the importance of the different factors affecting the radiological impact can be analyzed.

Scenarios of two types were analyzed. In scenarios with a pathway via the Baltic Sea and with favourable hydrogeological conditions the role of near-field phenomena is of minor importance and the requirements for a repository system are modest. In case of very unfavourable hydrogeological conditions the near-field is important, the repository system should provide release duration of 10^5 years at least. In scenarios having a pathway via a well the role of the near-field phenomena is essential and the release duration should be at least 10^6 years, unless very favourable geosphere conditions exist. The possibilities to meet the requirements for a repository system seem to be rather good.

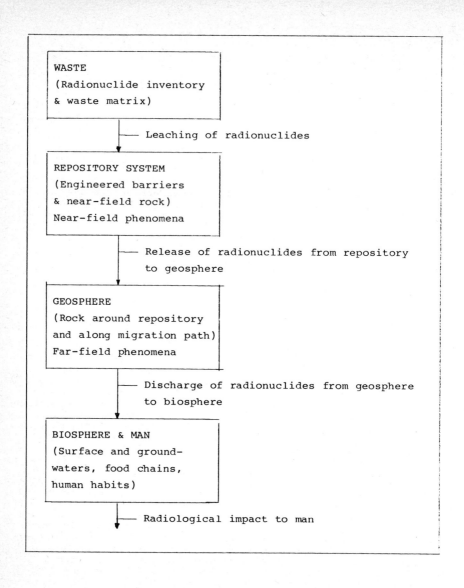

Figure 1. A schematic model of the main components affecting the
safety of nuclear waste disposal

1. INTRODUCTION

A study aiming at clarifying the role of the near-field phenomena in the overall safety of the final disposal of long-lived and high-level waste (Fig. 1) is currently being carried out. The important items to be assessed include e.g.: What functional capabilities are required from the near-field repository system in case of different conditions of far-field geosphere and biosphere? Which factors of the engineered barriers and the surrounding rock in the near-field are decisive for slowing down the release of the radionuclides? Can the repository system meet the possible performance requirements that are set in different cases? This paper deals mainly with the first question, since the analyses carried out earlier must be recomputed due to the changes introduced in the dose conversion factors [1]. The effect of the change specially in case of ^{237}Np and ^{231}Pa can be essential

The disposal concept being analyzed in this study consists of a repository located in hard Precambrian rock at a depth of about 500 metres. The repository consists of disposal tunnels with vertical disposal holes in the floor. The structure of the canister and the possible backfill were not fixed. The waste was assumed to be in the form of spent fuel (UO_2), the amount being 1200 tU.

2. METHODOLOGY

2.1 Principle

The principle of the methodology used in analyzing the role of the near-field phenomena and furthermore in estimating acceptable release rates is illustrated in Fig. 2. With certain limits for the radiological impact to man a preliminary generic screening of the acceptable disposal systems can be accomplished, keeping in mind the uncertainty of the models and the data used.

The criteria and limits for acceptable radiological impact are an issue of continuous reconsideration and development. For this study only one quantity was chosen to represent radiological impact: the effective dose equivalent rate $\dot{H}_E$(Sv/a). Two alternatives for the acceptable value of this quantity were employed.

One limit was determined as

$$\dot{H}_E^{(1)} = H_{50} \tag{2-1}$$

H_{50} is committed effective dose equivalent for 50 years due to a single intake.

In case of continuous exposure the formula (2-1) can be utilized. The value of H_{50} was chosen to be 10^{-4} Sv/a (10^{-2} rem/a), consistent with the regulation applied to reactor operation in Finland.

The other limit was determined by formula

$$\dot{H}_E^{(2)} = \frac{S_{500}^C \cdot E}{T_C \cdot N} \tag{2-2}$$

S_{500}^C is the limit of accepted collective dose equivalent commitment during 500 years (manSv/MW·a)
E is installed nuclear power capacity x operation time (MW·a)
N is population exposed to radiation
T_C is time period of exposure, 500 a

The value applied in Finland for S_{500}^C is 10^{-2} manSv/MW·a concerning the whole fuel cycle, recommended in [2]. For waste dis-

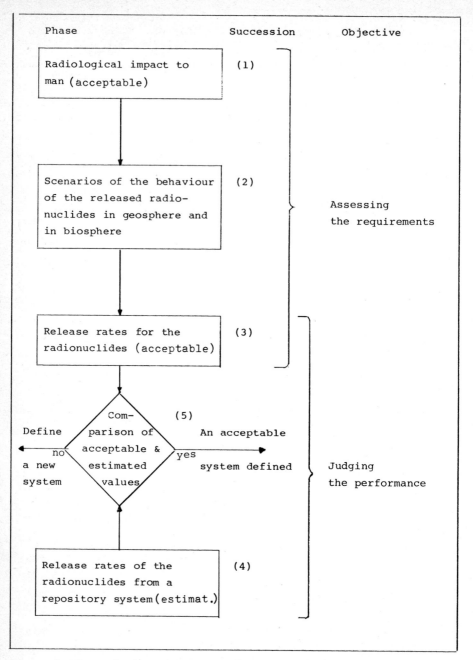

Figure 2. The principle of the methodology for assessing the requirements and judging the performance of a repository system.

posal a fourth of that was chosen (a subjective choice), since half of the limit is already reserved for reactor operation. By using the values $E = 4 \cdot 10^4$ MW$\cdot$a and $N = 2 \cdot 10^5$ persons, we got $\dot{H}_E^{(2)} = 10^{-6}$ Sv/a. The value of $\dot{H}_E^{(2)}$ is sensitive to the ratio of E/N which may, however, be quite constant.

2.2 Practice

For the analyses the methodology had to be modified. The succession of the phases presented in Fig. 2 was 2,3,1 instead of 1,2,3. The reason for using the more elaborate iterative way was the fact that no model with capability for backwards calculations exists today.

The iterative way employed is time and money consuming, especially when a comprehensive code like GETOUT or similar is utilized. A simplified, analytical closed-form model would be more suitable for preliminary sensitivity and screening analysis.

3.   THE ROLE OF NEAR-FIELD PHENOMENA

3.1 Choice of scenarios and parameter values

The importance of near-field versus far-field effects was studied varying the factors belonging to the repository and to the geosphere and employing two sets of biosphere factors connected with two pathways, viz. via Baltic Sea and via a well. The scenarios cover a wide range of cases connected with parameter values from "optimistic" through "reasonable" to "pessimistic" existing in situations considered "normal" as well as "accidental". The parameter values used are shown in Table I. In all the cases the discharge of the radionuclides from the repository to biosphere was thought to occur along with groundwater flow.

The parameters belonging to the near-field were:
- Time of the initiation of the release from the
  repository, T_i
- Duration of the release, T_d

The values of T_i and T_d were chosen to cover the range implied by the use of different construction and materials (e.g. SS and Cu as regards canister, bentonite or no backfill) as well as implied by accidental situations. Furthermore the public opinion is likely to consider the consequences of the release to be more serious the earlier in the future they might occur, regardless of their mathematical expectation value. Thus the successive time periods, e.g. $1 \ldots 10^3$, $10^3 \ldots 10^4$, $10^4 \ldots 10^5$ and $10^5 \ldots 10^6$ years, call for analyses having decreasing public interest and possibly decreasing need of accuracy. If only the parameter values regarded reasonable in case of sophisticated design and materials were used (e.g. Cu-canister of 20 cm wall thickness), we would have no estimates for "what if" questions pertinent to the most interesting future (before the year 10^5 or even 10^4).

The parameters belonging to the far-field were:
- Groundwater transport time (T_{gw}) connected with a
  constant value of path length (L) and hydraulic gradient
  (I) and with different, consistent values of hydraulic
  conductivity (K_p), effective porosity (ε_f), fracture
  spacing (S) and fracture aperture (2b).
- Sorption equilibrium coefficients (K_a) and related
  retardation coefficients (K_i)
- Coefficient of longitudinal dispersion (D_L) related to
  dispersion length (l_D) and groundwater flow velocity

The values of T_{gw} were chosen on the basis of previous groundwater flow calculations carried out employing as input data K_p

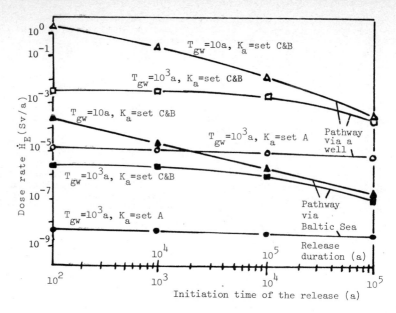

Fig. 3. Effect of different parameters on the maximum
individual dose equivalent rate.

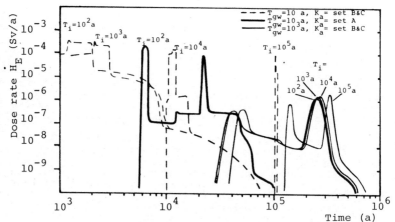

Fig.4. Dose rates from accidental releases of radionuclides. Release duration is 10^3 years, pathway via Baltic Sea.

values mainly based on literature [3] and choosing the consistent values of S and 2b in the way to give conservative values for T_{gw}. Transport time of 1000 years connected with hydraulic conductivity of 10^{-9} m/s is considered to be quite realistic; transport time of 10 years connected with $K_p = 10^{-7}$ m/s might be conservative by a factor of 5 (by using different values of S and 2b transport time could be about 50 years). The reason for using such a short transport time was the objective to analyze, how extreme requirements would be called for engineered barriers in case of a very unfavourable site.

The three sets of K_a values were chosen to correspond to oxidizing conditions (set B), conservative reducing conditions (set B) and optimistic reducing conditions (set A), see Table II.

The discharge rates of the radionuclides into the Baltic Sea and into a well were converted into dose rates by applying conversion factors that were derived from the results calculated with the BIOPATH code and presented in [4]. These factors were modified to correspond the new values of dose conversion factors published by ICRP [1]. The dose conversion factors employed are presented in Table III.

3.2 Effect of different parameters

The acceptable dose rate limit $\dot{H}_E^{(2)}$ (10^{-6} Sv/a) was applied to the Baltic Sea scenarios. The effect of the different parameters is illustrated in Fig. 3. Regarding cases of succesful site selection the hydrogeological conditions alone seem to be able to guarantee dose rates low enough. The effect of release duration is almost neglible, unless the duration is considerably more than 10^5 years. This is due to the combined effect of hydrological dispersion and the chemical retention for the most important nuclides. Even in the cases of accidental release ($T_d \approx 10^3$ years) the dose rate is approximately low enough, as pointed out in Fig. 4. The time of initiation of the accidental release has no influence.

In case of very unfavourable site ($T_{gw} \approx 10$ a) the influence of the release duration T_d is essential, Fig. 3. To meet the limit of acceptable dose rate $\dot{H}_E^{(2)}$, T_d should be more than 10^5 years, the dominant nuclide being ^{237}Np. In cases of accidental release the initiation time less than 10^6 years does not affect at all. The nuclides giving highest contributions are ^{237}Np, ^{239}Pu, ^{226}Ra and ^{231}Pa.

The accepted dose rate limit $\dot{H}_E^{(1)}$ (10^{-4} Sv/a) was applied to the well scenarios. The effects of parameter variations are presented in Fig. 3 and 4. If the dose rate limit applied is $\dot{H}_E^{(1)}$, there are obvious difficulties in meeting this limit. Either very favourable geosphere conditions or very long release duration are needed. Only in the case of the best combination of the geosphere parameters ($T_{gw} = 10^3$ years, sorption coefficients of set A) the dose rate is low enough regardless of the release period. In the other geosphere conditions the release duration T_d should be more than 10^6 years. The important nuclides are first of all ^{237}Np and ^{231}Pa but also ^{239}Pu and ^{240}Pu in case of relatively low values for T_{gw} and K_a. In case of short release duration (T_d 10^4 years) also ^{129}I is important.

3.3 Functional requirements for repository system

According to the sea scenarios the requirements for a repository would be rather modest, if the hydrogeological conditions are relatively favourable. In case of very unfavourable conditions the release duration should be more than 10^5 years. According to the well-scenarios and deterministic approach the release duration should be more than 10^6 years, unless very favourable geosphere conditions exist.

However, the limit of $\dot{H}_E^{(1)}$ applied for the well scenarios was determined as a dose value, not as a risk value, see formula 2-1. The accepted dose rate limit could be defined as a mathematical expectation value

$$\dot{H}_E = P_W \times \dot{H}_E^{(1)} \qquad\qquad (3\text{-}1)$$

where P_W is the probability of the well scenario to occur. Furthermore the share of the nuclides reaching the well compared to the amount releasing from the repository should be taken into account. By these assumptions the repository requirements would be remarkably less severe.

The method of formula 3-1 could be generalized to

$$\dot{H}_E = P \times \dot{H}_E^{(i)} \qquad i = 1,2 \qquad\qquad (3\text{-}2)$$

where P is the probability of the scenario. In this study this was not yet done.

The release rates connected with the different release periods are presented in Fig. 5. As far as the doses caused by ^{237}Np are concerned, the release rate as ^{237}Np is totally dominant, since its precursors ^{241}Pu (half-life 15 a) and ^{241}Am (half-life 430 a) have in most cases decayed before release. The acceptable release rate of ^{237}Np is about 500 MBq/a ($1.4 \cdot 10^{-2}$ Ci/a) if 10^5 years release duration is required, and about 50 MBq/a if 10^6 years are required. In case of doses due to ^{231}Pa the most important release rate is that of ^{235}U, the precursor of ^{231}Pa, having a long-half life of $7 \cdot 10^8$ years, and a high initial inventory. The acceptable release rate of ^{235}U is about 1 MBq/a ($3 \cdot 10^{-5}$ Ci/a) in case of 10^6 years release duration.

4.   PERFORMANCE OF REPOSITORY

   4.1 Phenomena

The near-field phenomena which can affect the release of a radionuclide are various by character: thermal, chemical, mechanical and radiation effects influence the behaviour of the waste, canister, backfilling material, surrounding rock and the groundwater. The processes connected with the release can be separated as follows:
   - Degradation of the engineered barriers causing the initiation of the leaching.
   - Leaching of the waste matrix and radionuclides into the water in the disposal hole.
   - Transport of the leached radionuclides from the hole to the groundwater in the surrounding rock.

Greatest uncertainties and lack of proper models seem to concern the first two of the process mentioned. The last process is not equally complicated and adequate models are available.

   4.2 Release rates

The results regarding the estimated release rates are for the present only preliminary, the proper analysis having not yet been completed.

The release rate presented in 3.3 for ^{235}U (1 MBq/a) is equal to total uranium release of 1.4 kg/a if no backfilling exists and corresponds to a concentration of at least $6 \cdot 10^{-4}$ M in the disposal hole, equal to 0.14 g/l. The maximum solubility in water is

10^{-8} M equal to $2.4 \cdot 10^{-6}$ g/l in case of $U^{(IV)}$ and about $4 \cdot 10^{-3}$ M equal to 1 g/l in case of $U^{(VI)}$ [3]. In natural groundwaters in the Precambrian rock maximum concentrations of the order of $1 \cdot 10^{-3}$ g/l have been detected [3].

The release rates of ^{237}Np, 50 MBq/a and 500 MBq/a (see 3.3) are equal to 2 g/a and 20 g/a. The rates correspond to concentrations of at least 10^{-6} M equal to $2 \cdot 10^{-4}$ g/l and 10^{-5} M equal to $2 \cdot 10^{-3}$ g/l consequently. The maximum solubility of artificial Np is not very well known, but is evidently lower than 10^{-4} M. Such a low solubility would have a decisive effect to the release rate.

If an intact backfilling material exists the transport from the hole to the rock can occur only by diffusion. By applying diffusion in stationary state in cylindrical geometry, and giving no merit to retardation, the release rate would be approximately $500...10^{-3}$ gU/a consistent with the maximum concentration of 1 g/l $...2 \cdot 10^{-6}$ g/l. By applying the model used in the KBS-study [5], which in addition takes into account the so called film transfer effect, the release rate would be ca. $50...10^{-4}$ gU/a. The influence of the retardation in the backfilling material is not very essential, see Fig. 6.

5.   DISCUSSION

The performance requirements for repository system in general and in greater detail for engineered barriers, and furthermore the requirements for analyzing and modelling the near-field phenomena are essentially dependent on the following factors:
- Principle and limit of the acceptable radiological impact applied
- Hydrological and geochemical conditions around the repository and along the pathways to biosphere
- Behaviour of the radionuclides in the biosphere, or in other words, conversion from discharge into doses.

The principle of applying either a deterministic dose rate (like formulae 2-1, 2-2) or a probabilistic expectation value (like formulae 3-1, 3-2) has an essential influence on the acceptability of the whole disposal system, not only on the repository itself. To be more realistic, the probabilities of "accidents" should be included in the analysis.

The model applied in the two-dimensional code FEFLOW [6] is sufficient in its structure to estimate the groundwater pattern and transport time, except in cases where an artificial temperature distribution is of importance. The main inaccuracy is due to uncertainty of the input data, especially of the fracture system, spacing and aperture. Transport time connected to a fixed K_p value can easily vary by a factor of ten, e.g. 10^2 or 10^4 years instead of estimated 10^3 years.

The models of the employed codes, GETOUT [7,8] and MMT1D [9], are almost competent enough to estimate the release and discharge rates of the nuclides. The main shortage of GETOUT is the requirement of simultaneous release of all the nuclides with a constant rate. In certain cases this leads to erroneous total discharge rates, which can be misleading. Neither of the codes can handle processes like diffusion into rock matrix or time dependent phenomena like temporary saturation. MMT1D is under development to be able to overcome both of the shortages. Essential inaccuracy will be caused by the uncertainty of the input data, especially of sorption and dispersion coefficient values, but also of groundwater flow velocity and of fracture system characteristics.

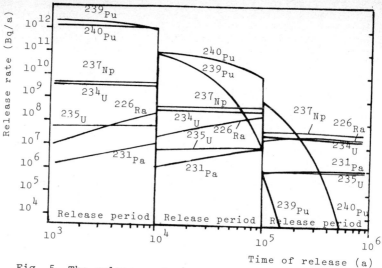

Fig. 5. The release rates of important nuclides from
a repository into groundwater in the surrounding
rock implied by different hypothetical release
periods.

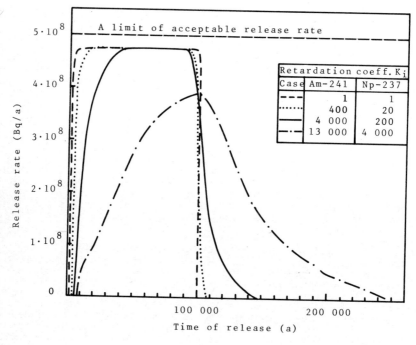

Fig. 6. Estimated release rate of Np-237 from repository
into groundwater in rock. Assumed thickness of
Bentonite backfill layer = 0,4 m, diffusion
coefficient $D = 4 \cdot 10^{-11}$ m^2/s, leaching period
$10^3 \ldots 10^5$ years.

The code BIOPATH [4], utilized but not at our disposal, includes many parameters which must be chosen almost arbitrarily in view of long time spans involved (pathways, human food habits, population etc.). A code called DETRA [10], currently under development, will not eliminate these difficulties but enables us to analyze their influence. They can be essential.

REFERENCES

1   Limits for intakes of radionuclides by workers, ICRP publication 30, Supplement to part 1, Annals of the ICRP, Vol. 3, No. 1-4.

2   Report on the Applicability of International Radiation Protection Recommendations in the Nordic Countries. The Radiation Protection Institutes in Denmark, Finland, Iceland, Norway and Sweden. Ed. B. Lindell, Stockholm, 1976.

3   Handling and final storage of unreprocessed spent nuclear fuel. Vol. II, Technical. KBS, Stockholm, 1978.

4   Bergman, R. et al., Dose and dose commitment from groundwater-borne radioactive elements in the final storage of spent nuclear fuel. KBS Tecnical Report No. 100. Stockholm, 1979.

5   Neretnieks, I., Transport of oxidants and radionuclides through a clay barrier. KBS Technical Report No. 79. Stockholm, 1978.

6   Puttonen, J. & Salo, J.-P., Groundwater flow calculations as a part of the safety analysis for the underground disposal of high-active waste (in Finnish), 12th Finnish Rock Mechanics Meeting. Helsinki, 1979.

7   Grundfelt, B. & Elert, M., GETOUT - a one-dimensional model for groundwater transport of radioactive decay chains. SKBF/KBS Technical Report 80-03. Stockholm, 1980.

8   DeMier, W.W. et al., GETOUT - A computer program for predicting radionuclide decay chain transport through geologic media, PNL-2970. Richland, Washington, 1979.

9   Washburn, J.F. et al., Multicomponent Mass Transport Model: A model for simulating migration of radionuclides in groundwater. PNL-3179. Richland, Washington, 1980.

10 Savolainen, I. & Hulmi, R., Transport in biosphere of radio-nuclides released from finally disposed nuclear waste - background information for transport and dose model (in Finnish). Report YJT-81-11. Nuclear Waste Commission of Finnish Power Companies, Helsinki, 1981.

11 Grundfelt, B., Nuclide migration from a rock repository for spent fuel (in Swedish). KBS Technical Report No. 77. Stockholm, 1978.

12 Hill, M.D. & Lawson, G., An assessment of the radiological consequences of disposal of high level waste in coastal geologic formations, NRPB-R108. Harwell, 1980.

Table I The parameter values employed in the sensitivity analysis

Parameter			Values
T_i	a	Time of the initiation of the release	$10^2, 10^3, 10^4, 10^5$
T_d	a	Duration time of the release	$10^3, 10^4, 10^5, 10^6$
T_{gw}	a	Groundwater transport time +)	$10^1, 10^2, 10^3$
K_p	m/s	Hydraulic conductivity +)	$10^{-7}, 10^{-8}, 10^{-9}$
2b	m	Fracture aperture +)	$10^{-4}, 3 \cdot 10^{-5}, 10^{-5}$
S	m	Fracture spacing +)	10, 3, 1
l_D	m	Dispersion length	0.2, 2, 20
K_a	m	Distribution coefficients *)	set C, set B, set A
Fixed parameter			Value
L	m	Length of groundwater path	5700
I	–	Hydraulic gradient	$2 \cdot 10^{-3}$
ε_f	–	Effective porosity of the rock	10^{-5}
D_L	m^2/s	Dispersion coefficient (longitudinal)	$3.6 \cdot 10^{-6}$

+) T_{gw} is a primary parameter, the values of K_p, 2b and S chosen to
be corresponding
*) See Table II

Table II   Area-based distribution coefficients
K_a (m^3/m^2)

Element	set A		set B		set C	
Zr	3.2-01	+)	2.5-02		4.2-02	
Tc	5.0-03		0		0	
I	0		0		0	
Cs	2.1-02		6.3-03		4.2-03	
Ra	2.5-01		6.3-03		3.3-03	
Th	2.4-01		1.4-03		2.6-02	
Pa	6.0-02	*)	5.0-03	*)	1.9-04	*)
U	1.2-01		1.0-02		2.1-04	
Np	1.1-01		1.0-02		1.3-03	
Pu	3.0-02		1.4-02		5.3-03	
Am	3.2+00		1.0-01		4.2-01	
Cm	1.6+00	*)	5.0-02	*)	2.0-01	*)

(+) 3.2-01 equal to $3.2 \cdot 10^{-1}$
(*) Own estimate based on data from ref. |11|
    Other data from ref. |3|

Set A: "optimistic" reducing conditions
Set B: "conservative" reducing conditions
Set C: "pessimistic" oxidizing conditions

Table III   Conversion factors from discharge rate to
            biosphere into effective dose equivalent rate

Nuclide	Dose conversion factor			
	via Baltic Sea		via a well	
	Sv/a	rem/a	Sv/a	rem/a
	Bq/a	Ci/a	Bq/a	Ci/a
Zr-93	1.6-18 +)	6.1-06	1.6-16	6.1-04
Tc-99	5.4-20	2.0-07	1.8-16	6.7-04
I-129	3.5-17	1.3-04	1.3-13	4.8-01
Cs-135	1.5-17	5.6-05	5.7-15	2.1-02
Ra-226	1.1-15	4.1-03	3.8-13	1.4+00
Th-229	3.5-15	1.3-03	1.1-12	3.9+00
Th-230	8.6-17	3.2-04	8.9-14	3.3-01
Pa-231	7.0-15	2.6-02	3.5-11	1.3+02
U-234	1.4-17	5.1-05	3.0-14	1.1-01
U-235	3.0-19	1.1-06	9.2-15	3.4-02
U-236	1.2-17	4.5-05	3.8-14	1.4-01
U-238	9.5-18	3.5-05	3.0-14	1.1-01
Np-237	1.8-15	6.5-03	5.4-12	2.0+01
Pu-239	7.8-18	2.9-05	2.7-13	5.0-01
Pu-240	7.8-18	2.9-05	2.7-13	5.0-01
Pu-242	7.8-18	2.9-05	2.7-13	5.0-01
Pu-244	7.8-18	2.9-05	2.7-13	5.0-01

(+) 1.6-18 equal to $1.6 \cdot 10^{-18}$
Data based on ref. |4|, |1| and |12|

DISCUSSION

M. MAKINO, Japan

How did you decide the acceptable dose rate levels both via the sea and via wells ?

E.K. PELTONEN, Finland

The values are subjective choices for the moment. The objective was that they were as far as possible consistent with the limits of acceptable radiological impact applied for nuclear power plant operation. The dose rate limit used in sea cases was derived backwards from collective dose commitment ; the limit applied in well cases is originally a committed effective dose for an individual belonging to a critical group of exposure. The details are presented in paragraph 2.1 in the paper.

NEAR-FIELD HEAT TRANSFER AT THE SPENT FUEL TEST-CLIMAX:
A COMPARISON OF MEASUREMENTS AND CALCULATIONS

W. C . Patrick, D. N. Montan, L. B. Ballou
Lawrence Livermore National Laboratory
Livermore, California, USA

ABSTRACT

The Spent Fuel Test in the Climax granitic stock at the DOE Nevada Test Site is a test of the feasibility of storage and retrieval of spent nuclear reactor fuel in a deep geologic environment. Eleven spent fuel elements, together with six thermally identical electrical resistance heaters and 20 peripheral guard heaters, are emplaced 420 m below surface in a three-drift test array. This array was designed to simulate the near-field effects of thousands of canisters of nuclear waste and to evaluate the effects of heat, alone, and heat plus ionizing radiation on the rock.

Thermal calculations and measurements are conducted to determine thermal transport from the spent fuel and electrical resistance heaters. We present calculations associated with the as-built Spent Fuel Test geometry and thermal source histories and compare these with thermocouple measurements made throughout the test array. Comparisons in space begin at the spent fuel canister and include the first few metres outside the test array. Comparisons in time begin at emplacement and progress through the first year of thermal loading in this multi-year test.

"Work performed under the auspices of the U.S. Department of Energy by the U.S. Department of Energy by the Lawrence Livermore National Laboratory under contract number W-7405-ENG-48."

INTRODUCTION

A test is currently in progress at the Department of Energy Nevada Test Site (NTS) to evaluate the feasibility of safe and reliable storage and retrieval of spent fuel assemblies in a deep geologic media [1]. The Spent Fuel Test-Climax (SFT-C) is funded through the Nevada Operations Office of the U.S. Department of Energy. Funding for the project was approved June 2, 1978, and emplacement of eleven spent fuel canisters was completed May 28, 1980.

A principal goal of the test is to demonstrate that spent fuel assemblies from commercial reactors can be safely and reliably packaged, transported, stored, and retrieved using existing technology. Instrumentation of the test facility will provide data to address technical questions as well. Technical issues are of two basic types: in situ response of the granitic media to combined mechanical, thermal, and radiation effects and evaluation of computational techniques for repository design.

This paper addresses thermal modeling that was conducted in support of the SFT-C. We discuss the role of thermal transport modeling in establishing the facility geometry and in locating thermocouples to measure temperatures in areas of interest. Calculational results are compared with field measurements to assess the adequacy of state-of-art thermal analysis models and codes.

EXPERIMENT GEOMETRY

Test Location and Geology

The SFT-C is located in the northeast quarter of the Nevada Test Site which is about 100 km northwest of Las Vegas, Nevada. The test is named for the Climax granitic stock in which it is located. The Climax stock outcrops over an area of about 4 km^2 on the surface and expands conically with depth. The SFT-C is located 420m below where a shaft had been sunk for nuclear weapons effects testing in the early 1960's [2,3].

The Climax stock consists of two intrusive units: a quartz monzonite and a granodiorite. The SFT-C is contained completely in the quartz monzonite unit. Laboratory properties of the rock are typical of other granitic rocks [4-8]. In situ testing has refined several of the laboratory values. Current best estimates of modulus and Poisson's ratio are 26 GPa and 0.246, respectively [9]. Heater test results indicate a thermal conductivity of 3.1 W/m-K and a thermal diffusivity of 1.26mm^2/sec [10].

Test Layout

Facility geometry was dictated to a large degree by the desire to simulate, within the central portion of the test, the early-time thermal and thermomechanical effects of a very large repository. Scoping calculations showed that spent fuel emplaced in the floor of a central storage drift flanked by two parallel drifts with electrical heaters in the floor could achieve this effect. The region in which this simulation is produced is shown in Figure 1 as a repository-model cell. Also shown are alternating canisters of spent fuel and thermally identical electrical simulators used to examine radiation effects in granite.

Detailed calculations were used to ascertain a power history for the electrical heaters which would produce the desired effects in the 15m x 15m repository-model cell. Calculations comparing temperatures at the center of this cell with those at the center of a 600m x 600m array of 8,000 spent fuel canisters show errors of less than 5% during the first 100 years after emplacement.

Thermal Sources

There are three principal sources of heat at the SFT-C, neglecting lights, transformers, small airconditioners, and similar devices. First, eleven Westinghouse 15 x 15 pressurized water reactor spent fuel assemblies, obtained from the Florida Power and Light Company Turkey Point Unit #3, are emplaced in the floor of the canister drift (Fig. 1). These units were emplaced at about 2.5 years after discharge when they had a thermal power output of 1500 + 50 W/assembly. Power levels were determined by calculations [11] normalized to calorimeter measurements which were performed on three assemblies just prior to spent fuel encapsulation [12]. The eleven spent fuel assemblies were emplaced between April 18 and May 28, 1980, at a rate of two assemblies per week [12].

The second source of heat is the set of six electrical simulators interspersed with the spent fuel canisters in the canister drift (Fig. 1). Each simulator consists of four tubular electrical resistance heating elements in a baffled and shrouded framework and is contained within a canister which is identical to those used for spent fuel encapsulation. These units are thus thermally identical to the spent fuel. Each simulator was energized at the time its corresponding fuel assembly was installed in its position. Adjustments to power were made subsequent to installation to match the decay of the spent fuel. Improved calculations and measurements of spent fuel power output resulted in a somewhat cooler decay curve, as discussed above. Simulator power, which was based on earlier calculations, is thus a few percent higher than that of the spent fuel. Further adjustments are planned to decrease the observed differences.

Ten guard heaters in each of two parallel heater drifts provide the third major source of heat. The power levels of these electrical resistance heaters are controlled to provide thermal boundaries on the repository model cell which closely approximate those of a large repository, as described above.

Ventilation air flowing through the SFT-C complex carries with it some of the heat generated by the three sources. The quantity of heat removed in this manner is determined by measuring flow rates, air temperature, and relative humidity at air inlet and exhaust locations [12]. Nominal flowrate has been 2.6m^3/sec for most of the test period. This flow is divided so that half flows through the canister drift and one quarter flows through each heater drift. Two planned decreases reduced the flowrate to about half its nominal value.

INSTRUMENTATION

Thermocouples have been installed throughout the test array. Placement is most concentrated near the canister emplacement holes where changes in temperature are most rapid and most pronounced. In this region, 18 thermocouples are associated with each hole: one at each of three axial and two azimuthal positions 180° apart on each canister, one at each of three axial and two azimuthal positions 90° apart on each liner, and one at each of three axial and two radial positions within the host rock.

Within a few metres of the canister are additional thermocouples associated with other geotechnical instrumentation. Further out, thermocouples measure temperature changes near one of the guard heaters and in a zone several metres outside the underground openings.

All thermocouples are Chromel-Alumel Type K and are encased in Inconel tubing with magnesium oxide packing. The thermocouples are grounded at the

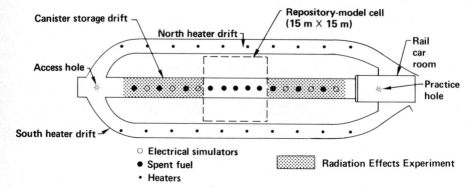

Fig. 1. Plan view of the Climax granite spent-fuel-storage test.

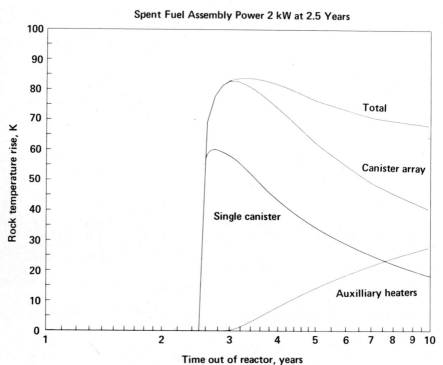

Fig. 2. Calculated temperature history at the storage-hole wall
at midheight of the center fuel assembly (see Fig. 1).

hot junctions. Temperatures at the cold junctions are measured with resistance temperatures devices [13]. Thermocouples were calibrated before emplacement at 0°C and 150°C. Subsequent calibrations are performed periodically on selected units and post-test calibration of all units is planned. All told, nearly 500 temperature measuring devices are in use at the SFT-C.

DATA ACQUISITION AND REDUCTION

Data acquisition is facilitated by twin Hewlett-Packard Model 2176A disc-based computer systems. These include HP1000 mainframes with necessary discs, magnetic tape drives, digital voltmeters, scanners, display terminals, line printers, modems, and related equipment to acquire data from nearly 850 channels of instrumentation. The system operates remotely and automatically at NTS. Acquired data is copied on magnetic tapes which are shipped to Livermore, California for processing, reduction, and analysis.

THERMAL CALCULATIONS

Analytical Models

The model used in the early phases of the thermal calculations was very simple. The granite was considered to be homogeneous, isotropic and of infinite extent. The model repository, or the spent-fuel test, was modeled as an array of parallel finite line sources whose length was that of the spent-fuel assembly ($\sim$4 m). The power vs time history of the sources was taken to be proportional to that for the PWR fuel with a burnup of 33000 MWD/MTU at a specific power of 37.5 MW/MTU, at that time the best estimate available [14].

This model, of course, ignores the existence of the mined drifts used to emplace the spent fuel and the drill holes in which they are stored. The neglect of the drifts was considered to be conservative in the sense that temperatures calculated would be higher than the "real world" since the openings would be better conductors than the rock they replaced and any ventilation would remove heat from the system. The use of line sources rather than cylindrical sources only affects early transients.

The calculations used in this phase were also simple, namely the superposition in space and time of the continuous point source solution of the diffusion equation.

$$\Delta T = \frac{Q}{4\pi kd} \text{ erfc} \sqrt{\frac{d^2}{4\kappa t}} .$$

The model repository, with which the calculations of the SFT-C geometry were compared is simple in design. It consists of a large number of long parallel drifts with the spent-fuel canisters placed in vertical holes drilled along the center lines of the drifts. The dimensions used in this model are 15 m, center-to-center, for the drifts with the waste canisters emplaced on 3-m centers along the drifts. The rationale for the 15-m drift spacing was an extraction ratio of 33% using a projected drift width of 5 m. Thermal calculations have been made for the center of such a repository. The parameters used in these calculations were k = 3 W/m-K, κ = 1.25 mm^2/s, Q= 2 kW @ 2.5 years out of reactor core, L = 3.66 m.

Calculations were performed with various spent fuel ages and test geometries to insure that three requirements were met. First, calculations of spent fuel aged 2.5, 5, and 10 years out of reactor core showed that 2.5-year-old fuel provided an early time peak that exceeded later time peaks associated with older fuel. This met the requirements of a slight thermal over test, for spent fuel storage, and a reasonably short test duration, less than 5 years. Second, calculations were used to establish the test layout, as mentioned earlier [Fig. 2]. These calculations showed that a single canister gives a peak rock temperature of 60°C, well below the 85°C calculated for the model repository. The 17-canister array peaks within a few degrees of the model repository, but drops to near half the value in 7 years. A number of calculations led to the design of the guard heater power history. When turned on at 0.3 years after the spent fuel is emplaced and maintained at a constant power of 1730 W each, they produced the contribution shown in the lower curve. When this is added to the canister array curve to give the top curve, the result is a peak rock temperature history that is remarkably close (within 1% for the first 7.5 years) to that of the model repository. Additional calculations have shown that the region between the canister row and the heaters experiences a temperature rise a few degrees higher than the region between rows in the repository design. Refinements in guard heater power have been made to produce even better agreement.

Numerical Models

In order to provide additional information for the final design, the decision was made to do more detailed thermal calculations beyond the capabilities of the calculational techniques used above. Finite difference models using the TRUMP code were employed [15]. These calculations were based on a unit cell of the repository or spent-fuel test. This cell contains one canister and is bounded by two parallel vertical planes perpendicular to the drifts and spaced half-way between canisters. For the model repository calculations, two additional vertical planes of symmetry parallel to and midway between the drifts also bound the unit-cell. The four-fold axis ofsymmetry at the canister center line provides two more vertical planes of symmetry, thus reducing the required calculations to a region comprising one-fourth of the unit-cell.

The mesh used contains three basic regions. The innermost region is 20 m wide by 40 m high and is divided into 1600 zones (0.5 by 1 m) to give the desired spatial resolution. To simulate an infinite medium for a long period of time, two more regions were added with 5- by 5-m and 20- by 20-m zones giving an overall size of 80 by 160 m. In the regions comprising the drifts, the regular zoning was replaced by single zones containing air. For three-dimensional modeling the region beneath the canister drift was three-dimensional with the zoning becoming progressively finer as the source was approached. Additional zones were used to model the canister, liner, shield plugs, railroad tracks, and other details. In all, more than 2500 zones were used.

Thermal Radiation and Convection

In some repository calculations that we have seen the drifts were considered as non-conductors (voids) and thus gave a large floor-to-ceiling temperature difference and isotherms that were perpendicular to the drift surfaces. This is quite incorrect. Either thermal radiation or convection will make the drifts better conductors than the rock that was removed to create them. When radiation and convection are properly treated, the drift surfaces are nearly isothermal and the nearby isotherms are nearly parallel to the surfaces. Inclusion of radiative transport does not grossly affect the temperature changes at a given point in the rock but it does have a major impact on thermal stresses, since isotherms change from being perpendicular to being parallel to the drift sidewalls.

Thermal radiation was modeled by straightforward application of the Stefan-Boltzmann equation. An emittance of 0.8 was normally used. One important consideration (often overlooked) is that in models having plane(s) of symmetry (such as occur in our model) the adiabatic boundary condition at such planes is equivalent to a mirror. Thus, the "half" main drift in our model with 16 radiating surfaces has 185 radiative connections.

Heat transfer from the drift surfaces to the air in the drift may be expressed by q = AHΔT, where H is the convection coefficient (typical units W/m^2-K). For the convection coefficient we have used a standard empirical correlation for natural convection in the turbulent regime relating the Nusselt number (N_{Nu}) to the Rayleigh number (N_{Ra}): $N_{Nu} = 0.13 \, N_{Ra}^{1/3}$, where the Rayleigh number is the product of the Grashof number and the Prandtl number.

Ventilation

The heat removed from a repository or the spent-fuel test by ventilation can amount to a significant fraction of the heat produced. Furthermore, the temperature rise of the air as it flows through the drift is nearly a linear function of the flow rate. Thus a careful approach to the calculation of thermal effects of ventilation seems prudent.

One approach is to hold the air, or some external boundary representing it, at the ambient air/rock temperature and then proceed with the calculation using convection coefficients. This ventilation model, while easily incorporated into the unit-cell model, has several serious drawbacks. In the spent-fuel test a fixed amount of air will enter one end of a drift and be removed at the other, picking up heat and rising in temperature as it passes through. The fixed air temperature will overestimate the heat removed, underestimate the rock temperatures (particularly the drift surfaces) and give no estimate of what the actual air temperature will be. On the other hand, physical implementation of the fixed air temperature model would require a very large amount of air.

In order to retain the unit-cell model and still do realistic ventilation effect calculations, we have adopted a "partial-flow" model in which the cell receives only a fraction of the total flow but that partial-flow enters the cell at ambient and leaves at the average air temperature of the cell. The heat removal may then be expressed as: $\Delta q = \rho c \dot{V}' \Delta x T$, where

ρ = air density (kg/m^3)
c = heat capacity (J/kg-K)
$\dot{V}'$ = flow rate per unit length of drift (m^2/s)
Δx = length of calculational cell (m)
T = temperature rise above ambient (K)

The question then arises as to what is the numerical value of $\dot{V}'$? The answer is to use several different values. They all represent different locations.

In the situation that we actually wish to model, a constant air flow, $\dot{V}_0$ (m^3/s), passes through the drift and at some point while traveling a distance Δx the temperature increases by an amount ΔT. The heat removal is then: $q = \rho c \dot{V}_0 \Delta T$. By equating this equation and the previous one and passing to the limit as Δx becomes infinitesimal we obtain:

$$\frac{dT}{dx} = \frac{\dot{V}'T}{\dot{V}_0}$$

Thus, by using different values of $\dot{V}'$ in the (complete) model and calculating the resulting air temperatures at a time or times of interest, a functional relationship between $\dot{V}'$ and T may be obtained. The above differential equation may then be integrated to give the desired result, air temperature rise as a function of distance along the drift. In practice we find that the product $\dot{V}'T$ is adequately approximated as a linear function of T.

Effects of Thermal Radiation, Convection, and Ventilation

The effects of thermal radiation, convection, and ventilation are shown in Figures 3-6. When the drift is treated as an insulator (no ventilation), a gradient from roof to floor of about 40°C results and floor temperatures are about 70°C, 4 years out of core. Including either convection or thermal

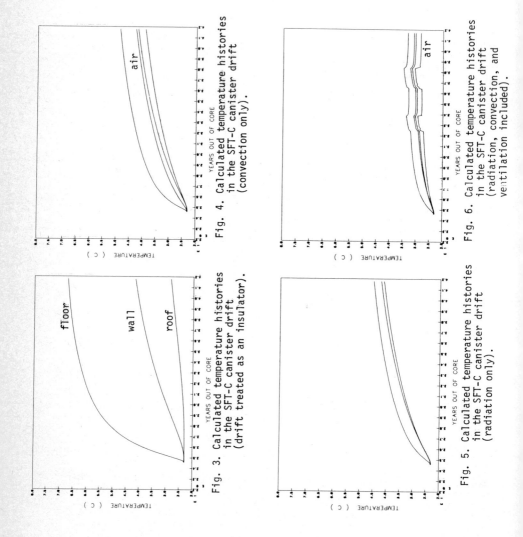

Fig. 3. Calculated temperature histories in the SFT-C canister drift (drift treated as an insulator).

Fig. 4. Calculated temperature histories in the SFT-C canister drift (convection only).

Fig. 5. Calculated temperature histories in the SFT-C canister drift (radiation only).

Fig. 6. Calculated temperature histories in the SFT-C canister drift (radiation, convection, and ventilation included).

radiation reduces the gradient to about 10°C or 2-4°C, respectively. Peak temperatures are also reduced throughout the test period. Introduction of ventilation results in markedly lower temperatures at all locations on the drift surfaces. For the example given, the reduction in temperature rise is about a factor of two at a time of 4 years out of core. Thermal gradients between roof and floor remain at about 2-4°C.

COMPARISON OF MEASUREMENTS & CALCULATIONS

Temperature data obtained during the first year of the test have been processed and are shown together with matching calculations in Figures 7-10. Individual calibration offset constants, which do not exceed 0.5°C, have not yet been applied to the data. The first of several points to be made is that agreement between calculational results and measurements is excellent. Differences in changes in temperature in the central region of the test array do not exceed 5% (Fig. 7) while those near the end of the array do not exceed 8% (Fig. 8 & 9). Second, plots of data and calculations are parallel to slightly converging in the central region, indicating that agreement should remain excellent. Near the end of the array, the plots are slightly diverging, due to end effects in this region of the test array. Third, the electrical simulators are closely matching the decay of the spent fuel, despite fluctuations in power levels due to outages and inaccuracies in control components. Note how the observed fluctuations are diminished as the thermal transient propagates through the rock. Fourth, hole liner temperatures are consistently lower than calculated. Several causes may be postulated for this difference. These include: incorrect total energy input, incorrect energy contribution due to gamma radiation, incorrect emittance, incorrect conductivity in the air gaps, and incorrect convection coefficients. The first two causes are quickly discounted by examining, respectively, the canister temperature and rock temperature curves, and the presence of low liner temperature in Figure 9 where no gamma radiation is present. Incorrect thermal properties are the likely cause of the observed difference in calculated and measured liner temperatures.

The calculated and measured temperatures associated with one half of the SFT-C array are shown in Figure 10. Once again, agreement is good with differences between calculated and measured temperatures generally not exceeding 2°C. Deviations near the guard heaters are larger since the guard heaters were modelled as strip sources rather than as individual assemblies.

CONCLUSIONS

Comparisons have been made between measured and calculated near-field temperatures at the Spent Fuel Test-Climax. Agreement in the region within 1m of the spent reactor fuel is excellent--within 5 to 8%. Agreement in the region extending several metres outside the test array is also good -- generally within 2°C. Data for these comparisions are based on temperatures recorded by thermocouples during the first year of testing. Individual calibration offset constants, which do not exceed 0.5°C, have not yet been applied to the data. Calculations with which data are compared, are based on the as-built SFT-C geometry and material properties. No adjustments in properties have been made to obtain the best possible fit of data and calculations. Comparisions indicate that some adjustment in thermal properties may be justified.

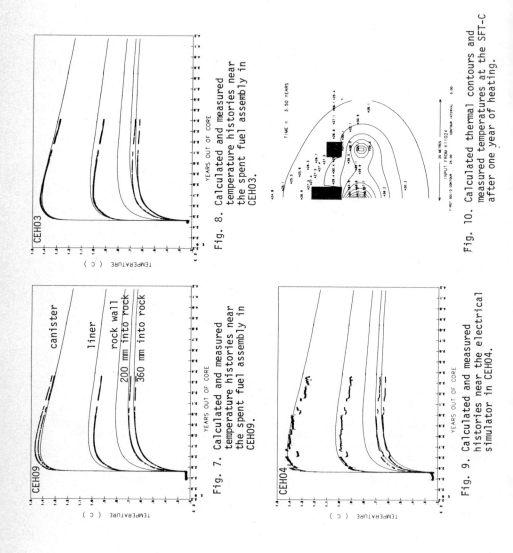

Fig. 7. Calculated and measured temperature histories near the spent fuel assembly in CEH09.

Fig. 8. Calculated and measured temperature histories near the spent fuel assembly in CEH03.

Fig. 9. Calculated and measured histories near the electrical simulator in CEH04.

Fig. 10. Calculated thermal contours and measured temperatures at the SFT-C after one year of heating.

REFERENCES

1. L. D. Ramspott, L. B. Ballou, R. C. Carlson, D. N. Montan, T. R. Butkovich, J. E. Duncan, W. C. Patrick, D. G. Wilder, W. G. Brough, M. C. Mayr, Technical Concept for a Test of Geologic Storage of Spent Reactor Fuel in the Climax Granite, Nevada Test Site, Lawrence Livermore National Laboratory, Livermore, CA, UCRL-52796 (1979).

2. D. G. Wilder and W. C. Patrick, Geotechnical Status Report for Test Storage of Spent Reactor Fuel in Climax Granite, NTS, Lawrence Livermore National Laboratory, Livermore, CA, UCRL-85096 (1980).

3. W. C. Patrick and M. C. Mayr, Excavation and Drilling Activities Associated with a Spent Fuel Test Facility in Granitic Rock, Lawrence Livermore National Laboratory, Livermore, CA, UCRL-in preparation.

4. F. Maldonado, Summary of the Geology and Physical Properties of the Climax Stock, Nevada Test Site, USGS Open-File Report 77-356, 1977.

5. L. Obert, Shot Hard Hat, Static Stress Determinations, DOD, 1963.

6. F. H. Wright, Shot Pile Driver, In Situ Rock Stress (U), DOE, 1967

7. D. P. Krynine and W. R. Judd, Principles of Engineering Geology and Geotechnics, McGraw-Hill Book Co., 1957.

8. K. Szechy, The Art of Tunnelling, Akademiai Kiado, Budapest, 1966.

9. F. E. Heuze, W. C. Patrick, R. V. Dela Cruz, and C. F. Voss, In Situ Geomechanics, Climax Granite, Nevada Test Site, Lawrence Livermore National Laboratory, Livermore, CA, UCRL-53076 (1981).

10. L. D. Ramspott, Climax Granite Test Results, Lawrence Livermore National Laboratory, Livermore, CA UCID-18502 (1980).

11. F. Schmittroth, G. J. Neely, and J. C. Krogness, A Comparison of Measured and Calculated Decay Heat for Spent Fuel Near 2.5 Years Cooling Time, Hanford Engineering Development Laboratory, Richland, WA, HEDL-TC-1759 (1980) (Preliminary -- Controlled Distribution).

12. R. C. Carlson, W. C. Patrick, D. G. Wilder, W. G. Brough, D. N. Montan, P. E. Harben, L. B. Ballou, H. C. Heard, SFT-C Technical Measurements Interim Report FY1980, Lawrence Livermore National Laboratory, Livermore, CA, UCRL-53064 (1980).

13. W. G. Brough and W. C. Patrick, Instrumentation Specification, Design, Installation, Maintenance, and Modification, Lawrence Livermore National Laboratory, Livermore, CA, UCRL-in preparation.

14. D. N. Montan, Thermal Analysis for a Spent Reactor Fuel Storage Test in Granite, Lawrence Livermore National Laboratory, Livermore, CA UCRL-83995 (1980).

15. A. L. Edwards, TRUMP, A Computer Program for Transient and Steady State Temperature Distribution in Multidimensional Systems, Lawrence Livermore National Laboratory, Livermore, CA UCRL-14754 Rev. 3 (1972).

DISCUSSION

K.K. TSUI, Canada

My question may be outside the scope of your paper since the paper deals with heat transfer only. How good is the comparison between measured and predicted stresses ?

W.C. PATRICK, United States

During the "mine-by" experiment, three IRAD vitrating-wire stressmeters were deployed in one pillar. These all showed *decreases* in vertical pillar stress whereas elastic continuum scoping calculations showed *increases* in vertical stress.

During the heated phase of the test, 18 IRAD gauges were deployed. Of these, 15 have failed. We are currently determining when and why they failed. In the meantime, we have begun replacement of these gauges.

K.R. SHULTZ, Canada

Are you measuring strain gauge transients and trying to develop earthquake models ?

W.C. PATRICK, United States

We are not specifically measuring the response of the SFT-C to long-period transients.

Instruments currently deployed do not appear to record such transients nor are the instruments adversely effected by known transients.

R. KOOPMANS, Canada

Last year, we had the opportunity of visiting the Climax test facility at the Nevada Test Site and observed that the concrete slab covering the spent fuel canisters was cracked. Could you please explain how this occurred and indicate whether or not your thermal models predicted this behavior ? Have the growth of the cracks stabilized since the time of our visit ?

W.C. PATRICK, United States

We believe that the tensile fracturing around the canister emplacement holes was in response to the temperature perturbations occurring beneath the floor. This region, located about 4-8 m below floor level, became compressed due to thermal expansion. The compressed zone created tension in the concrete above which fractured in a radial pattern around each emplacement hole. Since the concrete had several months to cure, the fractures are not believed to be due to shrinkage of the concrete.

The models did not predict the fractures because of their 2-dimensional geometry.

Fracture growth appears to have ceased.

Session 3

WATER MOVEMENT

Chairman - Président

N.A. CHAPMAN

(United Kingdom)

Séance 3

ECOULEMENT DES EAUX

CHAIRMAN'S SUMMARY OF THE SESSION 3

Two of the potential consequences of near-field temperature increases in a repository are the generation of thermal stresses in the host-rock and, in certain geological environments, the initiation of convective groundwater movements which may extend into the body of rock surrounding the repository. This Session examines both these phenomena, with papers describing convective flow in fractured media and presentations covering the thermomechanical behaviour of tuffs and potential thermal effects on a waste package emplaced in them. The near-field properties of plastic clay, which are being exhaustively studied in the Belgian programme, are also reviewed in detail.

The principal impression arising from the presentations and discussions is that there is currently a lack of data to enable the coupling of flow, transport and thermomechanical models. For example, it is difficult to assess whether near-field stress responses, particularly in fractured rocks, will have any effect on releases in the long-term. Whilst the pre-closure mechanical behaviour of a repository is clearly significant, especially in concepts requiring retrievability, the importance or otherwise of post-closure thermomechanical responses is difficult to assess.

Coupling of convective (bouyant) flow models with radio-nuclide transport codes has been more successful. Thermally driven groundwater movements, which are here described both by finite element models and empirical laboratory-scale experiments on single fissures, are increasingly being viewed as the principal long-term transport mechanism in fissured rocks. There is clearly scope for further work in this area, which is a fine example of the inter-relationship between near-field processes and far-field hydrogeology, and the need to consider both in unified assessments.

FREE CONVECTION IN FISSURED ROCKS

A. BARBREAU[x], A. COUDRAIN[xx], P. GOBLET[xx], J.M. HOSANSKI[xx], E. LEDOUX[xx]

[x] Commissariat à l'Energie Atomique, Fontenay-aux-Roses, France

[xx] Centre d'Informatique Géologique, Ecole des Mines de Paris, Fontainebleau, France

ABSTRACT

This work is a part of the evaluation of the thermal disturbance due to a deep radioactive waste repository, and of the radionuclide migration which might occur. The water movement due to buoyancy effects, as well as the movement of particles brought away by this movement, are studied. A mathematical model is proposed, and used in a computer code. The model results are compared with those of a laboratory experiment using a Hele-Shaw cell, and of a small scale migration test on a naturally fractured granite block.

RESUME

Ce travail s'insère dans l'étude de la perturbation thermique provoquée par un stock profond de déchets radioactifs sur la roche avoisinante et des transferts de radioéléments qui pourraient en résulter. On y étudie les circulations de fluide induites par l'effet de gravité dans des fractures verticales qui recoupent le stock, ainsi que le mouvement de particules entrainées par ces circulations. Un modèle mathématique est proposé, et mis en oeuvre dans un code de calcul informatique. Les résultats fournis par ce modèle sont comparés à ceux d'une expérience de laboratoire utilisant une cellule Hele-Shaw, et d'une expérience en miniature sur un bloc de granite fracturé naturellement.

INTRODUCTION

Burial into deep, low permeability geological formations such as cristalline rocks is one of the solutions presently considered for the disposal of long-lived radioactive wastes.

The extension of the repositories and the quasi general existence of major accidents in crystalline rocks make it likely that a fractured zone could intersect the storage zone.

The heat production from the radioactive wastes would then cause changes in the properties of the fluids in the fractures and lead to water movements from the repository towards the environment.

The purpose of the present work is to assess the importance of these convective movements, so as to evaluate the possible migration of radionuclide escaping from the repository.

1) THEORETICAL SCHEMES OF THERMAL TRANSFER (2),(3)

a) Modelling of the fissured medium

The fracturation is assumed to be sufficiently localized and constant in directions to be taken explicity into account.

In that case, a discontinuous approach of the medium seems possible and necessary. Fissures and matrix are therefore considered as two different media, related by boundary conditions along their interface.

b) Equations

The following assumptions are made:
- the fractures are plane, indeformable,
- the rock matrix is homogeneous, isotropic and perfectly impervious,
- temperatures vary sufficiently slowly so that continuity exists accross the boundary between fracture and matrix.

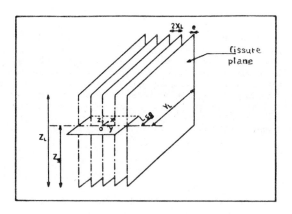

FIG.1 - SCHEMATIC REPRESENTATION OF A WASTE REPOSITORY
IN A GRANITE MASSIF

Equations in the fissure: The following equations are solved simultaneously:

- Mass balance: we neglect the variations of density (BOUSSINESQ's hypothesis):

$$\text{div } \vec{V} = 0$$

Velocity can therefore be expressed using a stream function

$$\vec{V} = \vec{\text{rot }} \psi$$

where ψ depends on x and z only (the velocity is supposed to be parallel to the fracture).

A momentum equation, which is a general form of Darcy's law:

$$\frac{\rho}{\varepsilon_f} \frac{\partial \vec{V}}{\partial t} = - \vec{\text{grad }} p + \rho \vec{g} - \frac{\mu}{gk_f} \vec{V}$$

Taking into account the definition of ψ, this equation can be rewritten:

$$(\frac{\rho}{\varepsilon_f} \frac{\partial}{\partial t} + \frac{\mu}{k_f}) \nabla^2_{xz} \psi - g \frac{\partial \rho}{\partial \theta} \frac{\partial \theta}{\partial x} = - \frac{1}{\varepsilon_f} \frac{\partial \rho}{\partial \theta} (\vec{\text{grad }} \theta \cdot \vec{\text{grad }} \frac{\partial \psi}{\partial t})$$

$$- \frac{1}{k_f} \frac{\partial \mu}{\partial \theta} (\vec{\text{grad }} \theta \cdot \vec{\text{grad }} \psi) \qquad [1]$$

which shows that the presence of a heat source of limited extension makes the configuration intrinsically unstable.

The variations of fluid density versus temperature come from literature. Those of the viscosity are given by BINGHAM's formula.

- Energy balance in an elementary volume $\varepsilon\delta x\delta z$:

$$\text{div } (\lambda^x \vec{\text{grad }} \theta) - \gamma_f \text{ div } (\theta \vec{\text{rot }} \psi) = \gamma^x \frac{\partial \theta}{\partial t} - \frac{2\lambda_g}{e} (\frac{\partial \theta_r}{\partial y})_{y=o} - \phi \qquad [2]$$

conductive flow   convective flow   flux to/from matrix

where $\phi(x,z)$ is the generated heat flux, distributed on the elementary volume
. the dispersive flux is neglected
. y=0 corresponds to the boundary between fissure and matrix.

Equations in the rock matrix:

- Energy balance for an elementary volume $\delta x . \delta y . \delta z$.

$$\text{div } (\lambda_g \vec{\text{grad }} \theta_r) = \gamma_g \frac{\partial \theta_r}{\partial t} - \frac{\lambda_g}{\partial y} \frac{\partial \theta_r}{\partial y} - \phi \qquad [3]$$

conductive flow   flux to/from fissure

where the generated flux $\phi(x,y,z)$ is distributed on the elementary volume.

The system [1], [2], [3] is solved numerically using finite differences.

2) STUDY OF THE THERMAL RELEASE OF A PUNCTUAL HEAT SOURCE IN A HELE-SHAW CELL (4)

a) Experimental device

The fissure is represented by the volume between two parallel sheets of plexiglas. The heat source is located in the middle of the fissure. The upper and lower boundaries are maintained at a fixed temperature. The other boundaries (x=+ x_L, y=+y_L) are considered adiabatic, owing to the low conductivity of the plexiglass.

The temperature field is visualized using cholesteric liquid crystals. The flowlines are visualized along three horizontal lines using the Baker method (modification of the fluid colour due to pH changes along the electrode) (1).

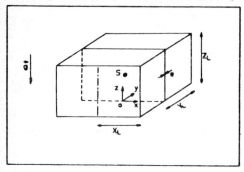

FIG.2 - CONFIGURATION STUDIED

b) <u>Numerical model</u>

The model represents 1/4 of the real configuration:

$$D = [0, X_L] \times ([-e, 0] + [0, y_L]) \times [0, z_L]$$

The discretization uses 16 vertical planes containing 21 x 21 meshes each. The heat transfer in the confining sheets is assumed to be one-dimensional in the y direction, and is solved explicitly at the end of each time step of the computation in the fissure. The flux in and outside the sheets is treated as a source-term of the heat equation in the fissure.

c) <u>Comparison of experimental and computed results</u>

The figure below shows a good qualitative and quantitative agreement between the measured and computed temperature fields.

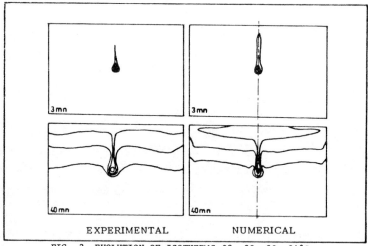

FIG. 3. EVOLUTION OF ISOTHERMS 28, 30, 32, 34°C

In the following figure, the velocity field along the 3 electrodes are compared for two different values of the injected heat flux.

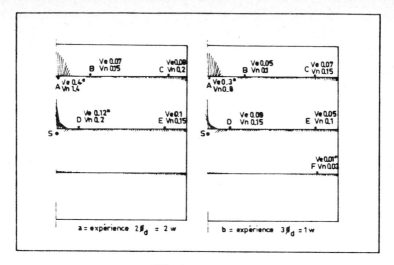

FIG. 4

We note a qualitative agreement between the sets of results: two contrarotative cells develop, and the velocities have the same order of magnitude. Measured values are systematically lower, probably due to the fact that the electrodes are not in the middle of the fracture, but against one of its walls, and that the velocity profile in the fissure is parabolic.

c) Conclusion

The equations used in the numerical model seem to fit satisfactorily with the experimental data. We have therefore developed a second experiment, closer to reality.

3) STUDY OF THE THERMAL RELEASE OF A LINEAR SOURCE IN A FISSURED GRANITE BLOC

a) Experimental device

A granite bloc from the Mayet de Montagne (France) is crossed by a natural, unclogged fissure. A linear electric resistor, perpendicular to the fracture plane, is used as a heat source. The lower boundary of the bloc is impervious and maintained at a fixed temperature. The upper boundary is in communication with circulating water, also at a fixed temperature. The lateral boundaries are insulated both hydraulically and thermically.

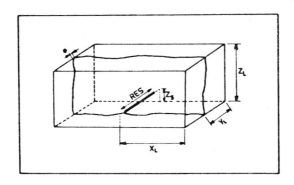

FIG. 5 - STUDIED CONFIGURATION

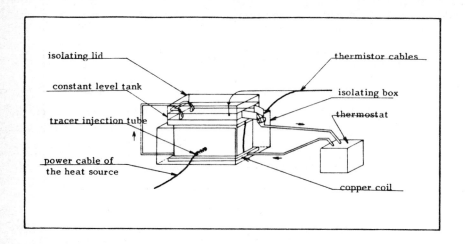

isolating lid

constant level tank

tracer injection tube

power cable of
the heat source

thermistor cables

isolating box

thermostat

copper coil

FIG. 6 - EXPERIMENTAL DEVICE FOR THERMAL EXPERIMENTS

Systematic measurements are possible in 14 points of the fissure and 4 of the matrix. A capillary tube allows the injection of a fluorescent tracer into the fracture. The measured values of the average aperture and permeability of the fissure are respectively 0,15 cm and 6,65.10^{-4} cm^2.

The thermal characteristics of the granite of the Mayet de Montagne are known from previous experiments:

$$g = 8.10^{-3} \ cal/cm.s.°C$$
$$g = 0,6 \ cal/cm^3.°C$$

b) Numerical model

The model represents 1/4 of the real configuration.

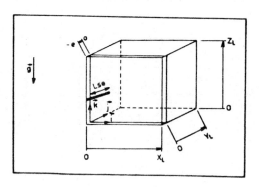

FIG. 7

The discretization uses 6 vertical planes of 20 x 20 meshes each. The complex thermal boundary conditions make the problem fully tri-dimensional. However, the solution in the matrix is computed solving the equation first in the xz plane, then in the y direction.

c) Comparison between experimental and computed results

The figure below shows the computed temperature and flow field, and the measured temperatures. A good agreement can be seen, except in the vicinity of the source where the discretization seems too coarse.

The thermal steady state is obtained, experimentally and numerically, after about 2 hours.

The stream functions show the development of two cells. The steady state is reached numerically after 1 hour.

The thermal flux between matrix and fissure goes first from the fissure to the matrix above the source and in the opposite direction under the source. At the end of the experiment, the flux goes from the matrix, where the thermal diffusivity is higher, towards the fissure in all the field.

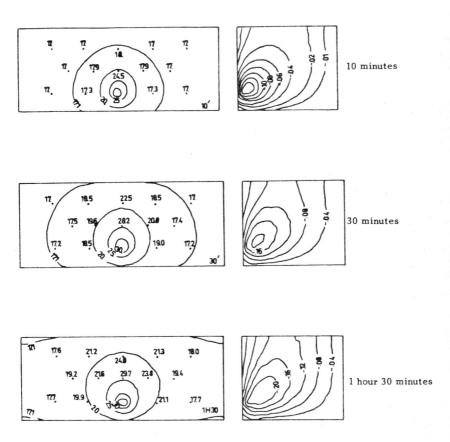

FIG. 8 - EVOLUTION OF THE THERMAL FIELD AND OF THE STREAM FUNCTION IN THE FISSURE.

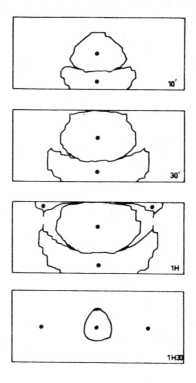

FIG.9 - DIRECTION OF THE FLUX BETWEEN FISSURE AND MATRIX
+ towards granite
- towards fissure

Tracer injection: After the thermal steady state has been reached, fluoresceine has been injected above the thermal source. The tracer appeared at the top of the fissure after 14 seconds. This result, which could be repeated several times, is in good agreement with the calculations of the model.

d) Conclusion:

The ability of the numerical model to reproduce experimental measurements has been demonstrated. An experiment is now planed in situ, with a heater at a shallow depth in a fractured crystalline massif, and an injection of tracer. Both temperatures and tracer concentrations would be monitored, in order to check the validity of the formulation at a larger scale and in real conditions.

Predictions of the influence of the thermal loading (coupled with stress variations and fracture aperture variations, for which laboratory experiments have been performed) would then be possible.

SYMBOLS

e	fissure aperture	L
g	gravity acceleration	LT^{-2}
k_f	intrinsic permeability of the fissure	L^2
Lse	length of the thermal source in the matrix	L
p	pressure	$ML^{-1}T^{-2}$
P	power	ML^2T^{-3}
RES	total length of the thermal source	L
t	time	T
V	filtration velocity	LT^{-1}
V_N	computed filtration velocity	LT^{-1}
V_E	measured filtration velocity	LT^{-1}
x_L, y_L, z_L	limits of the domain	L
z_s	elevation of the thermal source	L
γ_f	calorific capacity of the fluid	$ML^{-1}T^{-2}d°^{-1}$
γ	equivalent calorific capacity in the fissure	"
γ_g	equivalent calorific capacity in the matrix	"
ε_f	fissure porosity	
θ	temperature in the fissure	d°
θ_r	temperature in the matrix	d°

REFERENCES

(1) BAKER,J. (1966): A technique for the precise measurement of small fluid velocities
J. of Fluid Mechanics, vol. 26, part 3

(2) BORIES, S. (1970): Sur les mécanismes fondamentaux de la convection naturelle en milieu poreux.
Thèse, Université de Toulouse.

(3) COMBARNOUS, M. (1970): Convection naturelle et convection mixte en milieu poreux.
Thèse, Paris

(4) HOSANSKI, J.M.: Contribution à l'étude des transferts thermiques en milieu fissuré.
Thèse de Docteur-Ingénieur, ENSMP-Université Paris VI.

DISCUSSION

A.T. JAKUBICK, Canada

The fracture width used in the granite fracture experiments was 1.4 mm. This is 1 or 2 orders of magnitude larger than usually encountered under in situ conditions. What were the limiting conditions for darcyan flow in the model considered ? Did you check if your really had darcyan flow ?

P. GOBLET, France

We have no direct measurement of the velocity inside the fracture. However, the numerical model tests the validity of the darcyan flow hypothesis. As the model reproduces satisfactorily the observed arrival times of the tracer, we can reasonably assume that flow has remained darcyan along the experiment.

C. McCOMBIE, Switzerland

In the text of your paper, you state that you assume Boussinesq's hypothesis in the fissure - that is you neglect variations of density. In your presentation you appeared to imply that density variations were allowed for. What is the actual assumption made in your model ?

P. GOBLET, France

We use an extended version of Boussinesq's hypothesis, which is as follows :

- Neglect $\frac{\partial \rho}{\partial t}$ and $\vec{\text{grad}}\ \rho . \vec{V}$ in the mass balance equation, which allows us to continue using stream functions.

- Neglect the inertial term (of course we verify afterwards the order of magnitude of the neglected terms).

- Introduce laws of variation of density and viscosity versus temperature.

K.K. TSUI, Canada

If I understand correctly, there is no confining stress applied to your physical models in the laboratory. Can your physical models be modified to include the effects of confining stresses ?

P. GOBLET, France

Presently the models are not equiped for the application of confining stresses. However, as their geometry (fracture aperture) and boundary conditions can be modified, we could simulate the influence of such stresses during an experiment.

R.G. BACA, United States

How was the fracture permeability estimated ? By using the square of the aperture ?

P. GOBLET, France

We measured the permeability by means of two flow tests. The first one consisted in a linear flow obtained by imposition of a fixed head on the upper and lower limits of the fracture.

The second one was a "radial" test, with fixed head around the fracture and constant injection rate through the injection hole. The aperture of the equivalent plane fissure is smaller than the average measured aperture, which shows the influence of rugosity.

ANALYSIS OF HOST ROCK PERFORMANCE FOR A NUCLEAR WASTE REPOSITORY USING COUPLED FLOW AND TRANSPORT MODELS

R. G. Baca
Rockwell Hanford Operations
Richland, Washington

D. W. Langford and R. L. England
Boeing Computer Services Richland, Inc.
Richland, Washington

ABSTRACT

The performance of a nuclear waste repository in a deep geologic system will primarily depend on site-specific geohydrologic properties of the host rock and on basic design factors associated with the engineered barriers. The importance of these geotechnical aspects is clearly reflected in the recent technical criteria proposed by the United States' regulatory agencies for deep geologic repositories. The issue of compliance with the proposed criteria is one that can only be addressed using predictive models which describe the changes in the physical processes which occur as a result of the repository environment.

Numerical models were recently developed for simulation of coupled flow and transport processes in the near-field zone of a deep geologic repository. These models are currently being applied to a reference repository location in the Columbia River basalt at the Hanford Site in Washington state. This paper presents results from some preliminary computer model simulations carried out to evaluate host rock performance. Conclusions are drawn with regard to the most important physical processes and geohydrological properties determining the degree of waste isolation achieved by the repository system.

INTRODUCTION

Technical criteria for deep geologic repositories are currently being developed in the United States by the federal regulatory agencies. These criteria will ultimately form the basis for assessing the performance of candidate repository sites. Specific criteria proposed by the U.S. Nuclear Regulatory Commission [1], for example, would set such requirements as: the design of engineered barriers (waste package and underground facility) would assure containment of the waste for a minimum of 1,000 yr after closure, nuclide release rates from the repository would not exceed 10^{-5}/yr, and groundwater travel times from the repository to the accessible environment of 1,000 yr or longer. The U.S. Environmental Protection Agency [2] has proposed criteria which would limit the total activity of individual nuclides crossing a 1-mi (1.6 km) boundary from the repository, over a 10,000-yr period.

The Basalt Waste Isolation Project under way at Rockwell Hanford Operations is investigating the feasibility of constructing a deep geologic repository in the Columbia River basalt. The basalt formations which underlie the Hanford Site in southeastern Washington state are currently being characterized and studied. Accumulations of basalt in this area are particularly promising because of their large thickness, lateral extent, and relatively low permeability. Hydrologic modeling studies conducted to-date [3,4] for the candidate siting area indicate that groundwater travel times to the accessible environment are long; e.g., 10^5 yr, and that potential nuclide migration would be confined to the deep basalt formations.

In this paper, computer simulation results from a near-field analysis are presented which provide a perspective on the waste isolation performance of a reference repository site in a basalt geology. Transport calculations for a specific release scenario are presented and compared with the proposed criteria.

PERFORMANCE ANALYSIS SCENARIO

The release scenario considered in this performance analysis is the so-called no-disruption base case scenario. This scenario, in essence, represents the expected performance of the waste isolation system under "unperturbed" geologic conditions. In defining this particular scenario, the following general assumptions are made regarding the nature of the release sequence and the characteristics of the reference repository: (1) the repository contains the spent fuel inventory and is completely backfilled and the shafts are plugged and sealed; (2) the waste package and engineered barriers retain integrity for 100 yr after closure, after which the release is initiated; and (3) the radiocontaminants are released to the groundwater at a constant rate of 10^{-5}/yr.

The second assumption is conservative relative to the criteria proposed by the USNRC [1981]. Moreover, in the actual geologic system, the release rate would be controlled by the solubility limit of each nuclide. The basic objective of this analysis is to simulate the flow and transport processes over a suitable waste isolation period, namely, 10,000 yr.

The proposed repository in basalt would be designed to accommodate approximately one-half of the entire United States' inventory of 10-yr-old spent fuel projected for the year 2000, or about 47,000 t of heavy metals. The current conceptual design consists of two sections to the repository separated by about 830 m; together, these sections would span a total retangular region of 3,000 m long by 2,400 m wide. Located at a depth of about 1,000 m, the repository would be constructed in the central portion of a dense basalt flow. The rooms of the repository would be 6.5 m high, with canister boreholes in the floor of the room. Each canister would have a peak power output of about 1.5 kW, producing a maximum areal thermal load of 11 W/m^2. In this analysis, the performance of the host rock is assessed on the basis of the transport of a key radionuclide, namely ^{99}Tc, which has a large initial inventory and long half-life. The initial inventory assumed is 3.5×10^4 kg $(6.1 \times 10^5$ Ci). By virtue of its anionic form, ^{99}Tc has relatively low sorption in rocks. For conservatism, it is assumed that this key radionuclide is not sorbed in the rock mass.

Basic Features of Geologic System

The basalt rock beneath the Hanford Site represent a stratiform or tabular geology characterized by sequences of individual basalt flows or layers which, in certain locations, are separated by pervious water-bearing horizons. The candidate basalt flow located in the Grande Ronde Formation is referred to as the Umtanum basalt flow. This candidate flow is about 71 m thick and is laterally continuous across the basin.

For the purpose of evaluating host rock performance, a vertical cross section through the Grande Ronde Formation has been adopted in which the major stratigraphic features are grouped into 30 property types. Thus, the basic "geologic conceptual model" [5] consists of a multilayer system composed of an alternating sequence of dense basalt flows and flow contacts.

Unique aspects of the hydrologic features in the deep basalt flows are related to the large contrasts in their hydraulic properties. Flow contacts, for example, generally possess hydraulic conductivities of 10^{-6} to 10^{-9} m/sec, with porosities estimated to be on the order of a few percent; whereas, the confining basalt layers, have hydraulic conductivities of 10^{-10} m/sec, or less, and estimated porosity range of 10^{-2} to 10^{-4}.

Hydraulic head measurements taken at various deep boreholes within the Hanford Site indicate that the regional head gradients range between 1.0 and 0.1 m/km. Vertical head gradients across the candidate flow range from 25.0 to 1.0 m/km, with the lower values being more representative of the reference repository location. The inferred potentiometric fields in combination with the large permeability contrasts between the basalt flow and flow contacts strongly suggest a predominantly horizontal flow pattern across the central portion of the Hanford Site.

In this particular near-field analysis, the following assumptions are made with regard to the hydraulic properties of the geohydrologic systems:

1. Regional hydraulic head gradient is 1.5 m/km.

2. Vertical head gradient in the deep basalt formation is 2 m/km.

3. Horizontal hydraulic conductivity of the basalt flows is 10^{-11} m/sec, with anisotropy ratio (K_{zz}/K_{xx}) of 10.

4. Hydraulic conductivity of the flow contacts is 10^{-7} m/sec, with anisotropy ratio of 1.

5. Porosity of the basalt flows is 10^{-4} or 0.01%.

6. Porosity of the flow contacts is 10^{-2} or 1.0%.

The values assumed were chosen to be representative of nominal values, but also tending toward the conservative. The low porosities in particular produce conservative calculations for both groundwater and radionuclide travel times.

PERFORMANCE ANALYSIS APPROACH

The basic hydrologic variables which reflect the performance of a repository system consist of: groundwater pathlines, groundwater travel times, and radionuclide concentrations. Calculation of these variables requires the simulation of the coupled processes of groundwater flow, heat transfer, and nuclide transport. Mathematical models [6,7] of these processes have been developed from the basic conservation principles for mass and energy.

Mathematical Models

Groundwater flow in a nonisothermal regime is strongly dependent on the temperature of the fluid. This coupling is particularly important in the near-field by
virtue of the variations in fluid density and hydraulic conductivity of the
medium which occur with temperature changes. Dependence of these properties on
the thermal regime is made clear by considering the Darcy flow equation for the
nonisothermal case [8], which is written in indicial notation as:

$$q_i = -K_{ii}(\partial_i h + \delta_b \delta_{i3})$$ (1)

where the quantities are defined by: q is the Darcian flow velocity, m/sec; K is
the hydraulic conductivity, m/sec; h is the hydraulic head, m; δ_b is the density
disparity, unitless; and δ_{i3} is the Kronecker delta.

Recalling the basic relationship of hydraulic conductivity, K, and the fluid
properties, namely:

$$K = \frac{k\rho g}{\mu}$$ (2)

where the quantity k is the intrinsic permeability of the medium (m^2); ρ is the
fluid density (kg/m^3); g is the acceleration of gravity (m/sec^2); and μ is the
fluid viscosity (kg/m/sec). Over the temperature range expected in the near-field
zone, the fluid density would change by a few percent; however, the fluid viscosity will decrease by 20 to 30 times. As a result, the hydraulic conductivity will
increase by a significant amount. The density disparity term is a function of
fluid density and is computed from the equation:

$$\delta_b = \frac{\rho}{\rho_0} - 1$$ (3)

In this expression, ρ_0 is the fluid density (kg/m^3) at the reference temperature.

The density disparity, δ_b, represents the buoyancy driving force created by the
thermal regime. Evaluating the above equation for various temperature increments,
ΔT, one notes that for every 20°C the buoyancy driving force produced is equivalent to a hydraulic head gradient of 0.01 m/m. Thus, in the vicinity of a repository where the ΔT is relatively high, the buoyancy driving forces could be as
large as 0.05 m/m; such driving forces are significantly greater than the observed
natural hydraulic gradients by one order of magnitude or more.

By introducing the Darcian flow equation into the general equation for fluid
continuity, one obtains the governing equation for groundwater flow in an "equivalent porous continuum," namely:

$$S_s \frac{\partial h}{\partial t} = K_{ii}(\partial_{ii} h + \delta_{i3}\partial_i \delta_b) + \gamma \frac{\partial T}{\partial t}$$ (4)

where the variables and coefficients are defined by: S_s is the specific storage,
1/m; γ is the thermal coupling coefficient, 1/C°; t is the time, sec; and T is the
fluid temperature, °C. The repeated index implies summation.

A thermal energy balance on the combined water/rock system yields the governing equation for heat transport. In indicial notation, the expression is:

$$S_t \frac{\partial T}{\partial t} + \rho c_f q_i \partial_i T = D_t \partial_{ii} T - Q \qquad (5)$$

where S_t is the heat capacity, J/m^3; D_t is the thermal conductivity, J/sec-m-°C; and c_f is the specific heat capacity, J/kg-°C; and Q is the heat-generation rate, J/m^3-sec.

From the basic mass balance requirement for an arbitrary dissolved constituent or solute, one can derive the governing equation for radiocontaminant transport expressed by:

$$R_d \phi \frac{\partial C}{\partial t} + q_i \partial_i C = D_m \phi \partial_{ii} C - \lambda R_d \phi C + m_i \qquad (6)$$

where C is the nuclide concentration, mg/1; R_d is the retardation factor, unitless; ϕ is the effective porosity, unitless; D_m is the mass dispersion coefficient, m^2/sec; λ is the decay coefficient, 1/sec; and m_i is the mass source term, mg/1-sec.

The complete set of governing equations provides a systems description of the relevant hydrologic processes in terms of three state variables: hydraulic head, h; temperature, T; and concentration, C. From these state variables, groundwater pathlines and travel times are directly computed. From the flow patterns and concentration distributions, the total mass of a nuclide crossing an arbitrary boundary is computed over the time period of interest.

Numerical Models

A set of numerical models has been developed to implement various forms of the governing equations for flow and transport in a hardrock geology. These numerical models are specifically designed for repository analysis at two space scales: very near field (canister to room scale) and near field (repository scale). The numerical models fall into two basic categories: finite element and finite difference.

The primary finite element models used in near-field analyses are referred to as MAGNUM AND CHAINT, which are used conjunctively. Both numerical models are based on Galerkin finite element solution techniques [7]. The primary capability of these models is that they can describe relevant phenomena in a "dual porosity" continuum with discrete hydrogeologic features. The continuous portions of the rock mass are represented using triangular and quadrilateral finite elements; line elements are embedded along the sides of the two-dimensional elements to represent discrete macroscale (or microscale) features. Some of the basic features of MAGNUM AND CHAINT finite element models are:

- Models accommodate stratigraphic features and variable media properties.

- Computer codes provide options for simultaneous or sequential solution of governing equations.

- Transport code accommodates any combination of single or multicomponent sets of nuclides.

In the second category, the finite difference model, PORFLO, solves the governing equations as they apply to an "equivalent porous continuum." A nodal point integration method [9,10] is employed in conjunction with an alternating direction implicit algorithm in formulating the finite difference method. Major advantages of this numerical model are the ease of use and low cost of use. Other general features of this model include:

- Numerical method ensures energy and mass consistency at the subdomain level.

- A donor cell method is used to handle advection-dominated flow regimes.

- Model formulation is generalized to accommodate planar or axisymmetric coordinate systems.

Both the finite element and finite difference models have been extensively verified against known solutions for various boundary value problems as well as validated with experimental data.

RESULTS OF NEAR-FIELD ANALYSIS

In this performance analysis of a reference repository in basaltic rock, the finite difference model, PORFLO, was applied to simulate the coupled flow and transport processes in the near-field zone. The postulated release scenario assumes the release of a key radiocontaminant which initiates during the peak thermal period. The rate and extent of radiocontaminant movement are computed for a 10,000-yr period after closure. The principal simulation results are summarized below.

Groundwater Pathlines and Travel Times

Pathlines, which are the trajectories of particles moving with the groundwater, provide a detailed representation of the principal "pathways" of potential waste movement. The cumulative time of travel along any pathline is simply the associated groundwater travel time. By connecting points of equal time along a set of pathlines, a visualization of the flow field or flow fronts between the bounding pathlines is obtained.

For the no-disruption release scenario, the pathlines and travel times were computed for various starting points in the rock mass above the repository and assuming a release time of 100 yr after closure. These results are presented in Figure 1. The pathline patterns in the vicinity of the repository indicate a general upward direction; this trend is produced by the buoyancy driving force which is significant within the bounds of the thermal plume. The time lines in this region above the repository sections indicate a well-developed parabolic flow field.

Outside the zone of thermal influence, the pathline directions are determined by the natural hydraulic gradients and rock mass hydraulic properties. In this region, the pathline patterns exhibit a "stair step" pattern and predominantly horizontal direction. This particular trend is expected because of the large permeability contrasts between the flow contacts and the dense basalt flows; in essence, the groundwater seeks the path of least resistance which is along pervious flow contacts. The nominal upward groundwater movement through the basalt flows occurs as a result of the upward hydraulic gradient and the assumed anisotropy of the dense basalt flows.

With regard to the heat-transfer simulation, it was observed that the temperature field is symmetric around the repository sections. This pattern suggests that the heat transfer through the fluid phase is much smaller than the heat transfer through the rock mass. This observation is also reflected in the Peclet number which is very small because of the small flow rates and low porosities of the rock mass. Thus, the major component of heat transfer is by simple conduction through the geologic media. The average maximum temperature in the repository was computed to be 130°C and occurs at about 50 yr after closure (Figure 2). In addition, it is observed that significant temperature perturbations in the vicinity of the repository persist over the entire 10,000-yr period.

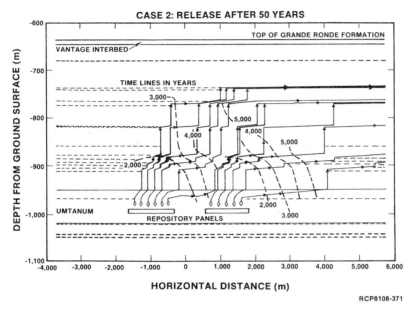

FIGURE 1. Groundwater Pathlines and Travel Times.

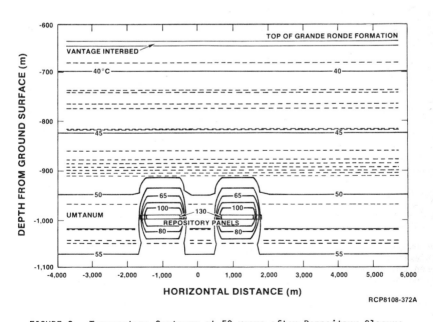

FIGURE 2. Temperature Contours at 50 years after Repository Closure.

Radiocontaminant Movement

The patterns of radiocontaminant movement computed for this release scenario clearly show a correlation to the groundwater pathlines and flow fronts. As illustrated in Figure 3, waste migration above the repository is primarily upward. Some advective transport along the flow contacts occurs and is reflected in the shape of the contours. By virtue of the wide separation between the bounding pathlines, the contaminant plume is spread over a significant portion of rock mass, rather than being confined to selected flow contacts. This characteristic enhances dilution which, for the most part, accounts for the low radiocontaminant concentrations. At the end of the waste-isolation period, the extent of the plume (defined by the boundary of the maximum permissible concentration, MPC, level) is limited to 200 m vertical movement and about 4 km horizontal movement.

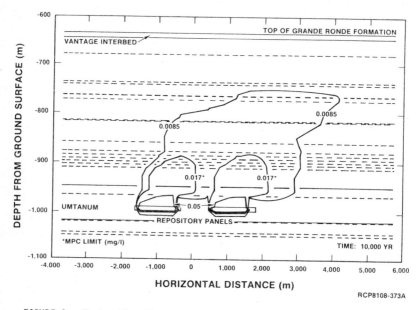

RCP8108-373A

FIGURE 3. Technetium-99 Concentration Contours at 10,000 years after Repository Closure.

The mechanics of waste migration through the host rock, the Umtanum basalt flow, is directly related to the interplay of the groundwater flow and heat transfer processes. High temperatures and large temperature gradients simultaneously increase the hydraulic conductivity of the geologic medium and generate buoyancy driving forces. The effect of the higher conductivity is manifested in a decrease in the isothermal (or natural) head gradient. This response is explained by the fact that the increased conductance produces a condition where less driving force is needed to accommodate the same flow.

The buoyancy driving force is a very significant factor, determining the radio-contaminant flux across the boundaries of the host rock. The relationship between temperature effects and the rate of release from the host rock is illustrated in computations tabulated below:

TABLE I. Radiocontaminant Fluxes for Host Rock
Assuming No-Disruption-Release Scenario.

Time (yr)	Repository temperature, ΔT (°C)	Integrated mass flux of ^{99}Tc (kg)
100	86	0.00
250	82	0.00
500	76	0.01
1,000	64	0.11
3,000	54	4.77
5,000	51	23.57
10,000	38	166.46

The total inventory released from the Umtanum basalt flow after a 10,000-yr period is about 170 kg or only 0.5% of the initial inventory.

Mass fluxes computed for a hypothetical buffer zone boundary at 1 mi (1.6 km) from the repository indicate that less than 12 Ci/1,000 t of ^{99}Tc would leave the buffer zone in 10,000 yr. This value is more than 2 orders of magnitude smaller than the current USEPA [2] criteria for technetium. For the more conservative case where the release rate is increased an order of magnitude; e.g., 10^{-4}/yr, less than 120 Ci/1,000 t would pass the 1-mi (1.6-km) boundary in 10,000 yr. Thus, compliance with the proposed criteria would still be achieved. Considering the properties of the key radiocontaminant and the conservative assumptions made in defining the scenario, these results suggest that the USEPA criteria for individual radiocontaminants could be easily met even if the proposed USNRC criteria (for waste-package design and the release rate) are not met. In turn, this could lead to the significant conclusion that there is little or no incentive for development of advanced waste packages with extended lifetimes.

In summary, the simulations of coupled flow and transport indicate that the heat transfer from the repository would have a major role in determining the rates and direction of potential waste movement. The rate of release of radiocontaminants from the host rock will be a direct function of the temperature gradients and, therefore, dependent on the repository heat load. Groundwater flow patterns in the near field are affected by the temperature gradients in a way which produces greater separation of the pathlines from the repository. In turn, this has the effect of spreading the radiocontaminant plume over a relatively large portion of low-permeability rock mass, which increases the radiocontaminant travel time. Over the period of waste isolation, the radiocontaminant plumes would be contained within the deep basalt formation.

CONCLUSIONS

A near-field analysis of a reference repository location was conducted to evaluate the long-term performance of the waste-isolation system. Numerical models were applied to simulate the process of coupled groundwater flow, heat transfer, and nuclide migration. The rate and extent of nuclide transport were calculated for a no-disruption-release scenario. Results of these computer simulations were compared to the criteria proposed for deep geologic repositories. The principal findings and conclusions of the analysis are: (1) groundwater-flow patterns around a repository are significantly altered for long-term periods by the temperature field, (2) heat transfer through the water/rock system occurs primarily by conduction through the rock mass, (3) the rate of nuclide transport through the host basalt flow will be strongly influenced by the buoyancy driving forces and the hydraulic conductivity of the rock mass.

With regard to waste-isolation effectiveness of the basalt medium, the results of the analysis indicate that:

1. Waste migration would be confined to the deep basalt formation.

2. Compliance with proposed USEPA criteria for ^{99}Tc could be achieved with an engineered barrier system characterized by 100-yr waste package, and release rate of 10^{-4}/yr.

3. Buoyancy effects created by the waste heat are more significant driving forces for potential waste movement than natural hydraulic gradients.

4. Any contaminant plume would be spread over a large portion of rock mass rather than channeled in any specific flow contact.

The latter observation is explained by the fact that the pathline patterns unique to a layered system are highly "refracted" by virtue of the permeability contrasts. The net result is that the groundwater pathlines bounding the contaminant plumes are widely separated; this spreading assures significant dilution of the dissolved radiocontaminants as well as increasing nuclide travel times.

REFERENCES

1. USNRC, 1981, "Disposal of High-Level Radioactive Wastes in Geologic Repositories - Technical Criteria, Proposed Rule Making," 10 CFR 60, Subpart E, 45 Federal Register 31393, U.S. Nuclear Regulatory Commission, Washington, D.C.

2. USEPA, 1981, "Environmental Standards and Federal Radiation Protection Guidance for Management and Disposal of Spent Nuclear Fuel, High-Level and Transuranic Radioactive Wastes," Draft, 40 CFR 191, 45 Federal Register 31393, U.S. Environmental Protection Agency, Washington, D.C.

3. King, I. P., D. B. McLaughlin, W. R. Norton, R. G. Baca, and R. C. Arnett, 1980, Parametric and Sensitivity Analysis of Waste Isolation in a Basalt Medium, RHO-BWI-C-94, Rockwell Hanford Operations, Richland, Washington.

4. Arnett, R. C., R. D. Mudd, R. G. Baca, M. D. Martin, W. R. Norton, and D. B. McLaughlin, 1981, Pasco Basin Hydrologic Modeling and Far-Field Radionuclide Migration Potential, RHO-BWI-LD-44, Rockwell Hanford Operations, Richland, Washington.

5. Myers, C. W., S. M. Price, J. A. Caggiano, M. P. Cochran, W. J. Czimer, N. J. Davidson, R. C. Edwards, K. R. Fecht, G. E. Holmes, M. G. Jones, J. R. Kunk, R. D. Landon, R. K. Ledgerwood, J. T. Lillie, P. E. Long, T. H. Mitchell, E. H. Price, S. P. Reidel, and A. M. Tallman, 1979, Geologic Studies of the Columbia Plateau: A Status Report, RHO-BWI-ST-4, Rockwell Hanford Operations, Richland, Washington.

6. Baca, R. G., J. B. Case, and J. G. Patricio, 1980, "Coupled Geomechanical/ Hydrological Modeling: An Overview of Basalt Waste Isolation Project Studies," Proceedings of Second Workshop on Thermomechanical - Hydrochemical Modeling for a Hardrock Waste Repository, LBL-11204, CONF-800751, Lawerence Berkeley Laboratory, Berkeley, California, also RHO-BWI-SA-82, Rockwell Hanford Operations, Richland, Washington.

7. Baca, R. G., R. C. Arnett, and I. P. King, 1981, "Numerical Modeling of Flow and Transport Processes in a Fractured-Porous Rock System," Proceedings of 22nd U.S. Symposium on Rock Mechanics, Massachusetts Institute of Technology, Cambridge, Massachusetts, also RHO-BWI-SA-113, Rockwell Hanford Operations, Richland, Washington.

8. Bear, J., 1972, Dynamics of Fluids in Porous Media, American Elsevier Publishing Company, New York, New York.

9. Pantankar, S. V., 1980, <u>Numerical Heat Transfer and Fluid Flow</u>, Hemisphere
    Publishing Corporation, McGraw-Hill Book Company, New York, New York.

10. Gosman, A. D., W. M. Pun, A. K. Runchal, D. B. Spalding, and M. Wolfstein,
    1969, <u>Heat and Mass Transfer in Recirculating Flows</u>, Academic Press, New York,
    New York.

ACKNOWLEDGMENTS

The authors wish to express their gratitude to various individuals who made direct
contributions and provided information used in this work. First of all, the
authors wish to thank Dr. R. A. Deju, Project Director, and Mr. R. E. Gephart for
their careful review and helpful comments which significantly improved the clarity
and focus of the paper. Special thanks are expressed to Dr. B. Sagar, Analytic
and Computational Research, Inc. for his assistance in the applications of the
numerical model.

The work summarized in this paper was conducted under the Basalt Waste Isolation
Project for the U.S. Department of Energy by Rockwell Hanford Operations under
contract DE-AC06-77RL01030.

DISCUSSION

P. UERPMANN, Federal Republic of Germany

How significant is the heat transport due to fluid flow in comparison to the thermal conductivity of the solid basalt in your model ?

R.G. BACA, United States

Heat transport in fluid is very small as compared to conduction in the rock mass. The reason for this is that the groundwater velocities are very small (so advection is small) ; and the porosity of the rock is small so that the thermal conductivity contributed by the fluid is small.

A. MULLER, United States

In your example of radionuclide transport, you use ^{99}Tc which is unretarded by the geologic medium. This is equivalent to simply modelling the water transport. How do you treat more realistic cases where the nuclide is retarded ? More specifically, do you have use retardation data for the specific layer in the model and how do you account for matrix retardation versus secondary mineral retardation (utilization factor) ?

R.G. BACA, United States

The retardation factor used in the model is computer from the equivalent sorption coefficient depending on whether it represents sorption in the "matrix" or sorption on fracture interfaces. In most model applications, however, conservative values of the retardation factor are used to account for the uncertainty generally associated with laboratory sorption data.

E.K. PELTONEN, Finland

You have employed a porisoty value of 10^{-2} ; is it a value of primary or secondary porosity ?

R.G. BACA, United States

The porosity value assumed for the flow contacts (interflow zones) is a primary porosity.

Session 4

CLAY AND TUFF

Chairman - Président

N.A. CHAPMAN

(United Kingdom)

Séance 4

FORMATIONS ARGILEUSES ET TUF VOLCANIQUE

EXPERIMENTATION ON AND EVALUATION OF NEAR-FIELD PHENOMENA IN CLAY. THE BELGIAN APPROACH

R. Heremans
A. Bonne, P. Manfroy
Centre d'Etude de l'Energie nucléaire CEN/SCK
Mol- BELGIUM

ABSTRACT

A deep tertiary plastic clay formation is investigated since 1975 for its accepta-bility as long term safe host medium for conditioned radioactive wastes.

Rock-waste interaction studies were started on samples taken during a core drilling on the potential site. The characterization of the clay material, including the interstitial water, was the main objective of the first laboratory researches. Simultaneously influence of temperature and γ irradiation on the clay properties were investigated, corrosion and leaching tests were performed and special atten-tion was given to sorption capacity of the clay material. In 1979 a simulated full scale heat transfer experiment was started at shallow depth in a clay quarry where also "in situ" corrosion experiments are still running. Final demonstration of near field phenomena will start in a couple of years from now. An underground facility at 220 m depth into the selected clay formation is already under construc-tion and will be operational early 1983.

Results from laboratory and "in situ" experiments will progressively be intro-duced in various models developed in the frame of a general deterministic system analysis.

Work performed in the frame of a contract between the European Community of Atomic Energy and the Centre d'Etude de l'Energie nucléaire of BELGIUM.

INTRODUCTION

In 1974 the decision was taken in Belgium to start R and D work on geological disposal of conditioned reprocessing wastes. At that time the construction of the first industrial nuclear power plants was already started. Sixteen hundred MW(e) were put into operation mid of the years seventy, this represented nearly 20 % of our total annual electricity production. Another three thousand eight hundred MW(e) are scheduled to start in the period 1982-1984. This means about 45 % of our electricity. Taking into account this nuclear power capacity and a possible slight increase in the future, one can postulate that the first repository in Belgium should have a capacity of about 1,000 m^3 of vitrified HLW corresponding to a total nuclear energy production of 300 GW(e) which is roughly the production foreseen in Belgium during 40 years.

An inventory of potential geological disposal host formations indicated that only clays and shales could be considered in our country. Seven potential areas were identified, one of them situated in the North-Eastern part of Belgium where the main Nuclear Research Centre and some industrial nuclear facilities are located. So, the field-work investigation was started on the C.E.N./S.C.K. site in Mol. A first cored bore-hole was performed to the top of the Cretaceous at -570 m. The tertiary clay formation known as "Boom clay" was met between approximatively -160 m and -270 m. It is a rather homogeneous, stiff but still plastic clay, containing up to 25 % of water, about 20 % of smectite and a variable carbonaceous percentage (to a few one). In order to characterize precisely the candidate host medium, the cored samples were submitted to extensive laboratory analyses and tests, mineralogical and chemical components were identified and physical and physico-chemical properties determined. It is obvious that a good knowledge of the various internal equilibria in the clay is essential for an accurate evaluation and a better understanding of the mutual interactions between waste, repository and host-rock. The result of the effects of these interactions will define the near-field scenario of the natural evolution of the whole repository. From the experimental point of view laboratory scale investigations and near-surface "in situ" tests are going on since several years. "In situ" experiments at real depth are planned to start early 1983. Along with these activities supporting desk studies are going on.

In the following pages an outline is given of our research activities on the near-field phenomena derived from nuclear waste emplacement into a man made repository in a deep clay formation.

LABORATORY INVESTIGATIONS

These are performed on core samples taken at real depths at the potential site. Alteration by oxygen (air) of some components of the clay material and dessication are two factors one has to deal with in the experiments and the interpretation of the results.

Gamma irradiation and heat impact

In order to identify the effects of γ irradiation on natural clay, orientative tests were performed. The samples were sealed in quartz tubes in absence of air and put in the centre of spent BR2 fuel assemblies. For an integrated dose of 10^8 rads, hydrogen and carbone dioxide gas releases were detected together with small amounts of methane in some cases. Following reactions are supposed to take place :

$$2H_2O \rightarrow 2H_2 + O_2$$
$$\text{organic } 2(CH) + 2O_2 \rightarrow H_2 + 2CO_2.$$

Additional laboratory investigations are going on in order to confirm and quantify the various phenomena and also to feature the effects which could result from these reactions : for instance the physico-chemical behaviour of H_2 and CO_2 in natural clay and the consequences for material corrosion and sorption capacity. For the actual repository concept, calculations indicated that the affected zone in the vicinity of the buried waste could be 1.5 m thick.

High irradiation field also means heat generation. Although the influence of heating during long periods of time (several centuries) on the mineralogical composition and fabric of the Boom clay were not investigated on a global experimental basis, one can however expect that no important modifications will occur at temperatures below 400° taking into account the lithostatic pressure of ± 40 bars. To be sure, however, a correct assessment of the influence of temperature should be obtained from tests of hydrothermal alteration carried out at the maximum values of temperature and pressure to envisage in a repository.

Aside the clay minerals other typical components are found in Boom clay. Pyrite, for example, which under the action of heat and particularly in the near-field, where oxygen is present and diffuses, may be oxidized to products as SO_2 of paramount importance in the corrosion of metals. Organic materials decompose in oxidizing and even neutral atmospheres. Studies indicate that, with time, decomposition could well be virtually complete at 250 °C. These are all reasons, why in the actual disposal concept a temperature of 150 °C was considered as a maximum acceptable at waste rock interface.

Besides heat transfer experiments the most relevant research works performed today concern :

- the influence of temperature on material corrosion and clay sorption (items treated in the corresponding sections) ;

- the influence of temperature on the geomechanical properties of the clay material.

In this last field a large laboratory tests programme on undisturbed clay samples is now starting. The samples were taken at shallow depth in a clay pit excavated in the area where the Boom clay formation crops out (about 50 km SW of Mol). It is however expected that the results can be extrapolated to the Mol case, taking into account the observed analogies in the results obtained on clay samples coming from both site. The aim of the running programme is to determine the influence of temperature variations from -25° up to +80 °C on compressibility, swelling and permeability coefficients measured by oedometric tests and also on the cohesion and the angle of shearing resistance measured by triaxial tests. Temperature cycling between 20 °C and 200 °C will also be applied to samples tested for compression. Today the number of results is limited, however some trends already come out for the lower temperature range :

- increase of shear stress between 0 and -25 °C, this increase is a function both of consolidation pressure and temperature (the lower the temperature the higher the shear stress) ;

- while at normal temperature and at -25 °C the elongation at failure is around 8 %, this value reaches to a maximum of 15 to 20 % at -5 °C ;

- repeated cycles of temperature between ambient temperature and -25 °C has no influence on elongation and failure strength.

A complete set of results is expected before the end of this year. The tests at temperatures below 0 °C received priority due to the fact that the freezing technique is applied for the digging of the experimental gallery.

When looking at the heat impact it is of course important to know the temperature distribution in the rock mass as a function of time and to determine the physical parameters involved.

Laboratory experiments on clay cores indicated a thermal conductivity of 1.5 W/m °C for natural clay containing up to 25 % of water and at an average temperature around 35 °C. The laboratory set up used and the diagram obtained are presented on Figures 1 and 2. This conductivity decreases very rapidly to a value of 0.3 W/m °C at 100 °C and as a function of decreasing water content. Dehydratation of confined clay under an hydrostatic pressure of 20 bars is however theoretically impossible even at temperatures up to 200 °C. This was qualitatively confirmed during a one year laboratory test performed on a clay lump 30 cm in diameter, 50 cm high at 6 atm. and 160 °C.

As far as water convection is concerned and taking into account the local known conditions, a preliminary mathematical assessment has shown that such a phenomenon might be neglected. So the heat transfer model developed at C.E.N./S.C.K. considers that the thermal flow is only transmitted by conduction. On this basis a finite element calculation allowing to assess the mechanical stresses in the near-field due to thermal expansion of the massif has been performed. It has to be mentioned, however, that the clay was considered as an homogeneous, isotropic and perfectly elastic medium. The results of these calculations indicated that

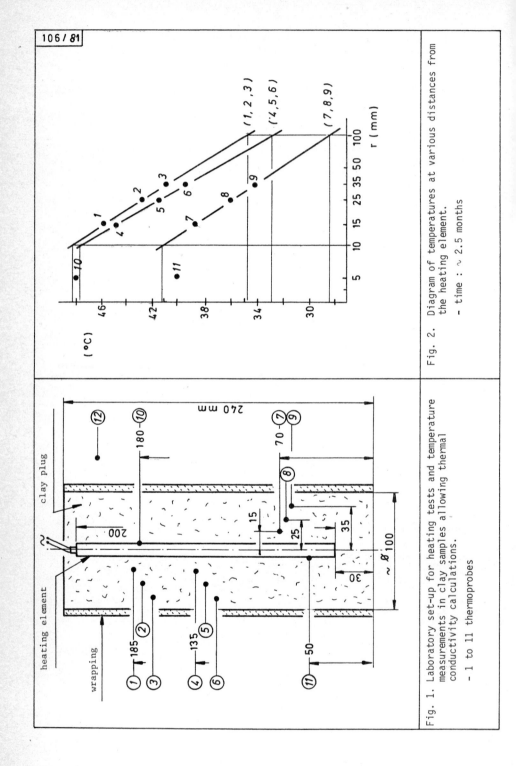

106 / 81

(°C)

r (mm)

Fig. 2. Diagram of temperatures at various distances from
the heating element.
- time : ∼ 2.5 months

240 mm

clay plug

heating element

wrapping

Fig. 1. Laboratory set-up for heating tests and temperature
measurements in clay samples allowing thermal
conductivity calculations.

- 1 to 11 thermoprobes

in this case the thermal expansion stresses represent only a small fraction of those due to the lithostatic and swelling pressures.

In order to compare small scale conductivity results to actual conditions, a near-surface experiment with a full size simulated HLW canister (30 cm in diameter and 150 cm length) was performed as described more in detail later on in this paper.

Corrosion and leaching

No decision is taken concerning the final conception of the waste package at disposal time. However, one can postulate that, depending upon the nature of the package components, the waste container will be at some time in contact with natural clay or at least with interstitial clay water or vapors. In our option the present objective is to select a metal or alloy which could resist to clay corrosion for several centuries (700 to 1,000 years). This material, if not suitable at the same time as container material for vitrified waste, can then be used as overpack for the classical stainless steel container. The materials to be tested, have been selected on the basis of a bibliographic screening of corrosion resistant materials in such media as, geothermal brines, groundwaters, agressive soils charged with solutes of sulphur, chlorine, organics a.s.o..

These materials fall into 5 categories :

a) aluminium and its alloys ;

b) carbon and chromized steels ;

c) austenitic and ferritic steels ;

d) nickel alloys and super alloys ;

e) titanium and titanium alloys.

As cast iron is a potential construction material for underground gallery lining it was also included in the corrosion tests. The samples are tested in the annealed and heat treated conditions while some of them are welded or stressed. The laboratory corrosion conditions were chosen to simulate as well as possible two different final "in situ" conditions : direct contact with natural clay and contact with an atmosphere prevailing in the direct vicinity of heated clay. The testing temperatures ranged from ambient up to 300 °C. Laboratory corrosion tests in the presence of a γ source are planned to start in the near future, and beside these investigations, attention will also be given to bacterial corrosion in clay medium.

As it will be mentioned later on, near-surface corrosion tests completed our programme in this field, the general trends of the already obtained results will be given there.

In addition to detailed examination of the metallic samples after regular periods of time (every 6 months), analyses of corrosion products are performed. It is indeed important to include the soluble corrosion products in the sorption studies and to evaluate their impact on the radionuclide migration in the direct vicinity of the leached source. The importance of this phenomenon is illustrated in Table I.

Table I. K_D values for Cs and Eu in the absence and presence of corrosion products (pH 3)

Solution type	K_D - Cs	K_D - Eu
A	254	7,827
B	115	545
C	15	2

A : Groundwater Cs : 5 mg.ℓ^{-1} Eu : 20 mg.ℓ^{-1}

B : Groundwater +   Fe : 100   Cr : 26   Ni : 14   Mn : 1.5
   in mg.ℓ^{-1} Sr : 13.3   Zr : 30   Mo : 20   Cs : 5.8   Eu : 30.6

C : Groundwater + concentr. sol. B x 10

Corrosion or leaching of various waste matrices is of course also one of our main points of concern. A laboratory for dynamic and static tests on actinide containing materials is now fully equipped. Experiments are performed with simulated interstitial clay water as well as in direct contact with natural clay. The main objectives of these researches are to improve the corrosion and the leaching resistance of the matrices by modifying the nature and the quantities of some constituents, and to develop a representative waste acceptance test. The leachates, are also used as feed solutions for migration tests.

Retardation to radionuclide migration

The presence of vermiculite or vermiculite-like and smectite clay minerals in the Boom clay makes it an excellent sorbent for various radionuclides.

The average cation exchange capacity measured for natural samples is 0.30 ± 0.05 meq/g. Extensive laboratory work was devoted to measurements of distribution coefficients under a wide variety of conditions or pre-treatments, types and concentrations of radionuclides, pH, temperature and γ irradiation. Exchanging ability of clay is selective, and taking into account the numerous radionuclides and corrosion products of concern, selectivity should be studied. This can be expressed by the selectivity coefficient :

$$K_c = \frac{Z_1 \cdot m_2}{Z_2 \cdot m_1}$$

where Z = the equilibrium equivalent fraction of an ion in the clay

m = the equilibrium equivalent fraction of an ion in solution.

Homo-ionic Boom clay is very selective for example for Cs and Eu, especially for lower concentration levels, while for Sr it is not. Competition between radio-nuclides, corrosion products and other leached ions from waste conditioning matrices were only partially studied, but intensive experimentation is planned, some K_c results are already presented in Table II.

Table II. K_c values for Sr, Cs and Eu of homo-ionic Boom clay (Ca-form) (absence of sulphate)

m_{Sr}	K_c (Sr)	m_{Cs}	K_c (Cs)	m_{Eu}	K_c (Eu)
$0.03 \ 10^{-2}$	0.93	$2.3 \ 10^{-4}$	369	$1.7 \ 10^{-5}$	74
$0.10 \ 10^{-2}$	1.02	$1.1 \ 10^{-3}$	281	$1.9 \ 10^{-4}$	23
$0.16 \ 10^{-2}$	1.03	$2.2 \ 10^{-3}$	216	$5.0 \ 10^{-4}$	25
$0.25 \ 10^{-2}$	1.00	$4.0 \ 10^{-3}$	103	$1.2 \ 10^{-3}$	13
$0.33 \ 10^{-2}$	1.00	$5.4 \ 10^{-3}$	83		
$N_{tot} = 10^{-2}$		$N_{tot} = 10^{-2}$		$N_{tot} = 10^{-1}$	

Taking into account that calculations indicated that migration of the majority of the radionuclides is limited in clay, it is expected that in the very near-field "cocktails" of different competing soluble specimens will occur.

The behaviour of radionuclides (and other ions) in a clay formation is also determined by other conditions, for instance solubility, complexation and so on. It is known that the interstitial water solution in argillaceous formations is generally of high ionic concentration, the high concentration of some anions for instance $SO_4^=$ (which can be explained by oxydation of sulphur components in the near-field) can rule the physico-chemical conditions of the interstitial medium for instance for Sr. In this case the behaviour of Sr in the near-field will be determined by the solubility of $SrSO_4$ in the interstitial solution. At ionic strength of 1 the solubility product for $SrSO_4$ is about 10^{-4}. It appears therefore that precipitation of Sr will occur and the solute Sr concentration will reach a limit of about $2 \ 10^{-3}$ M or 175 ppm.

A typical example of complex forming is that one of some Pu ions with organic substances or HCO_3 and subsequent colloid formation as Pu hydroxides. Non fixed species originating by reactions as just described will however be quasi immobilized by the absence of any noticeable water movement in the clay due to its low permeability combined in the Mol case with a small hydraulic gradient. Permeability of undisturbed Boom clay samples ranges between 1.2×10^{-7} and $7.1 \times 10^{-10} cm.s^{-1}$, while the present hydraulic gradient is about 1.5 cm H_2O/m.

In order to simulate "in situ" conditions at laboratory scale, a diffusion cell was developed which allows to perform diffusion experiments on reconsolidated clay plugs at various temperatures between +5 °C and +90 °C. One complementary particularity of the instrument is that it allows to calculate the permeability of the samples at the same time. The device is schematically presented in Figure 3 while Figure 4 gives two photographies of it. As far as influence of heat is concerned it was stated that distribution coefficients tend to decrease when exchange experiments are carried out above room temperature (up to 80 °C), but research work is not yet completed. When clay is heated in air and then rehydrated, it is generally found that the decrease of the distribution coefficients is more pronounced for samples having initially high values. Also, the K_D tends to decrease as the temperature of the treatments is increased. Exchange experiments on Boom clay show that the heat treatments limited to 300 °C decrease the adsorption of Sr, Cs (mainly) and Eu only when these elements are considered at very low concentrations. Possible, irreversible alterations in the organic fraction are responsible for this fact. Table III gives some values for distribution coefficients (K_D) at various temperatures.

Preliminary results obtained for diffusion coefficients indicate values for Cs of 1×10^{-8} and Sr 4×10^{-7} $cm^2.s^{-1}$ at room temperature.

Table III. Influence of temperature treatment on the
distribution coefficient of clay (pH = 6 to 8)

Nuclide Initial concentr. 0.1 mg ℓ^{-1}	T °C	K_d $cm^3 g^{-1}$
	100	527
Sr	300	208
	500	101
	100	4,900
Cs	300	1,229
	500	2,414
	100	3,667
Eu	300	2,418
	500	229

NEAR-SURFACE "IN SITU" TESTS

As already mentioned heat transfer and corrosion experiments are performed "in situ" in an area where the Boom clay formation crops out. The test site is situated about 50 km southwest of Mol, the tests are running at approximatively 20 m below the mid level of a quarry.

Heat transfer test

The simulated HLW canister is a stainless steel cylinder 30 cm in diameter and 150 cm long. Heat output is obtained by electrical wiring. The heater was buried at a maximum depth of 650 cm and the space between clay and cylinder filled up with a sandy material having nearly the same thermal conductivity as the clay. In total 36 thermoprobes were installed at various distances from the heat source as shown in Figure 5. The experiment ran for one year at a 1,000 W(e) output and

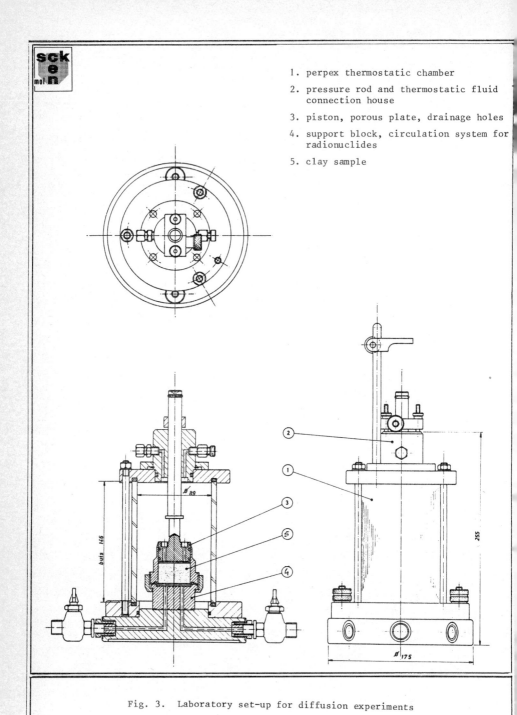

1. perpex thermostatic chamber
2. pressure rod and thermostatic fluid connection house
3. piston, porous plate, drainage holes
4. support block, circulation system for radionuclides
5. clay sample

Fig. 3.   Laboratory set-up for diffusion experiments

Fig. 4. Close-up photographies of the cell used for diffusion experiments

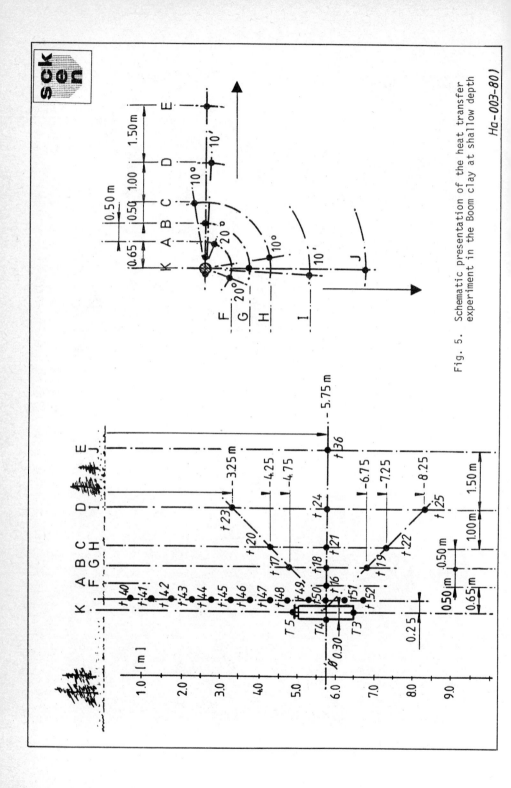

Fig. 5. Schematic presentation of the heat transfer experiment in the Boom clay at shallow depth

Ha-003-80)

for another 6 months at 1,500 W(e). Due to ground surface weathering phenomena and quarrying works, water infiltrations near to the heat source could occur. It was possible to take into account these water movements in the mathematical model. To illustrate the output of this experiment a sample is given on Figure 6 of the recorded temperature evolution as a function of time for one set of thermoprobes. The corresponding calculated temperatures are given in the same figure. The average values found for "in situ" heat conductivity is 1.69 W/m °C and for heat diffusivity 18.8m^2/yr.

Corrosion loops

The first 3 operating loops are "open" systems in which dry air skims along natural clay, picks up volatiles and is swept over metallic samples at the same temperature as the clay. The contact temperatures are for each loop respectively, ambient temperature (± 13 °C), ± 50 °C and ± 150 °C. The system is schematically presented in Figure 7 which also gives a view of the installation in the quarry. Some other tests, in which metallic samples are in direct contact with clay at different temperatures are now under development.

As far as general results on corrosion are concerned, some trends observed can be summarized here :

- the first comparisons between laboratory and "in situ" tests indicated good analogy ;

- Ti and Ti alloys, Hastelloy C, Inconcel 625 and UHB 904L gave the best resistance to local and preferential corrosion in clay medium ;

- heat pre-treatment of the samples generally improves their corrosion rate ;

- the presence of sulphur and some soluble sulphur components is one of the main reasons for the corrosion detected.

"IN SITU" TESTS AT REAL DEPTH

Near-field phenomena will be studied "in situ" at real depth as a last experimental phase in the R and D programme. Therefore the C.E.N./S.C.K. decided during the year 1978 to start the conceptual design of an underground laboratory located at a depth of approximately 220 m in the Boom clay. At the end of 1979 the administrative authorizations were obtained and the project scientifically and technically sufficiently advanced to start field works.

The facility which is represented in Figure 8 includes a shaft with a final diameter of 265 cm and its utilities, a connection room 610 cm diameter and 700 cm high and an horizontal gallery 350 cm in diameter and about 2,500 cm long. Shaft and connection room will be lined with concrete while the horizontal gallery will be lined with bolted cast iron segments. Holes equipped with flanges will allow to reach the clay for emplacement of various experiments. The construction of the shaft reached the top of the Boom clay last June, and is now beneath the 200 m level, the facility ought to be fully operational at the end of 1982. The foreseen underground experiments will concern the near-field phenomena already studied in laboratory and near-surface. Another capital question to be answered is the behaviour of the clay massif from a geomechanical and hydrogeological point of view in the vicinity of shafts, galleries and disposal holes. This can only be done in conditions where engineering works are performed at real depth.

Therefore once the top of the transition zone between the upper aquifer and the compact clay was reached (this transition zone composed of alternating clay, silty clay and clayey sand layers has a thickness of about 20 m) various pressure cells were installed in the frozen ground at 4 different levels in the transition zone and in the compact clay. These cells will allow the measurements of vertical and radial total ground pressures. Piezometric cells are to be installed at the same levels in order to record continuously pore water pressure during and after defreezing. Similar instrumentation will also be installed along the horizontal gallery where at the same time influence of increasing temperature, due to the heating experiment, can be studied. More details concerning the "in place" geomechanical programme was presented by one of the authors (PM) during the "Rockstore 80 Conference" in Stockholm.

An important basic experiment also to start will be in connection with heat transfer and heat and irradiation impact. It is foreseen to bury a simulated VHLW canister in a hole of 2,000 cm length connected with the main gallery, this source will be equipped with electric heaters and later on supplied with a γ

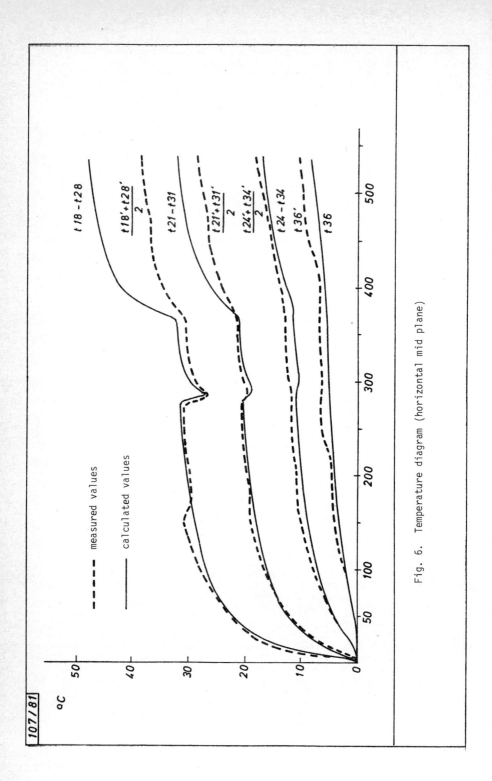

Fig. 6. Temperature diagram (horizontal mid plane)

107/81

°C

measured values

calculated values

$t18 - t28$

$\dfrac{t18 + t28'}{2}$

$t21 - t31$

$\dfrac{t21' + t31'}{2}$

$\dfrac{t24' + t34'}{2}$

$t24 - t34$

$t36'$

$t36$

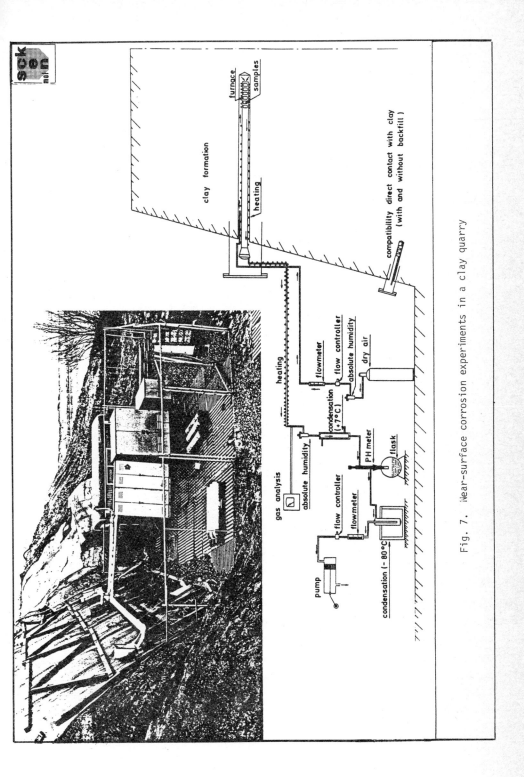

Fig. 7. Near-surface corrosion experiments in a clay quarry

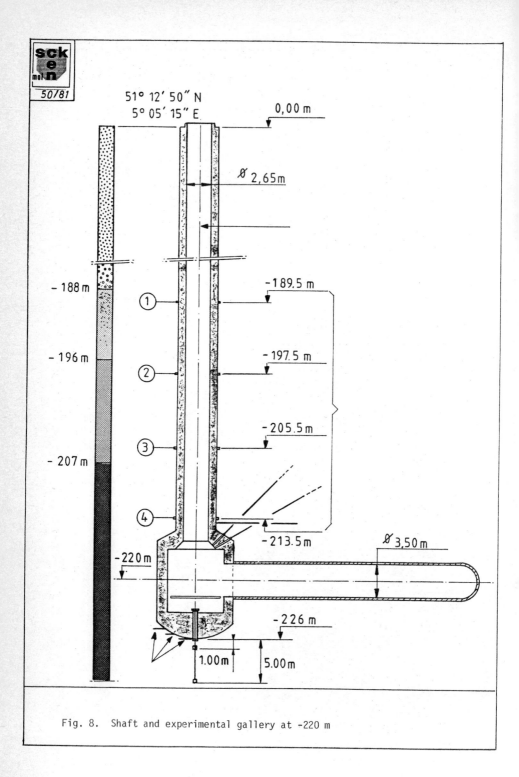

Fig. 8. Shaft and experimental gallery at -220 m

irradiation source not yet defined today. This experiment is foreseen to run several years and will provide informations not only on the general behaviour of the host rock and engineered structures under thermal stresses but at the same time on heat impact on corrosion phenomena. Other experiments concerning corrosion, leaching, diffusion etc. are of course also foreseen in the coming years but not yet planned in detail.

It is a matter of fact that beside these "in situ" tests at depth, laboratory research will continue with the objective to explain the observed phenomena on a scientific basis.

CONCLUSIONS

In the preceeding pages the studies of near-field phenomena in clay, as they are tackled in Belgium in the frame of the R and D programme on geological disposal of conditioned radioactive waste, were briefly described. An overlook is presented in Table IV. In the paper no research on backfilling materials was mentioned. In the Belgian R and D programme this repository component was not yet investigated i.r.t. the near-field phenomena. Indeed the most promising back-filling materials are also clays or clay mixtures and it may be expected that the near-field effects in backfilling materials will be very similar to those encoun-tered in the host rock itself. The results of our extensive and various experiments incorporate important basic data needed for the definition of the source term in the repository evolution analysis which is part of the site performance assessment study. Numerous years of experimentation and more particularly confir-mation tests "in situ" at real depth are necessary. The underground laboratory we actually build and which will be in operation early 1983 is an essential tool on the way for better knowledge of near-field phenomena.

Table IV. C.E.N./S.C.K. experimental programme on near-field phenomena

Near-field phenomena due to	Laboratory	In situ near-surface	In situ at depth
1. Excavation and mining			
- geomechanical	θ	×	θ
- chemical	θ	×	+
- biological	+	×	−
2. Waste disposal			
- thermal	θ	θ	+
- radiological	θ	×	+
3. Interactions waste-repos-itory immediate host rock			
- corrosion	θ	θ	θ
- leaching	θ	θ	θ
- migration	θ	×	+

θ = performed or running
+ = planned
− = not planned immadiately
× = not foreseen

- 201 -

DISCUSSION

R.H. KOSTER, Federal Republic of Germany

A comment on your corrosion phenomena : On the basis of knowledge of corrosion resistance against for instance geothermal brines, you come up with 5 material categories. From our corrosion experiments with several metal alloys in concentrated salt brines, we can conclude that specially Hastelloy C-4 - a Ni alloy - and cast iron should be suitable as construction material for containers or waste package. Astenic, ferritic steel and even titanium alloys like Ti-Pd alloy (99.8 % Ti) we have to exclude as candidate canister material in case of salt brine attack due to selective corrosion processes such as stress corrosion.

R.H. HEREMANS, Belgium

Your comments concerning the behaviour of Ti and Ti alloys are particularly interesting ; we certainly will give special attention to these materials in our analyses of their corrosion in clay media.

J.L. CAMBON, France

Do you think the influence of corrosion products on k_d values, as shown in your paper, is site specific of Boom clay, or can it be extended to other clay materials or other types of rock ?

R.H. HEREMANS, Belgium

The influence of corrosion products on k_d values found for Cs and Eu is certainly not specific to Boom clay. One can assume that competition between various ions will exist for most ion exchange processes.

H.S. RADHAKRISHNA, Canada

Concerning the decrease in thermal conductivity of clay from 1.5 W/m°C at 20°C ; to 0.3 W/m°C at 100°C ; is this a true effect of ambient temperature or is it due to the fact that the soil dries out at 100°C ? In otherwords, were the measurements made in a closed system or an open system for moisture migration ?

R.H. HEREMANS, Belgium

The value of 1.5 W/m°C for the thermal conductivity of Boom clay was found for a core at a temperature of about 35°C in a system where water could not escape. The value of 0.3 W/m°C at 100°C was found for a clay sample where the moisture could escape, in otherwords on a dry sample at that temperature.

H.S. RADHAKRISHNA, Canada

What was the mineral composition of the clay, and did it contain any organic compounds ?

R.H. HEREMANS, Belgium

The main clay minerals present in the Boom clay are, vermiculite and vermiculite like minerals, illite, smectite, chlorite

and various interstratifieds. The total amount of organic materials
varies between 2.3 % (1.14 % of C) and 5.5 % (3.25 % of C).

W.R. FISCHLE, Federal Republic of Germany

Where is the ground water level in your in-situ experiment ?

R.H. HEREMANS, Belgium

The top of the Boom clay formation at the experimental site
in Mol is situated at about -190 m, between this level and ground
surface, aquiferous sands are present.

TEMPERATURES AND STRESSES IN THE VICINITY OF A

NUCLEAR-WASTE REPOSITORY IN WELDED TUFF

D. K. Parrish, H. Waldman and J. D. Osnes
RE/SPEC INC.
Rapid City, SD/USA   57709

ABSTRACT

   Finite element thermomechanical models were used to assess the feasibility of a nuclear waste repository located in welded tuff 800 m below the ground surface and 330 m below the groundwater table. These models include the effects of porewater vaporization and joints in the rock mass. Porewater vaporization (boiling) initially decreases the rate of temperature rise, but the change in thermal properties after dehydration causes the boiling model to heat faster, reach higher temperatures and cool faster than the models without porewater vaporization (no boiling). Plane-strain finite-element models incorporating Drucker-Prager failure criteria for both intact rock and rock joints were used to assess the mechanical effects of initial excavation and subsequent heating. The regions in which stresses satisfy one of the failure criteria depend on the temperature history, which, in turn, is a function of the gross thermal load and the porewater vaporization. The models predict that redistribution of stresses during excavation causes a small region of failure at the room wall. This region expands only slightly during subsequent heating.

1. INTRODUCTION

The Nevada Nuclear Waste Storage Investigations (NNWSI) are investigating the suitability of rocks on the Nevada Test Site and contiguous federal lands for storing radioactive waste. Large volumes of tuff occur in many parts of the western U.S. and, in particular, in southern Nevada. Moreover, tuff contains zeolite and clay minerals which have high sorptive coefficients for large-radius cations typical of radioactive wastes. Consequently, tuffs are being evaluated as a potential repository host rock. As part of this evaluation process, the Mine Design Study (MIDES) Working Group at Sandia National Laboratories is performing calculations to access the feasibility of placing a nuclear waste repository in tuff. RE/SPEC has developed a series of three-dimensional and two-dimensional finite element models to analyze the thermomechanical response of welded tuff surrounding repository rooms 800 m below the ground surface and 330 m below the water table.

A repository located below the water table in a welded tuff contained within a series of other welded and nonwelded tuff units may induce several thermal and mechanical phenomena which pose unique modeling problems. In this paper, we focus on the effects that porewater vaporization, joint slip and intact rock failure may have on the thermomechanical response of the tuff to a repository.

2. GEOLOGICAL SETTING

The NNWSI are focused on the tuffaceous media of the Yucca Mountain area near the southwestern corner of the Nevada Test Site; thus, the geological setting described here is specific to that area. A geologic cross-section (Figure 1) shows that Yucca Mountain is a series of horsts and grabens containing a sequence of thick, late Tertiary, gently dipping, tuff deposits. Tuff is a volcanic rock composed of particles ejected during volcanic eruption. The mineralogy of tuff is highly variable. Freshly deposited tuff consists of fragmented particles of volcanic glass and rock fragments. The high temperatures and rapidly developed lithostatic loads may "weld" the glass shards into a compact matrix. Later, the glass devitrifies and alters to quartz, K-rich alkali feldspar, zeolites, and clay minerals. The rock fragments are quartz, K-rich alkali feldspars, and pumice fragments.

Tuff porosity may vary from almost nil, in welded, unfractured tuffs deposited by ash flows, to 50% in ash tuffs deposited by air fall. Joints of variable spacing present throughout the stratigraphic column add to the porosity and permeability of tuff. The water table lies at a depth of 470 m. Detailed descriptions of the geology are presented by Spengler et al (1), Sykes et al (2), and Heiken and Bevier (3).

The models in this study assume a repository located within the Bullfrog member of the Crater Flat Tuff at a depth of 800 m. The Bullfrog member consists of devitrified welded tuff composed primarily of quartz and K-rich alkali feldspar with zeolite alteration products (1). The ambient rock temperature of the Bullfrog tuff is assumed to be 35°C.

3. CONCEPTUAL REPOSITORY MODELS

The conceptual repository used for the models described in this paper consists of a series of long parallel rooms with two rows of waste canister emplacement holes drilled parallel to the room centerline (Figure 2). After emplacement, waste canisters filled with spent unreprocessed fuel (spent fuel) 10 years after discharge from a UO_2 fueled pressurized power reactor, will heat to a maximum temperature, then cool. The exact temperatures, the temperature distribution and the thermomechanical consequences of the heating depend on the size of the room and intervening pillars (extraction ratio), the areal density of the spent fuel (gross thermal load), and the thermal response of the rock. All the models presented here assume that the room extraction ratio (room width/(room width + pillar width)) is 0.20, and that the gross thermal load (total initial energy stored/total repository area) is 25 W/m^2.

RSI-PUBL. NO. 81-09

The thermal response of the tuff is complicated by the assumption that the repository is below the water table, because the temperatures in the saturated tuff may exceed the boiling temperature of the porewater. A realistic model of the dehydration (boiling of pore water) and subsequent resaturation of tuff requires a fully coupled thermohydrological mechanical model including the system of repeated welded and nonwelded tuffs subject to the regional hydraulic gradient and flow system. Instead two simplified models were constructed to obtain the temperature fields near a spent fuel repository. These models are recognized as being artificial, but they have served as useful baseline models for comparing the effects of separate parameters and for obtaining an insight to the thermal response of tuff to a repository.

3.1. No-Boiling Models

One set of models, (no-boiling models) consider no boiling or vaporization of the porewater. These models make use of the best known thermal properties (those for saturated tuff, Table I). In some cases, boiling temperatures will never be approached because, in a sealed repository, the hydrostatic porefluid pressure may cause the boiling temperature of water to be higher than the induced temperatures around the repository. Even without sealing, the boiling temperatures of the bulk of the rock will be greater than 100°C. A no-boiling model is reasonable if other "drying" mechanisms are neglected (for example, evaporation and diffusion).

3.2. Boiling Models

The boiling models assume that the porewater boils when the rock temperature reaches 100°C. These may be considered "worst" case models because the porewater boiling temperatures in the bulk of the rock are likely to be greater than 100°C so that these models produce the maximum dehydration zone. The boiling models are further simplified by the following assumptions:

(1) no flow of porewater into the dehydrated zone,
(2) superheating of the steam is negligible,
(3) vaporized water escapes immediately and irreversibly from the system, even after backfilling,
(4) the liquid/vapor porewater interface exactly coincides with the 100°C isotherm.

The boiling phenomenon was simulated by assuming a phase change takes place over a 20° range in temperature (90°C to 110°C). At temperatures below 90°C the thermal properties for saturated tuff were used, and above 110°C the thermal properties for dry tuff were used (Figure 3, Table I). In the temperature interval 90°C to 110°C, the thermal conductivity was taken to be the mean of the saturated and dry values. The volumetric heat capacity was taken to be the mean of the saturated and dry values plus the appropriate value for the latent heat of vaporization of the porewater (Figure 3).

3.3. Elastic-Plastic Models of Jointed Tuff

The volcanic ash flows at the Nevada Test Site are faulted and jointed. The repository will be contained within a major fault-bounded block, so discrete faults are not included in the models. On the other hand, excavation and heating may induce stresses which can cause slip on joints as well as fracture of intact rock. An elastic-plastic model was developed to assess the effect of joint slip and intact-rock fracture.

The spacing and orientation of joints in the Bullfrog tuff are not known. Rock surrounding test tunnels at the Nevada Test Site contain at least three sets of nearly vertical joints (1). As a conservative approximation, the repository was envisioned as emplaced in a rock mass transected by a set of ubiquitous (that is, infinitely closely spaced) vertical joints. The deformation by slip on such a set of joints was modeled as deformation of a transversely isotropic element.

A Drucker-Prager yield criterion (5) was chosen as the criterion for determining whether the intact rock had fractured and whether the joints had yielded. An associated flow rule for the Drucker-Prager criterion was used to calculate the plastic strain after the criterion was satisfied. The Drucker-Prager criterion was selected because it predicts stress at a lower deviatoric stress than other available criteria (6). Therefore, the detrimental effects that joint slip and intact-rock failure might have on the repository are maximized by these models.

Two parameters, the angle of internal friction and the cohesion, are required to define a Drucker-Prager yield criterion. These parameters were specified for each of four criteria which were used in the model: (1) intact rock prior to failure (peak, Table II), (2) intact rock after failure (residual, Table II), (3) joints prior to slip (peak, Table 2), and (4) joints after slip (residual, Table 2). The properties for intact rock before and after failure were obtained from laboratory tests (7). The peak and residual properties of joints were assumed to be the same as fractured tuff.

The temperatures for both the boiling or no-boiling thermoelastic plastic models were calculated using a two-dimensional model of a typical room and pillar (plane YY'Z'Z in Figure 1). Two-dimensional models compute temperatures identical to the three-dimensional models, except within one hole diameter of the emplacement hole. Therefore, two-dimensional models are adequate for computing the temperatures in the room floor and pillar.

Plane-strain finite-element models computed the thermoelastic stresses at each of eight time steps. At each time step, the thermoelastic stresses were superimposed on the previous stress state, then each integration point was compared with a Drucker-Prager yield criterion. If the criterion was satisfied, at any point, then the stresses were redistributed by a technique involving repeated resolution of the stiffness equations with updated displacements according to an associated flow rule. This was done until the yield functions for both failure modes indicated no further yielding.

4. EFFECTS OF POREWATER VAPORIZATION ON TUFF TEMPERATURES

Temperatures calculated by three-dimensional finite element models assuming the gross thermal loading was 25 W/m^2 demonstrate the effect that the dehydration has on temperatures at various points around the repository (Figure 4). The boiling models differ from the no-boiling models in two ways. First, the boiling models remove the latent heat of vaporization from the system, thereby reducing the total heat deposited to the rock and tending to induce lower rock temperatures. Second, after the porewater boils, the thermal conductivity (K) and the volumetric heat capacity (ρCp) of the rock change. The thermal diffusivity ($\alpha = K/\rho Cp$) of unsaturated tuff is lower than the thermal diffusivity of saturated tuff. Neglecting other effects, changing the thermal properties after the rock reaches 110°C tends to induce higher rock temperatures. The net effect of the two basic phenomena in the boiling models combined with the decaying spent-fuel heat-generation produces a complicated thermal history.

The temperatures calculated by either the no-boiling model or the boiling model are identical until the boiling phenomena begins (90°C) (Figure 3). After the boiling temperatures are exceeded (110°C), the boiling model temperatures increase faster than the no-boiling-model temperatures (Note, for example, the change in slope of curves (A-D) in Figure 4). However, at a later time, the temperatures in the boiling model increase more rapidly than those in the no-boiling model. The increased heating rate occurs at about one year when the gross thermal load is 25 W/m^2. Eventually, in the vicinity of the emplacement hole, the boiling-model temperatures exceed the no-boiling-model temperatures. The peak temperatures in the boiling models are higher than those in the no-boiling models (210°C vs. 190°C) and the peak temperatures are reached sooner in the boiling models. For example, in the boiling model, the room centerline (point D in Figure 3b) reaches a maximum temperature of 175°C approximately 20 years after waste emplacement, whereas the same point in the no-boiling model reaches a maximum temperature of only 165°C approximately 25 years after emplacement.

Although the boiling models predict higher temperatures earlier, the boiling-model temperatures decrease faster than the temperatures in the no-boiling models. Consequently, at 100 years after waste emplacement, the vicinity of the waste canister is cooler in the boiling models than in the no-boiling models.

In the regions around the room and pillar (points F-H in Figure 3), where boiling temperatures are not reached, the boiling models predict lower temperatures than the no-boiling models during the 100 year duration of the model.

5. EFFECTS OF EXCAVATION AND HEATING ON ROOM STABILITY

Plane-strain finite-element models of an infinite array of spent fuel storage rooms were used to study the effects of excavation and canister heating on the stress state around each room. Initial excavation perturbs the in situ stress state and the models predict that the redistributed stresses around the room walls satisfy the Drucker-Prager yield criterion specified for the vertical joints (Figure 5). The region of predicted joint slip (x's in Figure 5) extends from the room wall and corners approximately 2.5 meters into the pillar. The temperatures calculated by either the boiling or the no-boiling thermal models caused very little additional deformation. In only one model, GTL = 25 W/m^2, no-boiling (Figure 5d) did joint yield occur further into the pillar than 2.5 m. In that case, the calculated stress at one integration point located 3.75 m from the room wall satisfied the yield criterion. Stresses in small regions near the corners of the room satisfied the Drucker-Prager yield criterion specified for intact rock (solid squares in Figure 6).

6. CONCLUSIONS

Placement of a spent fuel repository in saturated tuff is likely to induce a complex thermal response in the rock. Maximum rock temperatures may exceed the vaporization temperature of the water in the vicinity of the repository, but in the pillar regions removed from the repository the hydrostatic head pressures may be high enough to inhibit vaporization. Two sets of models provide some insight to the complex sequence of thermal events that may be induced, one model (boiling model) includes the heat losses due to water vaporization, and the change in thermal properties when the saturated rock changes to dry rock. The other model (no-boiling model) assumes the rock remains saturated throughout the time modeled. In the vicinity of the waste canister (the region of highest temperatures), the onset of boiling causes a lower rate of temperature increase than is observed in the no-boiling models. However, as the rock continues to heat, the altered thermal properties cause the dried rock to heat more rapidly to a higher temperature than is calculated for the no-boiling models. After the maximum temperature is reached, the canister regions in the boiling models cool faster than the same region in the no-boiling models.

The temperature history is simpler at points in the room and the pillar, where the thermal properties are not altered by heating. The boiling models calculate lower temperatures.

Plane-strain models of a representative room and pillar in the repository indicate that most deformation (indicated by yield according to a Drucker-Prager criterion) occurs when the room is excavated. The yield occurs because of slip on the ubiquitous vertical joints in the model. Subsequent heating induces slightly more deformation. After 100 years, slip on joints is restricted to a region within one-half a room width of the room (3.75 m) and intact rock fracture is restricted to within 1.25 meters on the room corners. These models indicate that if a repository room can be excavated from welded tuff, subsequent heating will not seriously affect the rock stability.

7. REFERENCES

(1) Spengler, R. W., D. C. Muller, and R. B. Livermore, 1979, Preliminary Report on the Geology and Geophysics of Drill Hole UE 25a-1, Yucca Mountain, Nevada Test Site, U.S.G.S. Open File Report 79-1244.

(2) Sykes, M. L., G. H. Heiken, and J. R. Smyth, 1979, Mineralogy and Petrology of Tuff Unit from the UE25a-1 Drill Site, Yucca Mountain, Nevada, Los Alamos Scientific Laboratory Informal Report, LA-8139-MS.

(3) Heiken, G. H. and M. L. Bevier, 1979, Petrology of Tuff Units from the J-13

Drill Site, Jackass Flats, Nevada, Los Alamos Scientific Laboratory, Informal Report, <u>LA-7563-MS</u>.

(4) Johnstone, J. K. and K. Wolfsburg, Eds: "Evaluation of Tuff as a Medium for a Nuclear Waste Repository: Interim Status Report on the Properties of Tuff", <u>SAND 80-1464</u>, Sandia National Laboratories, Albuquerque, NM, July, 1980.

(5) Desai, C. S. and J. F. Abel: <u>Introduction to the Finite Element Method, A Numerical Method for Engineering Analysis</u>, Van Nostrand Reinhold Co., 1972.

(6) Callahan, G. D. and A. F. Fossum, "Quality Assurance Study of the Special Purpose Program SPECTROM: Plasticity Problems", Topical Report RSI-0111, Prepared for the Office of Nuclear Waste Isolation, Battelle Memorial Institute, Columbus, OH, <u>ONWI-141</u>, July, 1980.

(7) Olsson, W. A. and A. K. Jones: "Rock Mechanics Properties of Volcanic Tuffs from the Nevada Test Site", <u>SAND 80-1453</u>, Sandia National Laboratories, Albuquerque, NM, November, 1980.

RSI DWG 001-81-450

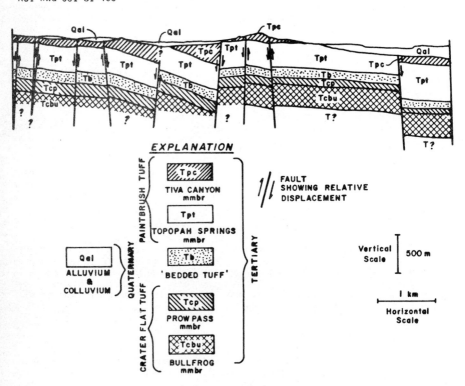

Figure 1.   Cross-Section of Tuff Geology (after Sykes et al, 1979).

RSI DWG 038-81-105

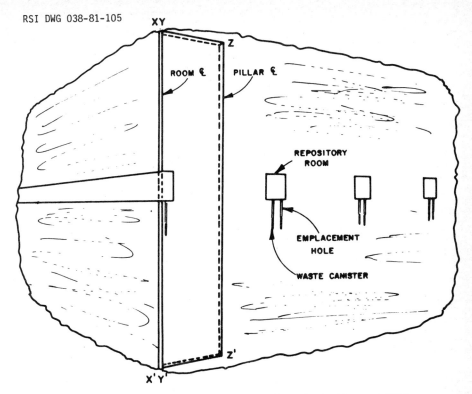

Figure 2. Conceptual Repository Showing the Room Centerline and Pillar
Centerlines Which Were Used as Symmetry Planes in the Two-
Dimensional Models.

TABLE I

GROSS THERMAL LOAD = 25 W/m^2

| | | TUFF (4) | | AIR | SF WASTE CANISTER |
		BOILING	NO-BOILING		
K(W/m-K)	<90°	2.40	2.4	61*	3.75
	90-110°	1.98	2.4	61*	3.75
	>110°	1.55	2.4	61*	3.75
Cp(J/kg-K)		1597	2280	1014	714
ρ(kg/m^3)		1597	2280	.90	4193

* The values given for the thermal conductivity of air represent an
equivalent thermal conductivity which includes the effects of radiation
and free convection between the room floor an ceiling/walls.

TABLE II

TUFF
PHYSICAL MATERIAL PROPERTIES

PROPERTY	UNITS	INTACT ROCK	JOINTS[†]
YOUNG'S MODULUS, E	GPa	20	---
POISSON'S RATIO, ν	---	0.25	---
DENSITY, ρ	Kg/m^3	2280	---
THERMAL EXPANSION COEFFICIENT	10^{-6} K^{-1}	7.5 10.3[*]	---
ANGLE OF FRICTION, PEAK	DEGREES	43.2	35.0
ANGLE OF FRICTION, RESIDUAL	DEGREES	35.0	35.0
COHESION, PEAK	MPa	8.5	0.0
COHESION, RESIDUAL	MPa	0.0	0.0

[*] THE GREATER VALUE GIVEN FOR THE THERMAL EXPANSION COEFFICIENT WAS USED FOR TEMPERATURES OVER 100° C.

[†] THE JOINT ANGLE IS 90° FROM HORIZONTAL.

RSI DWG 038-81-149

TUFF THERMAL MATERIAL PROPERTIES

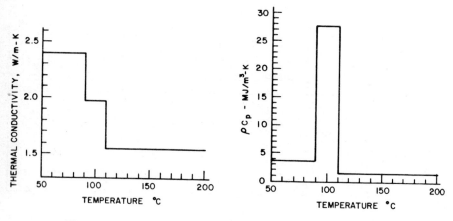

Figure 3.   Thermal Properties Used for Boiling Models of Tuff.

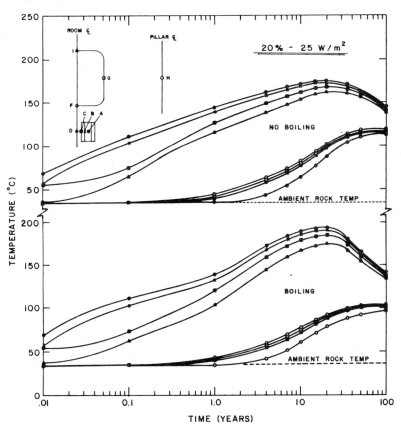

Figure 4. Temperatures Calculated for Boiling and No-Boiling Models of a Spent Fuel Repository Room in Welded Tuff. Gross Thermal Load = 25 W/m^2; Extraction Ratio = 20%. A. Canister Center, B. Canister Surface, C. Emplacement Hole Wall, D. Room Centerline, F. Room Floor, G. Room Wall, H. Pillar Centerline, I. Room Roof.

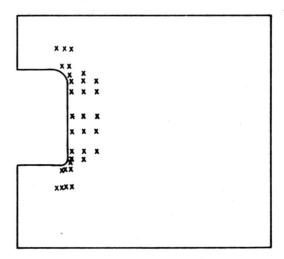

Figure 5. Joint Yield After Excavation of Spent Fuel
Repository Room. X = Gaussian Integration
Point at Which Drucker-Prager Yield
Criterion for Joint Slip is Satisfied.

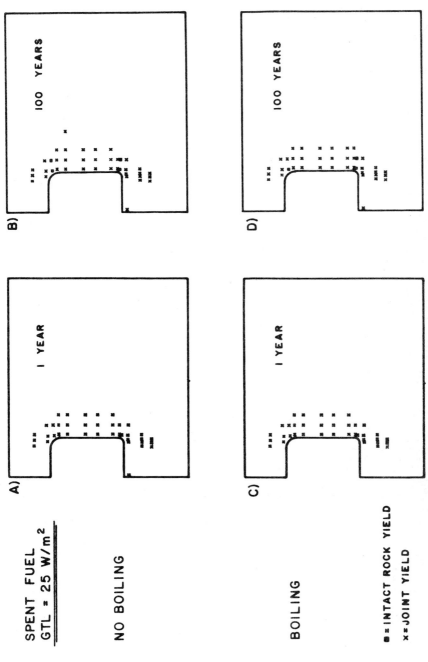

Figure 6. Effect of Heating on Jointed Tuff Surrounding Spent Fuel Repository Room. Symbols Denote Gaussian Integration Points that Satisfy Drucker-Prager Yield Criteria.

RSI DWG 038-81-152

DISCUSSION

M. WALLNER, Federal Republic of Germany

I have a question concerning your joint modelling. What was you joint orientation ? When you have inclined joints not dipping perpendicularly towards the drift it seems to me that the modelling in a plane-strain model cannot give correct results. In this case you have to consider the 3D effects, which would give different results.

D.K. PARRISH, United States

With the orientation of the plane of isotropy used in the model, the joints are vertical in all these models. We have run excavation models in which the joints in each element were oriented within ± 15° of vertical. The distribution of points where yield occurred in those models are essentially the same as the distribution shown in Figure 5. That is, in these two-dimensional models, a variation of 15° from vertical does not cause yield further into the pillar than with vertical joints.

These 2-D models do not consider joints which have a component of dip parallel to the long axis of the drift. A three dimensional model incorporating such joints would produce different results but I think these 2-D models are a worst case.

F. GERA, Italy

If we assume that the repository has been backfilled and sealed all fluids would be under hydrostatic pressure. Under these conditions is it conceivable that the high temperature might create sufficient fluid pressure to fracture the formation ? If the scenario is possible what temperature might be required to induce failure in the rock ?

D.K. PARRISH, United States

The scenario you propose is possible. If the room is back-filled after storage, it is likely that the room backfill may saturate, thus raising the hydrostatic pressure on the pore water, thus raising the boiling temperature. The No-boiling models are as close as we come to modelling that scenario, we model the phenomena only insofar as we use laboratory data for the thermal expansion of saturated tuff. I have not made the calculation to determine whether we should expect excessive pore pressures because of the differences in the thermal expansion of water and tuff mineral components.

H.S. RADHAKRISHNA, Canada

In your "Boiling Model" you have not considered the pressure dependence of the boiling point of pore water. Do you consider that as a conservative assumption with respect to the rock stresses in the near-field ?

D.K. PARRISH, United States

Technical conservatism is complex when considering a complex system such as a nuclear waste repository. Each parameter and consequence must be considered before a degree of conservatism can be determined. I hope, but I am not certain, that these simplified models represent special extreme cases that provide bounds on the temperatures

we might expect. The "no-boiling" models permit no boiling at all, so
they over compensate for the pressure dependence of the boiling phenomenon. As you have seen these models produce higher temperatures in the
room and pillar than the boiling models. On the other hand, the boiling
models remove more heat than would be removed if the pressure dependence
were modeled. I think these models provide a lower bound for temperature. Assessing the maximum or minimum effects that these two assumptions have on stress is more complicated. I do not think I can predict
the degree of conservatism or the bounds without actually modelling
all the phenomena we might expect.

A PARAMETRIC STUDY OF THE EFFECTS OF THERMAL ENVIRONMENT ON A WASTE PACKAGE FOR A TUFF REPOSITORY

J. K. Johnstone, W. D. Sundberg and J. L. Krumhansl
Sandia National Laboratories
Albuquerque, New Mexico, USA

ABSTRACT

We have modeled the thermal environment in a simple reference waste package in a tuff repository for a variety of variables.
The waste package was composed of the waste form, canister, overpack and backfill. The emplacement hole was 122 cm dia. Waste
forms used in the calculations were commercial high level waste
(CHLW) and spent fuel (SF). Canister loadings varied from 50 to
100 kW/acre. Primary attention was focused on the backfill
behavior in the thermal and chemical environment. Results are
related to the maximum temperature calculated for the backfill.
These calculations raise serious concerns about the effectiveness
of the backfill within the context of the total waste package.

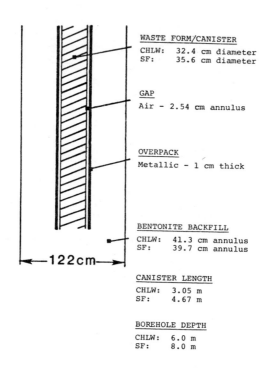

WASTE FORM/CANISTER

CHLW: 32.4 cm diameter
SF: 35.6 cm diameter

GAP

Air - 2.54 cm annulus

OVERPACK

Metallic - 1 cm thick

BENTONITE BACKFILL

CHLW: 41.3 cm annulus
SF: 39.7 cm annulus

CANISTER LENGTH

CHLW: 3.05 m
SF: 4.67 m

BOREHOLE DEPTH

CHLW: 6.0 m
SF: 8.0 m

Figure 1.   Reference Waste Package Used For
Thermal Calculations

INTRODUCTION

The nuclear waste package is a major component within the National Waste Terminal Storage (NWTS) program. Within the past year, an activity was initiated in the Nevada Nuclear Waste Storage Investigations (NNWSI) project to evaluate the waste package concept for a repository in tuff. Because of the potential complexity of waste packages (1,2), we undertook a systems analysis of a simplified reference waste package in a tuff host rock. The studies reported here focus on the thermal environment experienced by the waste package for both spent fuel (SF) and reprocessed commercial high level waste (CHLW) for periods to 1000 years. The purpose of the initial studies was to evaluate the package component stability, identify critical areas for research, assess the impact of the waste waste package on the nearfield repository conditions and provide a basis for extension of the systems analysis into other functional aspects of package performance.

REFERENCE WASTE PACKAGE

At the time this study was initiated, there were no conceptual waste packages upon which to base dimensions and thermal/physical properties. There were, however, several lists of potential waste package components (1-3). In addition, the Reference Repository Conditions Working Group had standardized waste form/canister configurations and properties for thermal calculations in the various host rocks. We began with those configurations and constructed a simple reference waste package composed of a waste form/canister, an overpack, and a backfill. The overpack was separated from the canister by a narrow air gap. The overpack was assumed to be an iron alloy and the backfill was assumed to be Na-montmorillonite based bentonite. The dimensions associated with the reference waste packages are shown in Figure 1. Since the main purpose of this study was to calculate the thermal environment caused by the emplacement of nuclear waste and evaluate its effect on the package, the package was purposely kept conceptually simple.

During the course of this study, the Office of Nuclear Waste Isolation recommended a baseline spent fuel canister loading of 1.6 kW and an areal power density of 60 kW/acre (4,5). We incorporated these values into our calculations.

THE MODEL

The thermal analysis was carried out using COYOTE (6), a finite element, heat conduction code. We assumed vertical emplacement of the waste packages in boreholes in a single row in the drift floor and two-dimensional axisymmetric geometry. A rectangular unit cell containing one waste package, part of the room and part of the pillar, was converted to a cylindrical effective radius configuration by conserving the planar area. The room was left open (not backfilled) during the entire calculation. Heat transfer across the room accounted for both conduction and radiation. Convection was not considered.

Throughout the calculation, temperatures were monitored at four locations radially along the centerplane of the waste package. The four locations were the waste form centerline, the canister outside diameter, the overpack/backfill interface, and the backfill/rock wall interface.

PARAMETER VALUES

The calculations reported here were performed using time and temperature independent properties, an initial ambient temperature of 35°C, a 20% extraction ratio, and a 122 cm diameter waste

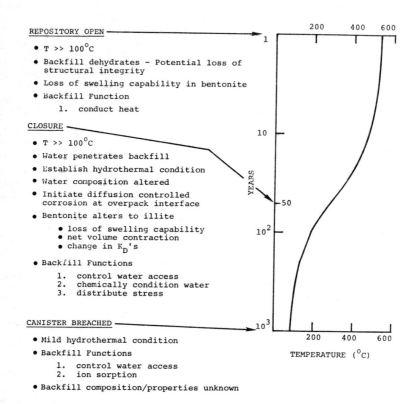

REPOSITORY OPEN

- T >> 100°C
- Backfill dehydrates - Potential loss of structural integrity
- Loss of swelling capability in bentonite
- Backfill Function
    1. conduct heat

CLOSURE

- T >> 100°C
- Water penetrates backfill
- Establish hydrothermal condition
- Water composition altered
- Initiate diffusion controlled corrosion at overpack interface
- Bentonite alters to illite
    - loss of swelling capability
    - net volume contraction
    - change in K_D's
- Backfill Functions
    1. control water access
    2. chemically condition water
    3. distribute stress

CANISTER BREACHED

- Mild hydrothermal condition
- Backfill Functions
    1. control water access
    2. ion sorption
- Backfill composition/properties unknown

Figure 2.   Backfill Maximum Calculated Temperature History for CHLW, 2.16 kW/canister, and 100 kW/acre

package. Parametric variables included, waste form (decay rate), canister thermal loading, areal power density, backfill thermal conductivity, and rock thermal conductivity. The values used are shown in Table I. Properties not listed in Table I were taken from handbooks and were representative for the component in question. All calculations were carried out to at least 100 years and most were carried to 1000 years.

Another parameter central to our study is the representative groundwater composition. While we have not measured this at locations of interest for a repository, it has been determined for water from a well which is located in the vicinity of the site of interest to the NNWSI project and penetrates many of the same geological units. The reference composition is given in Table II. The water contains moderately low dissolved solids. Of main concern, is the potassium to sodium ratio (~0.1) which is high enough to cause concern regarding the stability of Na-montmorillonite in a hydrothermal environment (7).

RESULTS

The results are summarized in Figures 2 through 4. In particular, we have concentrated on the potential response of the backfill and we show the calculated maximum temperature of the backfill located at the overpack/backfill interface as a function of time.

Figure 2 shows the maximum temperatures calculated for the backfill using CHLW with a canister loading of 2.16 kW and areal power density of 100 kW/acre. These conditions were acceptable for tuff in the absence of the overpack and backfill (8). During the period that the repository is open, the backfill maximum temperature reaches about 535°C. Such high temperatures would be expected to cause dehydration of the bentonite. For a highly compacted bentonite, this would cause the liberation of nearly 1000 liters of water and loss of the structural integrity of the backfill resulting in a decrease in thermal conductivity. In addition, dehydration at at temperatures in excess of about 390°C results in irreversible loss of the swelling capability of the bentonite (9). The only function of the backfill during this period is to conduct heat while, in fact, it is behaving more like a thermal insulator.

At closure of the repository (assumed to occur at 50 years), the calculated temperature is still very high, ~ 285°C. We believe that water would penetrate the backfill as rapidly as is consistent with the temperature gradient and the ability of the rock to deliver the water, thereby establishing hydrothermal conditions. The water composition would alter as a consequence of the temperature, radiation, backfill, and any buffers (conditioners) present in the backfill. Upon complete saturation of the backfill, diffusion controlled corrosion of the overpack would be initiated. The potassium/sodium ratio in tuff groundwater (0.1) would promote conversion of the bentonite to illite (7) resulting in loss of swelling capability (already lost during dehydration), net volume contraction, and a general decrease in ion exchange capacity. The function of the backfill during this period would be to control the access of water to the overpack (only in the sense of diffusion controlled transport), chemically condition the water (assuming that conditioners have survived the previous temperatures), and redistribute stress around the overpack.

At 1000 years, the time at which the package may be breached and radionuclides released (10), mild hydrothermal conditions, T > 90°C, would still be present. The backfill functions are to limit nuclide transport to diffusional processes and provide for ion sorption. Based on the foregoing discussion, backfill composition and properties are unknown.

Table I

Parameter Values for Thermal Model Calculations

A.

Waste Form	Canister Power kW/can	Power Density kW/acre
Commercial High Level Waste (CHLW)	2.16 1.08	100,50 100,50
Spent Fuel (SF)	0.55 1.60	100,50 60

B.

Component	Thermal Conductivity W/m°C
Backfill	0.30(dry), 0.75, 1.10(saturated)
Tuff(host rock)	1.55(dry), 2.40(saturated)

Table II

Reference Groundwater Composition for Tuff Based on
Composition of Jackass Flats Well J-13 at the
Nevada Test Site

Constituent	Concentration mg/liter	Constituent	Concentration mg/liter
Silica	61	Lithium	0.05
Aluminum	0.03	Bicarbonate	120
Iron	0.04	Carbonate	---
Calcium	14	Sulfate	22
Magnesium	2.1	Chloride	7.5
Strontium	0.05	Fluoride	2.2
Barium	0.003	Nitrate	5.6
Sodium	51	Phosphate	0.12
Potassium	4.9		

pH   7.1
Eh undetermined

Figure 3 is a similar plot for spent fuel with a canister loading of 0.55 kW and areal power density of 50 kW/acre, conditions which were also considered acceptable for tuff in the absence of the overpack and backfill. Compared to the previous case, the conditions are mild. The only concern arises after closure of the repository when conditions again exist for conversion of the bentonite to illite but at much lower temperatures.

Figure 4 is also for spent fuel and reflects the recent recommendation (4,5) that fuel bundles be disassembled and fuel rods close packed in the canisters. The calculation was carried out for 1.6 kW/canister and areal power density of 60 kW/acre (4). With one exception, all of the comments concerning CHLW (Figure 2) also apply in this case. The exception is that irreversible dehydration of the backfill does not appear likely while the repository is open.

CONCLUSIONS

These and additional calculations, raise serious concerns about the effectiveness of the backfill within the context of the total waste package. We have concluded that 1) the repository thermal load design limit may be fixed by the waste package, not the host rock, 2) for a specific thermal loading, thermal conductivity of the backfill is the most important factor governing thermal profiles in the waste package, and 3) the material selection for the waste package may be site specific. The results of these studies are being used to define future research and development activities for a waste package for a repository in tuff.

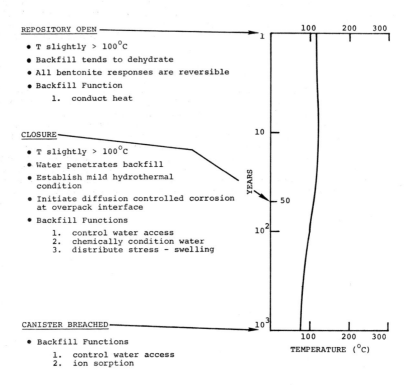

REPOSITORY OPEN

- T slightly > 100°C
- Backfill tends to dehydrate
- All bentonite responses are reversible
- Backfill Function
    1. conduct heat

CLOSURE

- T slightly > 100°C
- Water penetrates backfill
- Establish mild hydrothermal condition
- Initiate diffusion controlled corrosion at overpack interface
- Backfill Functions
    1. control water access
    2. chemically condition water
    3. distribute stress - swelling

CANISTER BREACHED

- Backfill Functions
    1. control water access
    2. ion sorption

Figure 3. Backfill Maximum Calculated Temperature
History for Spent Fuel, 0.55 kW/
Canister, and 50 kW/acre

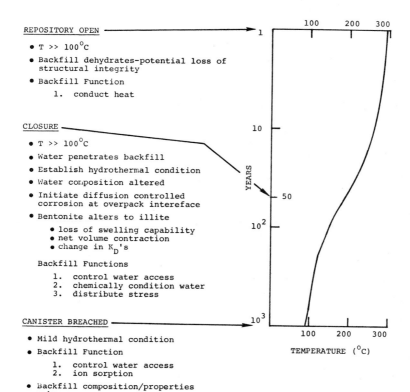

REPOSITORY OPEN

- T >> 100°C
- Backfill dehydrates-potential loss of structural integrity
- Backfill Function
    1. conduct heat

CLOSURE

- T >> 100°C
- Water penetrates backfill
- Establish hydrothermal condition
- Water composition altered
- Initiate diffusion controlled corrosion at overpack intereface
- Bentonite alters to illite
    - loss of swelling capability
    - net volume contraction
    - change in K_D's

  Backfill Functions
    1. control water access
    2. chemically condition water
    3. distribute stress

CANISTER BREACHED

- Mild hydrothermal condition
- Backfill Function
    1. control water access
    2. ion sorption
- Backfill composition/properties unknown

TEMPERATURE (°C)

Figure 4. Backfill Maximum Calculated Temperature History for Spent Fuel, 1.6 kW/canister, and 60 kW/acre

REFERENCES

1. Office of NWTS Integration, "NWTS Waste Package Program Plan, Volume 1: Program Strategy, Description and Schedule (draft)," NWTS-96 (Battelle Memorial Institute, Columbus, OH), June 1981.

2. W. E. Coons, E. L. Moore, M. J. Smith and J. D. Kaser, "The Functions of an Engineered Barrier System for a Nuclear Waste Repository in Basalt," RHO-BWI-LD-23 (Rockwell Hanford Operations, Richland, WA), January 1980.

3. Office of NWTS Integration, "NWTS Program Criteria for Mined Geologic Disposal of Nuclear Waste: Waste Package Performance Criteria (working draft)," NWTS-33(4) (Battelle Memorial Institute, Columbus, OH), January 1981.

4. E. R. Johnson Associates, Inc., "Systems Impacts on Spent Fuel Disassembly Alternatives (draft)," JAI-173, Reston, VA, May 1981.

5. Letter from S. Goldsmith, Manager Office of Nuclear Waste Isolation to J. O. Neff, Program Manager, National Program Office, "Spent Fuel Waste Form Recommendation," June 11, 1981.

6. D. K. Gartling, "COYOTE - A Finite Element Computer Program for Nonlinear Heat Conduction Problems," SAND77-1332 (Sandia National Laboratories, Albuquerque, NM), June 1978.

7. D. Eberl and J. Hower, "Kinetics of Illite Formation," Geo. Soc. Am. Bull., **87**, p.1326-1330, 1976.

8. Office of NWTS Integration: Reference Repository Conditions Interface Working Group, "Interim Reference Repository Conditions for Spent Fuel and Commercial High-Level Nuclear Waste Repositories in Tuff (draft)," NWTS 12 (Battelle Memorial Institute, Columbus, OH), December 1980.

9. R. E. Grim, _Clay Mineralogy_, Second Edition, p.313, McGraw-Hill Book Company, New York.

10. Nuclear Regulatory Commission, "Disposal of High-Level Radioactive Wastes in Geologic Repositories, Proposed Rule," 10CFR60, Federal Register, **48** [130], p.35280-35296, July 8, 1981.

DISCUSSION

R.V. MATALUCCI, United States

In your parametric study, did you vary the thickness of the backfill and study its affect on temperature rise at the canister/clay interface and clay/host rock interface ?

Do you intend to do this in the future ?

J.K. JOHNSTONE, United States

We have done calculations for an emplacement borehole diameter of 61 cm and backfill thickness of approximately 10 cm. We varied many of the same parameters as in the case of the thick backfill (122 cm diameter emplacement hole) and monitored both overpack/backfill and backfill/rock wall temperatures among others. In general, the thin backfill causes lower temperatures at the overpack/backfill interface and higher temperatures at the backfill/rock interface when compared to the results for thick backfill exposed to identical thermal conditions. The magnitude of the variations depends, of course, on the canister thermal loading, area. power density and thermal properties of the components.

W.C. PATRICK, United States

At early times the power levels of individual canisters is the controlling factor - not gross thermal load. The latter becomes important after 10-100 years.

J.K. JOHNSTONE, United States

Dr. Patrick is essentially correct. However, the degree of influence of gross thermal loading is a function of the waste form (reprocessed commercial high level waste vs spent fuel) and the canister power. Generally the influence of gross thermal loading on thermal profiles is greater for spent fuel and low canister power.

Session 5

SALT FORMATIONS

Chairman - Président

R.H. KOSTER
(Federal Republic of Germany)

Séance 5

FORMATIONS SALINES

THE MODELLING OF BARRIER-EFFECTS FROM OPEN AND CLOSED BORE-HOLES
IN SAFETY STUDIES FOR SALT DOME REPOSITORIES

R. Storck, R. Brüggemann, S. Hossain and T. Podtschaske
Technische Universität Berlin, FRG

ABSTRACT

Based on analytical solutions simple models have been developed to study the barrier-effects of open and sealed bore-holes towards the transport of radio-nuclides. The open bore-holes are shown to offer no significant barrier-effect. The barrier-effects of the sealed bore-holes have been investigated with the help of simple leach models. At the end, a parametric study has been carried out to show the effects of solubility limits, life time and permeability.

1.   INTRODUCTION

Safety studies for salt dome repositories in Germany are concerned mainly with the consequence analysis of a water inrush into the open or backfilled mine. The project "Safety-Studies-Entsorgung (PSE)" [1] has been developing a methodology for modelling the radio-nuclide transport through all barriers between waste and man. This methodology has been demonstrated recently for the szenario "Water Inrush into the Open Mine During the Operational Phase". It has been done to demonstrate the methodology rather than to produce results.

As a first approach the following items are considered as barriers: waste forms, bore-holes, drifts and shafts. For each item a submodel for the barrier-effect has been developed and all these have been combined together to obtain the release of radio-nuclides into the overlying strata. The transport through the overlying strata and the biosphere have been modelled separately to complete the demonstration for the radiation exposure to an individual.

This paper deals with the modelling of the bore-holes for high level and medium level heat producing wastes, which could be open or sealed by a plug of salt concrete. Applying simple leach models, preliminary results for the release of radio-nuclides into the drifts will be given and conclusions for future work will be drawn.

2. DESCRIPTION OF THE REPOSITORY

The repository, planned for the Gorleben site [2] , will be a one-level system situated 800 m below surface in homogeneous rock salt. Heat producing wastes, which are vitrified high level waste (HLW) and partly the solidified medium level waste (MLW), will be disposed off in bore-holes of 310 m depth in one wing of the repository. Bore-hole diameters will be 40 cm for the HLW and 80 cm for the MLW. The bore-holes shall be sealed with salt concrete directly after use; drifts shall be backfilled with crushed salt and sealed with salt concrete, when turning to the next disposal-drift. Hence, in the case of a water inrush, only one bore-hole for each of the HLW and the MLW could remain open.

3. BARRIER SYSTEM

The inflowing water on its way to the waste could be retarded by the sealed and backfilled drifts, the bore-hole sealings and the waste canisters. The sealings of drifts and bore-holes will not be designed for a hydrostatic pressure difference of about 100 bar. The canisters are only designed for transportation loads and not for loads due to rock pressure. Hence, any retardation and also any delay of water flow to the waste will be neglected.

The transport of leached radio-nuclides to the drifts will be retarded by the waste form itself, the containers and the bore-hole sealing. Neglecting special effects from the cracked canisters, the sealed bore-holes should be modelled with respect to the waste form and the sealing, and the open bore-holes only with respect to the waste form.

4. RADIO-NUCLIDE TRANSPORT

Leached radio-nuclides in the bore-holes can be transported by forced convection due to creeping of the rock, by diffusion and by natural convection due to the heat production, the salt dissolution and the gas production by radiolysis. Leached radio-nuclides can be retarded by sorption and precipitation.

As a first approach, sorption and precipitation are neglected in order to get simpler barrier models. Also sorption and precipitation data are not easily available for saturated brine conditions and the model neglecting these effects is a conservative one. Further diffusive transport is neglected against convective transport, and salt dissolution and gas production are neglected as driving mechanism for natural convection against heat production.

So we are left with the mechanism of temperature driven natural convection and forced convection caused by the creeping salt.

The temperature driven natural convection can be described exactly by the equations of hydrodynamics. In order to obtain first results for the velocities of radio-nuclide transport we have been successful in getting analytical solutions (as in [3] and [4]) under the following conditions:

(a) Underground cavities in nuclear repositories are wide-spread in only one direction and this direction is also the main direction of fluid flow.

(b) Flow pattern in such cases can be divided into a core-region and two end-regions.

(c) The core-region is characterized by low transverse
    velocities (and hence can be neglected) as compared
    to longitudinal velocities of fluid flow.

4.1 Open Bore-Hole

In order to ensure a safe disposal procedure, the bore-hole
diameter is larger than the diameter of the waste canister. Within a
flooded annular gap between the waste form and the rim of the borehole, a flow pattern will develop due to the temperature difference
across the gap. The annular gap will be enlarged by dissolution processes of rock salt and will be reduced by the creeping on the rock.

Analytical solutions of Navier-Stokes-equations for the core
region of a plane, vertical cavity under a horizontal temperature
gradient produce velocities of 880 m/h for MLW and 2200 m/h for HLW
(see Figure 1). Based on these velocities and with the following
assumptions, an analytical solution [5] can also be obtained for the
equation of the salt component in the brine:

(a) The inflowing brine at the top of the bore-hole to
    be saturated at the entrance temperature

(b) saturation of the brine at the rim of the bore-hole
    at the temperature of 5 K above entrance temperature

(c) convective transport of the salt only in the vertical
    direction and diffusive transport only in the hori-
    zontal direction.

Results are given in the table of Figure 1 as excavation velocity for the shift of the bore-hole rim (u_w) due to salt dissolution
and as the relative variation of bore-hole volume (r_A). Since the
creeping of the rock will be conpensated by the salt dissolution, the
open bore-hole is assumed to remain open for all time; and because of
the high velocities of fluid flow, the barrier-effect of open bore-
holes will be neglected.

4.2 Sealed Bore-Hole

After the disposal of 250 canisters into a bore-hole of 310 m
depth, the upper 10 m of the bore-hole will be filled with a sealing
of salt concrete. Although the gap around the canisters will be closed soon by creeping of the rock, there will be a remaining void
volume between the different waste forms and in the cracked waste.
The vitrified waste has to be assumed as a cracked medium firstly
because of the cooling down period and secondly due to the rock pressure. In case of a flooded mine, the remaining void volume will be
filled with fluid and a natural convection will take place within it.
Assuming the waste form as a porous medium, analytical solutions of
the Darcy-equations can be derived for the core region of a long
vertical tube filled with porous medium and an internal heat source
within it. As shown in Figure 2, the solution describes a flow pattern with a downcast flow at the rim of the bore-hole and an upcast
flow at the center of the bore-hole, where the temperatures are
higher. The average upcast velocity can be calculated as 7.6 m/a for
the HLW and as 1.3 m/a for the MLW considering the heat productions
at the disposal time respectively.

4.2.1 Barrier Model

Based on the calculated Darcy-velocities, a barrier model for
the sealed bore-hole can be constructed, assuming the radio-nuclides

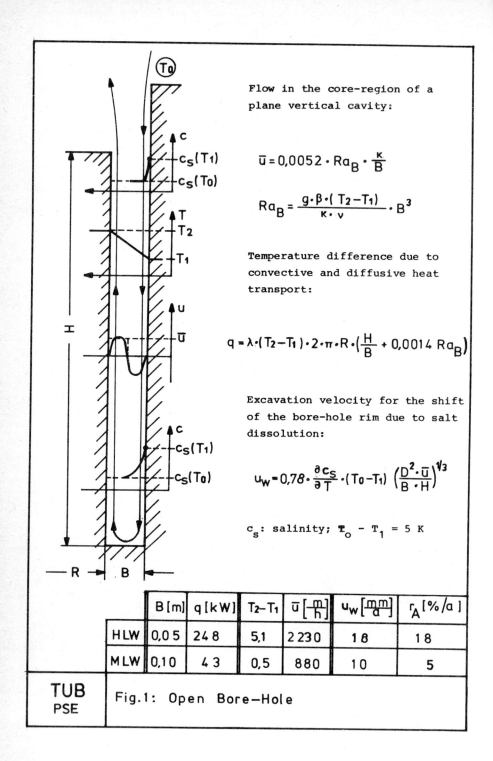

Flow in the core-region of a plane vertical cavity:

$$\bar{u} = 0{,}0052 \cdot Ra_B \cdot \frac{\kappa}{B}$$

$$Ra_B = \frac{g \cdot \beta \cdot (T_2 - T_1)}{\kappa \cdot \nu} \cdot B^3$$

Temperature difference due to convective and diffusive heat transport:

$$q = \lambda \cdot (T_2 - T_1) \cdot 2 \cdot \pi \cdot R \cdot \left(\frac{H}{B} + 0{,}0014\, Ra_B \right)$$

Excavation velocity for the shift of the bore-hole rim due to salt dissolution:

$$u_w = 0{,}78 \cdot \frac{\partial c_S}{\partial T} \cdot (T_0 - T_1) \left(\frac{D^2 \cdot \bar{u}}{B \cdot H} \right)^{1/3}$$

c_S: salinity; $T_0 - T_1 = 5$ K

	B [m]	q [kW]	$T_2 - T_1$	$\bar{u} \left[\frac{m}{h}\right]$	$u_w \left[\frac{mm}{a}\right]$	r_A [%/a]
HLW	0,05	248	5,1	2230	18	18
MLW	0,10	43	0,5	880	10	5

TUB
PSE

Fig.1: Open Bore—Hole

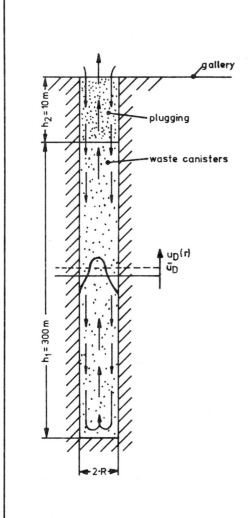

Darcy flow in the core-region of a cylindrical vertical porous medium with internal heat source

$$\bar{u}_D = 0.063 \cdot Ra_R \cdot \frac{\kappa_s}{R}$$

$$Ra_R = \frac{g \cdot \beta \cdot \kappa \cdot q'''}{\kappa_s \cdot \nu \cdot \lambda} \cdot R^3$$

$$k = 10^{-12}\ m^2$$

	R [m]	$q''' \left[\frac{kW}{m^3}\right]$	$\bar{u}_D \left[\frac{m}{a}\right]$
HLW	0,15	11,7	7,6
MLW	0,35	0,4	1,3

Fig. 2: Sealed Bore-Hole

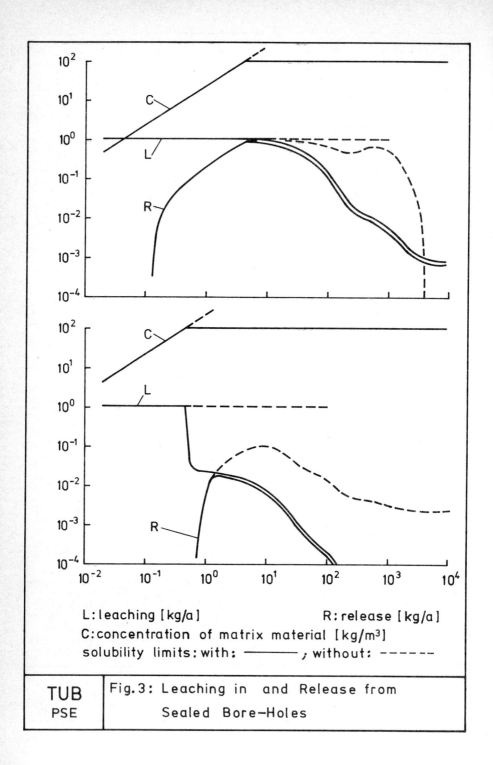

L: leaching [kg/a]   R: release [kg/a]
C: concentration of matrix material [kg/m³]
solubility limits: with: ————— ; without: ------

TUB
PSE

Fig.3: Leaching in and Release from
       Sealed Bore-Holes

to be transported upcast at every place in the bore-hole even in the sealing with a velocity of Darcy-velocity divided by the porosity of the medium.

4.2.2 Leach Model

For demonstrating the barrier model of the sealed bore-hole, simplified leach models (based on [6] , [7]) have been used which can be characterized by a time period over which a constant leaching will mobilize the whole waste. These life times are assumed to be 1000 years for vitrified HLW and 100 years for solidified MLW.

The void volume in sealed bore-holes and the replacement of the fluid by natural convection could be low enough, so that solubility limits become important. To consider solubility controlled leaching we introduced an upper limit for the concentration of matrix material. Since the simultaneous dissolution of matrix material is not necessarily the only controlling process for leaching of radio-nuclides, we arbitrarily took as upper limit 100 kg/m^3 instead of the real solubility limit for amorphous SiO_2 which is below 1 kg/m^3 [8]. With this solubility limit the dissolution of matrix material and proportionately the dissolution of radio-nuclides have been restricted.

As a result, the dissolution of radio-nuclides is not only a function of a pure leach model but also a function of the properties of the sealed bore-hole (feedback effect). Hence a coupled bore-hole and leach model has to be solved.

5.   RESULTS

The leach models and the bore-hole models have been coupled together with regard to the feedback effects from the solubility limit for HLW as well as for MLW sealed bore-holes. For each of them, real numbers for the amount of matrix material and for the heat production have been used, but normalized nuclide-inventories, which result in a unit leach rate of 1 kg/a, have been used to show the bore-hole effect in particular.

The time-dependent heat production rates have been approximated by a sum of five exponential functions with different half-life values to consider the special nuclide spectrum. Heat production rates related to the waste canisters at disposal times are 990 W for HLW and 158 W for MLW.

The calculated results in Figure 3 show the mobilization rate and the release rate for the sealed HLW and MLW bore-holes in two cases: with and without solubility limits. There is a short delay period of 0.1 year for the HLW-bore-hole and, due to the lower heat production a larger delay period of about 1 year for the MLW-bore-hole. Without any solubility limit, there is no significant retardation effect for the HLW but a significant one for the MLW-bore-hole. Using solubility limits, there will be an important effect for both, because the concentration limits for the matrix material will be reached already after 5 years for HLW or 0.5 years for MLW, and subsequently the mobilization rate and, belonging to the delay time, the release rate drops down. The cumulated amount of released material over 10,000 years is about 5 % for the HLW and 0.25 % for the MLW bore-hole.

These results are not yet able to demonstrate the safety or the risk of a repository in rock salt but they show the important parameters and effects, which have to be investigated in more detail. Parametric studies have been carried out for the permeability, the

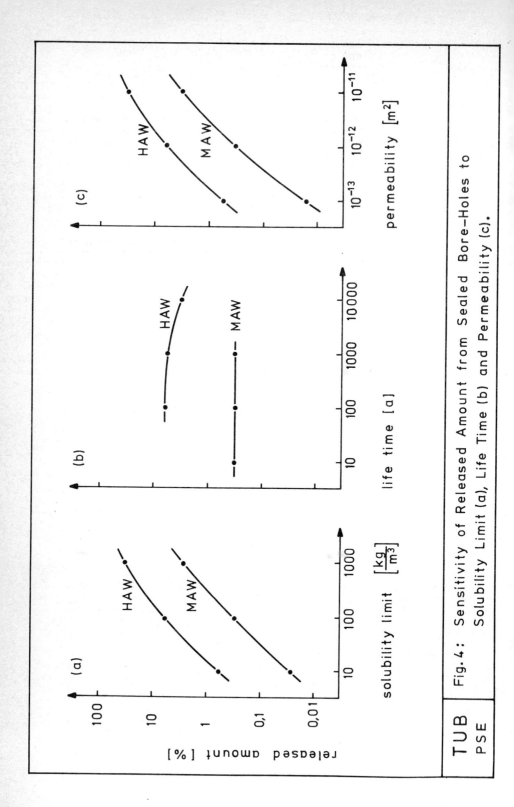

Fig. 4: Sensitivity of Released Amount from Sealed Bore–Holes to Solubility Limit (a), Life Time (b) and Permeability (c).

TUB
PSE

- 240 -

solubility limit and the life times of the waste form. As demonstrated in Figure 4, the permeabilities are as important as the solubility limits, whereas the life time of the waste is less important. Such investigations will help to divert research activities for essential effects.

6.   FUTURE WORK

Beside the demonstrated sub-model for the barrier effect of the sealed bore-hole, other barrier models have also been established for open drifts, sealed chambers and backfilled shafts, and the overall release from the salt dome has been assessed [9]. The subsequent hydro-geological calculations show, that the overlying strata is the main barrier, but if future work could reduce the conservative assumptions for the radio-nuclide transport in the flooded mine, it might be possible to get a second main barrier for the disposal system and to improve its redundancy.

For this reason we will continue the modelling of the mentioned and also of other barriers by considering several other effects such as solubility limits for specific radio-nuclides under saturated brine conditions, nuclide retention by sorption and precipitation under the same conditions, natural convection from salt dissolution and gas production by radiolysis.

REFERENCES

1.   Levi, H.W.: "Project 'Safety-Studies Entsorgung' in the Federal Republic of Germany", IAEA-SM-243/17, Proc. Intern. Symp. on the Underground Disposal of Radioactive Wastes, Helsinki, 1979.

2.   Röthemeyer, H.: "Site Investigations and Conceptual Design for the Mined Repository in the German Nuclear 'Entsorgungs-Zentrum'", IAEA-SM-243/48, Proc. Intern. Symp. on the Underground Disposal of Radioactive Wastes, Helsinki, 1979.

3.   Ostrach, S.: "Natural Convection in Enclosures", Advances in Heat Transfer, Vol. 8, p. 161 - 226, 1972.

4.   Gershuni, G.Z. and Zhukhovitskii, E.M.: Convective Stability of Incompressible Fluids, Israel Program for Scientific Translations, Keterpress Enterprises, Jerusalem, 1976.

5.   Bird, R.B., Stewart, W.E. and Lightfoot, E.N.: Transport Phenomena, John Wiley and Sons, New York, p. 551 - 552, 1960.

6.   Lutze, W., Hahn-Meitner-Institut für Kernforschung Berlin GmbH, Berlin, private communications.

7.   Köster, M., Gesellschaft für Kernforschung mbH, Karlsruhe, private communications.

8.   Petzold, A. and Hinz, W.: Silikatchemie, Ferdinand Enke Verlag, Stuttgart, p. 183, 1979.

9.   Abschlußbericht PSE-I, in preparation.

CRITICAL EXAMINATION OF CONDITIONS
FOR DUCTILE FRACTURE IN ROCK SALT

M. Wallner

Bundesanstalt für Geowissenschaften und Rohstoffe
Hannover, F. R. Germany

Abstract:

Rock salt because of its ductility in general prevents the
presence of open cracks. This favorable property is derived from
the capability of the material to reduce deviatoric stresses by
means of creep deformations. Ductile fracture of halite under
in situ stress conditions therefore is only possible under certain
circumstance. The condition to initiate and propagate fractures
in a salt formation on one hand and the fracture healing process
on the other are discussed.

An extensive experimental programme was performed to
ascertain the fracture strength of rock salt. A summery of results
from these tests is given and a proposed criterion for creep
fracture in compression is explained.

INTRODUCTION

The safe disposal of radioactive wastes is one of the most important problems of modern technology. In the Federal Republic of Germany R+D work was early placed upon a final storage in the diapiric salt structures of Northern Germany. The storage concept by means of a system of sequential or interconnected natural and technical barriers (multiple barrier system) should provide for a safe permanent disposal so that the wastes remain isolated from the biosphere until the activity of the various radionuclides has subsided to an acceptable level.

Technical barriers are the type of fixation and the packaging of the waste products as well as the sealing and backfilling mine works. The natural barrier, as the main barrier for the disposal in geological formations, is the rock mass itself.

In case of rock salt this barrier has the following favorable properties [1] :

- Due to its very low porosity rock salt is practically impermeable, even over long periods, for liquids and gases.

- Its ductility in general prevents the presence of open joints and therefore circulating water.

- Rock salt allows the excalation of large cavities without lining.

- Rock salt has a somewhat greater heat conductivity than other rocks.

- Most of the salt domes of Northern Germany have completed the phase of diapirism and are now in the stage of only very small later residual movements.

- The presence of tectonic active faults in the salt can be excluded because of the characteristic visco-plastic behavior of rock salt.

Despite these briefly mentioned advantages the final storage
of radioactive wastes in rock salt is not without problems and
requires a comprehensive safety analysis [1] .

This paper will only deal with the engineering geological
and rock mechanical conditions; physical and geochemical processes,
as a result of the interrelation between the different barriers,
which also play an important role, will not be considered.

STABILITY PROBLEMS

From a geomechanical point of view we have to consider at
least 3 types of stability problems:

Stability of drifts and cavities or of the pillars between drifts
or cavities respectively

In any case the excavation procedure represents a disturbance
of the initial state of equilibrium. Along with the excavation the
formerly almost hydrostatic state of stress will change into a
secondary state of stress where more or less high differential
stresses occur and have to be maintained by the rock mass.

Due to the ductility of rock salt these stress concentrations
are reduced by means of creep deformations. This on one hand
favorable property on the other hand causes time dependent
convergencies which might injure the accessebility of drifts and
cavities.

Long term integrity of the rock mass

The heat dissipation resulting from the radioactive decay of
high level wastes will increase the temperature in the storage area
and the surrounding rock. The rock mass will expand due to the
temperature rise. The deformation properties of the rock mass allow
a certain amount of thermal expansion, however, as a result of the
rigidity of the rock mass a resistance to deformation arises and
thermally induced stresses will be built up in the rock mass.

While the decay heat production decreases with time the temperature in the storage area drops after a certain time and consequently a contraction of the rock mass arises, which also leads to differential stresses and possibly to tensile stresses.

These thermomechanical effects can induce a stability problem especially in the neighbourhood of a layered salt formation with different stiff layers.

Dynamic loading due to earthquakes

From mining experiences it is known that the seismic loading to underground openings from an earthquake at a certain distance is uncritical.

The occurance of an active fault in the salt formation is not possible since the viscoplastic material properies prevent the rock mass from accumulation of differential stresses.

However, at least as a hypothetical case an active fault beneath the salt formation propagation through the rock salt mass during a heavy earthquake could be considered as a further stability problem.

FRACTURE AND HEALING PROCESSES

In all the above mentioned stability cases a fracture will be produced when the strength of the rock salt is exceeded. This will be especially the case if tensile stresses occur since rock salt like other rocks has a very small tensile strength.

In the case of compressive stresses the failure condition seems to be a little more complicated. As a phenomenological result one can establish that e.g. depending on the confining pressure the bearing capacity of a rock salt sample can be limited by a shear discontinuity but also by reaching a maximum differential stress without producing any fracture (cp. Fig. 1). A more precise definition of the strength will be explained in the following.

Besides that the question arises whether fractures under a triaxial state of stress cause a permanent damage to the integrity or not. It is well known that rock salt has a relatively large rate of creep under differential compressive loading, even at low temperatures. Thus, in rock salt, tensile stresses that created the fracture can relax rapidly and allow the compressive stresses in the other directions to close the crack [2] .

A similar procedure is possible for the healing of shear discontinuities.

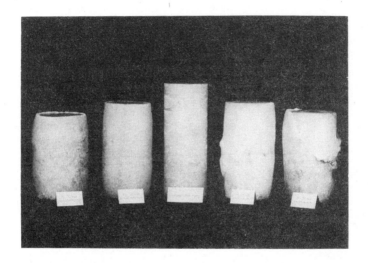

Fig. 1: Tested rock salt samples

LABORATORY TESTING PROGRAM

The question under what conditions fractures in rock salt occur and when not was examined in an extensive laboratory testing program in order to derive a reliable failure criterion and to delimit failure against creep phenomena.

Most of the tests were done on the triaxial testing apparatus shown in Fig. 2. In this test equipment the vertical load trans- mitted by a piston to the pressure cell and the confining pressure in the pressure cell itself are servo controlled. The equipment allows for load and displacement controlled tests on cylindrical samples (100 mm in diameter and 250 mm in length).

Some of the triaxial creep tests and a small number of creep failure tests where the tertiary creep phase was reached were also carried out on this machine.

Fig. 2: Triaxial testing equipment

The load controlled tests were run with a constant load rate of 1 MPa/min. The confining pressure was varied between 0 and 30 MPa.

In the displacement controlled tests the displacement rate was varied between 0.001 mm/min and 10 mm/min, the confining pressure between 0 and 20 MPa respectively.

Some typical test results are illustrated in Fig. 1. The sample in the middle shows the original specimen size before testing. The salt samples on the right side show a typical shear failure whereas the bearing capacity of the samples on the left side are limited by a maximum differential stress. In spite of large deformations these samples do not show any failure.

PRELIMINARY RESULTS

The test results of the triaxial compression tests were first plotted in a diagram showing the maximum octrahedral shear stress $\tau_{o\ max}$ at failure on the vertical axis and the octrahedral normal stress σ_o on the horizontal axis (Fig. 3).

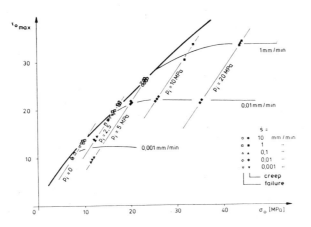

Fig. 3: $\tau_{o\ max}/\sigma_o$-diagram for strength test results

In Fig. 3 the open symbols indicate test results where failure has taken place whereas the full marked symbols stand for test results where only large deformation with a maximum differential stress occurs. From this diagram it is obvious that a failure criterion for rock salt only based on stresses is insufficient.

Analogously to the presentation of the results for steady state creep the strength test results were then plotted in a diagram showing the vertical strain rate $\dot\varepsilon$ on the vertical axis and the differential stress $\sigma_1-\sigma_3$ on the horizontal axis (Fig. 4).

For every confining pressure p_i a typical curve can be drawn which shows that the strength besides the dependence on the confining pressure itself also increases with increasing strain rate.

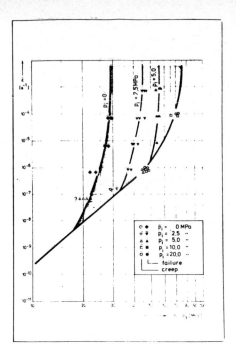

Fig. 4: $\dot{\epsilon}/\sigma_1 - \sigma_3$ -diagram for strength test results

Another result which is not shown here is that the temperature also influences the strength in a manner that the strength decreases with increasing temperatures [3] .

In comparison with the results for steady state creep [3,4] it can be concluded from the diagram in Fig. 4 that the displacement controlled compression tests where no failure was observed correlate to creep tests for evaluating steady state creep. As reciprocal tests they give the same result.

Proceeding from the lower curve in Fig. 4 the hatched area indicates the range where failure takes place and has to be specified by the state of stress, the deformation rate and the temperature.

In order to improve this result different test procedures were compared with each other as shown systematically in Fig. 5. The characteristic pathes of the stress rate controlled test, the strain rate controlled test and the creep test at constant differential stress are plotted.

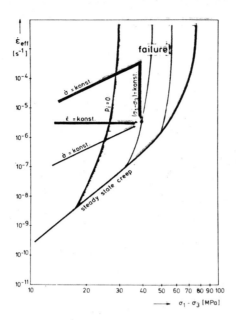

Fig. 5: Creep failure diagram, comparison of different test
procedures

As could be stated from the different tests all tests under
comparable test conditions lead to the same failure point. Therefore a generalized failure criterion could be derived based on the
state variables: state of stress, deformation rate or their
invariants respectively and the temperature.

From the present data a preliminary first approximation for
the failure criterion (identical with a creep fracture criterion)
was formulated:

$$f = \tau_o - C_1 \sigma_o^{\alpha} \, e^{\, -C_2 \sigma_o (\vartheta - \vartheta_o)^2} \, (1 + C_3 \ln \frac{II_{\dot{e}}}{II_{\dot{e}}^o}) = 0$$

In this empirical formula the state of stress is given by:

$$\sigma_o = \frac{1}{3} I_\sigma$$

$$\tau_o = \sqrt{\frac{2}{3} II_s}$$

where I_σ is the first stress invariant and II_s is the second deviatoric stress invariant respectively.

C_1, C_2 and C_3 as well as the stress exponent α are material constants. ϑ indicates the temperature and $II_{\dot{e}}$ stands for the second invariant of the deviatoric strain rate. ϑ_o and $II_{\dot{e}}^o$ are reference values.

A graphical representation of the failure criterion is shown in Fig. 6. This first approximation should be improved by necessary further test results.

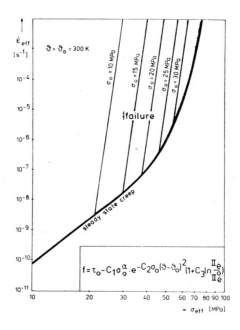

Fig. 6: Creep fracture criterion

ACKNOLEDGEMENT

This research work was done by the Bundesanstalt für Geo-wissenschaften und Rohstoffe (BGR), Hannover, within the frame of a subcontract for the Gesellschaft für Strahlen- und Umwelt-forschung (GST), Munich, on salt samples of the Asse mine. It is included in the research contract 130-80-7 WASD in the scope of the indirect action programs of the European Atomic Energy Community: Management and Storage of Radioactive Waste, action no 7: disposal of radioactive waste in geological formations.

The author acknowledges the permission to publish these results.

REFERENCES

(1) Albrecht, H., Langer, M. and Wallner, M.:"Thermomechanical Effects and Stability Problems Due to Nuclear Waste Disposal on Salt Rock", Proc. Rockstore 80, Vol. 2, pp. 801-809, Stock-holm, 1980

(2) Costin, L.S. and Wawersik, W.R.:"Creep Heeling of Fractures in Rock Salt", SAND 80-0392, Sandia National Laboratories, Albuquerque NA, 1980

(3) Langer, M., Delisle, G. and Wallner, M.:"Thermal Mechanical Modeling", Proc. of the U.S./FRG Bilateral Workshop, pp. 248-319, Berlin, 1979

(4) Langer, M.:"Rheological Behaviour of Rock Masses", Proc. 4. Int. Congr. Rock Mechanics, Vol. 3, pp. 29-62, Montreux, 1979

DISCUSSION

P. UERPMANN, Federal Republic of Germany

Would you say, that with the formula presented in your paper, you might have a tool which enables you to find a repository design where no rock salt failure occures ?

M. WALLNER, Federal Republic of Germany

Yes, with the restriction that up to now we have only a first approach for the strength or the creep failure respectively. But I think we now understand better the basic relationship. Within the on-going laboratory testing program we will improve this formula so that we than really have the basis for designing a safe repository mine from a rock mechanics point of view.

THE DYNAMIC NETWORK MODEL (DNET): A MODEL FOR
DETERMINING SALT DISSOLUTION RATES AND INCORPORATING
FEEDBACK EFFECTS IN SALT DISSOLUTION PROCESSES

R. M. Cranwell
Sandia National Laboratories
Albuquerque, New Mexico, USA   87185

J. E. Campbell
INTERA Environmental Consultants
Lakewood, Colorado, USA   80215

ABSTRACT

For nuclear waste isolation in deep, geologic formations,
transport in groundwater appears to be one of the more likely means
for radioactive waste to migrate from the repository to the biosphere. With respect to a repository in bedded salt, transport in
groundwater would, for most breachment scenarios, have to be preceded
by dissolution of all or portions of the salt layers surrounding the
repository. The Dynamic Network (DNET) model provides a capability
for investigating the rate of salt dissolution associated with a
variety of disruptive events and processes and also provides a capability for investigating the effects of feedback mechanisms such as
thermal expansion, subsidence, fracture formation and salt creep.

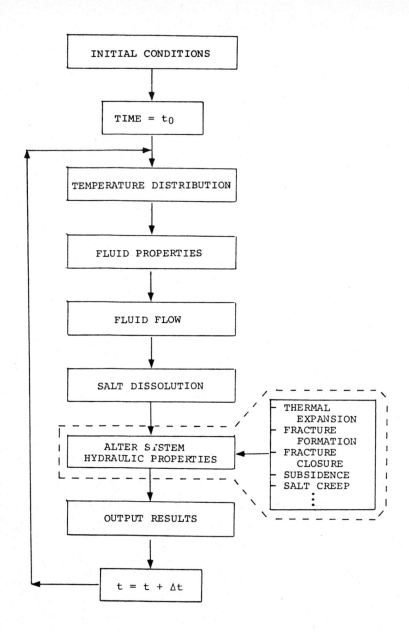

Figure 1. Computational Sequence in DNET

INTRODUCTION

A methodology for assessing the risk from geologic disposal of radioactive waste is being developed by the Fuel Cycle Risk Analysis Division at Sandia National Laboratories [1]. This methodology is to be demonstrated by application to a reference site -- a hypothetical waste repository in bedded salt. As part of this methodology, the Dynamic Network (DNET) model was developed to investigate processes near the repository such as salt dissolution and salt creep that could affect the release of radioactive waste to circulating groundwater. The DNET model also provides a systematic means for investigating the effects of feedback mechanisms such as thermal expansion, subsidence, fracture formation and fracture closure. These mechanisms can act to accelerate or decelerate the salt dissolution process and thus increase or decrease the potential for release of radioactive waste.

STRUCTURE OF DNET

DNET uses a network flow model similar to that used in the Network Flow and Transport (NWFT) model [2]. NWFT was developed to simulate far-field transport of contaminants dissolved in groundwater. Therefore, the flow system hydraulic properties in NWFT are assumed static. DNET, on the other hand, was developed for investigation of feedback mechanisms in the near vicinity of the depository. Thus in DNET, the hydraulic properties of the system are allowed to vary with time. DNET simulates several physical processes including the following: (1) fluid flow, (2) salt dissolution, (3) thermal expansion, (4) fracture formation and closure, and (5) salt creep. Because of the complexity involved in treating the several processes in DNET, the governing equations cannot be solved in an implicitly coupled fashion (i.e., simultaneously). Thus, the submodels which represent the various processes treated in DNET are applied sequentially.

COMPUTATIONAL SEQUENCE

The computational sequence in DNET is indicated in Figure 1. Computation begins at time t_0 after depository closure. The initial conditions input in DNET are assumed to be the conditions of the system at time t_0. However, for purposes of the thermal calculations, the radioactive waste heat source is decayed from depository closure at time $t = 0$ to time t_0 at which the analysis of DNET is begun. Fluid properties are functions of temperature and brine concentration. For the first time step, the brine concentration is input as an initial condition. In subsequent time steps, the brine concentration is calculated. Once fluid density and viscosity are determined (all other hydraulic properties are initialized in the input), fluid flow can be calculated. The salt solution model calculates salt removal by dissolution from appropriate portions of the flow system and determines brine concentrations throughout the system. Brine concentrations will be used in the following time step to determine fluid density and viscosity. System hydraulic properties are altered based on salt removal as well as several other processes as indicated in Figure 1. Once output information is printed, the time is incremented by Δt and DNET loops back as indicated for the next time step. The sequential application of the various submodels in DNET implies the assumption that the system is static over the time interval Δt.

THE NETWORK FLOW MODEL

The construction of the flow network used in DNET is loosely based on a hypothetical flow system which serves as a reference site for the risk methodology program. The reference site is discussed in

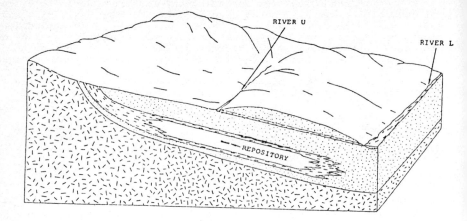

Figure 2. Physiographic Setting for Reference Site. One side of
the symmetric basin is shown. The upper end of the
valley is elliptic with the repository located on the
minor axis; the sides of the valley are parallel below
the repository.

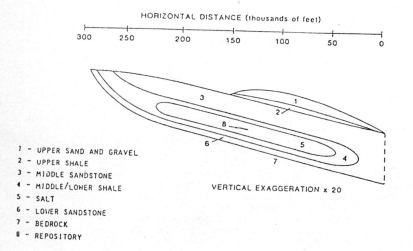

1 - UPPER SAND AND GRAVEL
2 - UPPER SHALE
3 - MIDDLE SANDSTONE
4 - MIDDLE/LOWER SHALE
5 - SALT
6 - LOWER SANDSTONE
7 - BEDROCK
8 - REPOSITORY

Figure 3. Geologic Cross Section at Reference Site.
This figure is an adaptation of Figure 1.2.2
in Campbell et al. [1].

detail in Campbell, et al. [1]. Figures 2 and 3 provide general representations of the site. Groundwater flow calculations have been performed for the reference site using the Sandia Waste Isolation Flow and Transport (SWIFT) model [3]. These calculations have shown that flow is essentially one-dimensional in the middle and lower sandstone aquifers. The SWIFT simulation of the reference site is shown in Figure 4.

The recommended geometry for the DNET network is represented by the darker lines in Figure 4. Figure 5 shows the leg and junction numbering system used for this network. The junction numbers are circled and the arrows represent the direction of positive flow. As DNET was developed to simulate salt dissolution and feedback effects near the repository, a smaller network representation could have been developed. However, DNET requires constant pressure boundary conditions at the aquifer inlets (Junctions 1 and 2) and at the discharge point to River L (Junction 3). These boundary conditions are valid if the aquifer inlet and discharge points are sufficiently far removed from the simulated disruption near the repository. The flow network in Figure 5, by representing the full (or nearly so) reference site flow system, assures that the disruptions near the repository have small effect on the boundary pressures.

Legs 1, 2 and 3 of the network are placed at the middle shale/sandstone interface and are used to represent the middle sandstone aquifer. Similarly, Legs 4, 5 and 6 are placed at the lower shale/sandstone interface and are used to represent the lower sandstone aquifer. Legs 15 and 17 are shown at the salt/middle shale and salt/ lower shale interfaces, respectively, in Figure 4. However, these legs have the flexibility of being placed at any desired location between Legs 2 and 5. Similarly, Leg 16, shown at the repository level in Figure 4, can be placed at any location between Legs 15 and 17. Legs 7, 9, 11 and 13, as well as Legs 8, 10, 12 and 14, represent vertical legs through the salt and shale and must maintain a total length of 1100 ft. These legs are used to represent various disruptive features which affect the salt and shale layers near the repository. Leg 18 represents discharge from the lower sandstone aquifer to River L.

FLOW CALCULATIONS

The following properties are assumed known for the flow calculations:

1. P_1, P_2, P_3 — Pressure boundary conditions for aquifer inlet and discharge points (Junction 1, 2, and 3)

2. k_1, k_2, ..., k_{18} — Permeability for Legs 1 to 18

3. A_1, A_2, ..., A_{18} — Cross-sectional area for Legs 1 to 18

4. L_1, L_2, ..., L_{18} — Lengths of Legs 1 to 18

5. D_1, D_2, ..., D_{14} — Elevations above datum for Junctions 1 to 14

6. ρ_1, ρ_2, ..., ρ_{18} — Average fluid density for Legs 1 to 18

7. ϕ_1, ϕ_2, ..., ϕ_{18} — Porosity for Legs 1 to 18

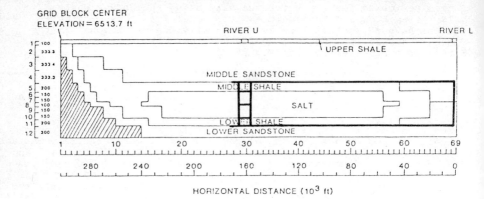

Figure 4. SWIFT Simulation of the Reference Site

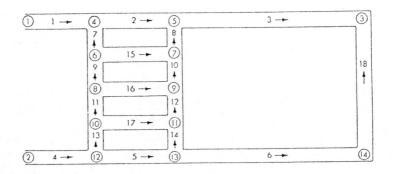

Figure 5. Network Flow Representation used in DNET
Positive flow directions are indicated by
arrows, junction numbers are circled.

Once input has been read, initial temperatures, fluid viscosities and fluid densities are calculated for each leg of the network. With these quantities determined, fluid discharge and interstitial velocities are calculated for each leg. Fluid discharge in Legs 1 through 18 is given by the following equations:

$$q_1 = \theta_1 \left[P_1 - P_4 + \rho_1 (D_1 - D_4) \right] \tag{1}$$

$$q_2 = \theta_2 \left[P_4 - P_5 + \rho_2 (D_4 - D_5) \right] \tag{2}$$

$$q_3 = \theta_3 \left[P_5 - P_3 + \rho_3 (D_5 - D_3) \right] \tag{3}$$

$$q_4 = \theta_4 \left[P_2 - P_{12} + \rho_4 (D_2 - D_{12}) \right] \tag{4}$$

$$q_5 = \theta_5 \left[P_{12} - P_{13} + \rho_5 (D_{12} - D_{13}) \right] \tag{5}$$

$$q_6 = \theta_6 \left[P_{13} - P_{14} + \rho_6 (D_{13} - D_{14}) \right] \tag{6}$$

$$q_7 = \theta_7 \left[P_6 - P_4 + \rho_7 (D_6 - D_4) \right] \tag{7}$$

$$q_8 = \theta_8 \left[P_7 - P_5 + \rho_8 (D_7 - D_5) \right] \tag{8}$$

$$q_9 = \theta_9 \left[P_8 - P_6 + \rho_9 (D_8 - D_6) \right] \tag{9}$$

$$q_{10} = \theta_{10} \left[P_9 - P_7 + \rho_{10} (D_9 - D_7) \right] \tag{10}$$

$$q_{11} = \theta_{11} \left[P_{10} - P_8 + \rho_{11} (D_{10} - D_8) \right] \tag{11}$$

$$q_{12} = \theta_{12} \left[P_{11} - P_9 + \rho_{12} (D_{11} - D_9) \right] \tag{12}$$

$$q_{13} = \theta_{13} \left[P_{12} - P_{10} + \rho_{13} (D_{12} - D_{10}) \right] \tag{13}$$

$$q_{14} = \theta_{14} \left[P_{13} - P_{11} + \rho_{14} (D_{13} - D_{11}) \right] \tag{14}$$

$$q_{15} = \theta_{15} \left[P_6 - P_7 + \rho_{15} (D_6 - D_7) \right] \tag{15}$$

$$q_{16} = \theta_{16} \left[P_8 - P_9 + \rho_{16} (D_8 - D_9) \right] \tag{16}$$

$$q_{17} = \theta_{17} \left[P_{10} - P_{11} + \rho_{17} (D_{10} - D_{11}) \right] \tag{17}$$

$$q_{18} = \theta_{18} \left[P_{14} - P_3 + \rho_{18} (D_{14} - D_3) \right] \tag{18}$$

where

$$\theta_i = \frac{k_i A_i}{\mu_i L_i} \quad , \quad \mu_i = \text{fluid viscosity of Leg } i.$$

The following conservation equations are applied at the leg junctions:

$$q_1 + q_7 = q_2 \qquad \text{Junction 4} \qquad (19)$$

$$q_2 + q_8 = q_3 \qquad \text{Junction 5} \qquad (20)$$

$$q_9 = q_7 + q_{15} \qquad \text{Junction 6} \qquad (21)$$

$$q_{10} + q_{15} = q_8 \qquad \text{Junction 7} \qquad (22)$$

$$q_{11} = q_9 + q_{16} \qquad \text{Junction 8} \qquad (23)$$

$$q_{16} + q_{12} = q_{10} \qquad \text{Junction 9} \qquad (24)$$

$$q_{13} = q_{11} + q_{17} \qquad \text{Junction 10} \qquad (25)$$

$$q_{17} + q_{14} = q_{12} \qquad \text{Junction 11} \qquad (26)$$

$$q_4 = q_{13} + q_5 \qquad \text{Junction 12} \qquad (27)$$

$$q_5 = q_{14} + q_6 \qquad \text{Junction 13} \qquad (28)$$

$$q_6 = q_{18} \qquad \text{Junction 14} \qquad (29)$$

Equations 19 through 29 are solved simultaneously to determine the unknown pressures P_4 through P_{14}. Fluid discharge by leg is calculated using Equations 1 through 18. Interstitial velocities, v_i, are calculated using the equation

$$v_i = \frac{q_i}{A_i \phi_i} \quad , \quad i = 1, 18 \qquad (30)$$

Pore volume changes and brine concentrations due to salt dissolution are then determined for appropriate portions of the flow system. Hydraulic properties are then altered based on salt removal as well as other processes such as thermal expansion, subsidence and salt creep.

SALT DISSOLUTION MODEL

Dissolution of salt is calculated using a simple, first order rate equation

$$\frac{dW}{dt} = K_s f_s (W_{sat} - W) \qquad (31)$$

where

W = weight percent dissolved salt

K_s = salt dissolution rate constant (day^{-1})

W_{sat} = weight percent of dissolved salt in saturated brine

t = time (day)

f_s = mass fraction of soluble material to total solids

Dividing Equation 31 by W_{sat} gives

$$\frac{dC}{dt} = K_s f_s (1 - C) \qquad (32)$$

Consider the application of Equation 32 to the flow network described in Figure 5. If a given leg in the network represents water movement through soluble material and if water enters the leg with brine concentration C_0, then the brine concentration will approach saturation asymptotically as water moves through the leg. If dispersion is ignored, then Equation 32 can be restated as

$$v \frac{dC}{dx} = K_s f_s (1 - C) \qquad (33)$$

which yields the solution

$$C(x) = 1 - e^{-K_s f_s x/v} + C_0 e^{-K_s f_s x/v} \qquad (34)$$

where

> x = distance measured from the inlet for the leg under consideration (ft)
>
> v = average interstitial fluid velocity (ft/day)

SALT CREEP

The formation and enlargement of solution cavities and the dissolution of salt along boreholes may be offset by the process of salt creep which can act to close solution cavities and boreholes through the salt. The process of salt creep is modeled in DNET as secondary or steady state creep using the empirical expression

$$\dot{\varepsilon} = A \exp{(-Q/R\theta)} (\sigma_1 - \sigma_3)^n \qquad (35)$$

where

> A, Q, n = parameters
>
> R = gas constant (1.9869×10^{-3} kcal/mole)
>
> θ = temperature °K
>
> $\sigma_1 - \sigma_3$ = deviatoric stress (psi)
>
> $\dot{\varepsilon}$ = strain rate (yr^{-1})

THERMAL TREATMENT

The heat source in DNET represents the emplacement of hot radioactive waste in the repository. The source is treated as a time-varying continuous planar source representing radioactive decay. Furthermore, for purposes of thermal analysis the region surrounding the repository is treated as a semi-infinite region with bounding plane being either the interface between the middle shale/middle sandstone aquifer or lower shale/lower sandstone aquifer.

The temperature field produced by a time-varying planar heat source in the semi-infinite body $z > 0$ is found by integrating the solution for the temperature field produced by an instantaneous heat source. The temperature field produced by such a source is found by solving the equation

$$\frac{\partial T}{\partial t} = \alpha \frac{\partial^2 T}{\partial z^2} \qquad (36)$$

subject to the boundary conditions

> $T = 0$ at $z = 0$ (a)
>
> and $T = f(z)$ when $t = 0$ (b)

where $f(z)$ is the initial temperature at z and α is the thermal diffusivity. Condition (a) is satisfied if we locate, in the infinite plane, a positive heat source in the plane $z = z_0$ and a negative source of the same strength in the plane $z = -z_0$.

If the instantaneous source is given by Q, the solution of Equation 36 is given by

$$T(R,t) = \frac{\sqrt{\alpha}\ Q}{2\sqrt{\pi t}\ K} \left[e^{-R^2/4\alpha t} - e^{-R'^2/4\alpha t} \right]$$ (37)

where

$$R = z_0 - z$$

$$R' = z_0 + z$$

$$K = \text{thermal conductivity}$$

This temperature field represents a change in the temperature at z and so must be added to the initial temperature f(z).

If the heat source Q(t) is a permanent source or varies in time, the temperature takes the form

$$T(R,t) = \frac{\sqrt{\alpha}}{2\sqrt{\pi}\ K} \left[\int_0^t \frac{Q(\tau)\exp(-R^2/4\alpha(t-\tau))}{(t-\tau)^{1/2}}\ d\tau \right.$$

(38)

$$\left. - \int_0^t \frac{Q(\tau)\exp(-R'^2/4\alpha(t-\tau))}{(t-\tau)^{1/2}}\ d\tau \right]$$

FRACTURE FORMATION AND CLOSURE

The generation of heat in the middle of the bedded salt will cause some thermal expansion in the salt layers before the thermal front reaches the overlying and underlying shale and causes thermal expansion in these layers. Furthermore, typical values of the thermal expansion coefficient of salt are generally higher than those for shale. Thus, in the initial stages of the analysis one would expect some extension and stretching of the shale layers caused by thermal expansion in the salt.

The formation of new cracks and the opening of existing cracks or fissures in the shale are treated as resulting from this extension and stretching. Actually, the effects of uplift from thermal expansion in the salt must be combined with the subsidence which could occur as salt is removed by dissolution along the salt/shale interface.

The distributions and orientations of the cracks are based on preliminary results of modeling of the maximum elastic stresses caused by thermal expansion. These results indicate that the loci of maximum tensile stresses are near the repository margins and are concentrated in the shale layers.

If we assume that, initially, some unit of the shale of length ℓ_0 is subjected to extension, then the net extension of the unit at time t is given by

$$\ell - \ell_0 = \sqrt{\ell_0^2 + h^2} - \ell_0 \qquad\qquad (39)$$

where

ℓ = length of shale after extension

h = net vertical displacement

Initially this extension can operate to open existing fractures or create new ones. However, as the thermal front reaches the shale layers, this extension can be offset by thermal expansion in the shale. Setting

T = temperature in shale at time t

T_A = ambient or initial temperature in shale

the effective extension at time t is given by

$$E = (\ell - \ell_0) - \ell_0 \cdot \alpha \cdot (T - T_A) \qquad\qquad (40)$$

where α = thermal expansion coefficient. If $E < 0$, the implications are that thermal expansion in the shale is more than enough to offset the extension caused by uplift or settling. In this case, the crack opening size should be reduced accordingly or set to some minimum value. For $E > 0$, no new cracks should be created unless E exceeds its maximum value from previous time steps. When the maximum is exceeded, the additional extension is used to create new cracks and to increase the average crack opening.

The formation of new cracks in the shale or the enlargement of the aperture of existing cracks can increase the permeability of the shale at these locations. DNET then calculates the additional rate of salt dissolution based on this increase in the permeability of the shale. These dissolution rates are used as input to the methodology to determine estimates of risk.

Additional information on the DNET model can be found in either [4] or [5].

REFERENCES

1. Campbell, J. E. et al.: "Risk Methodology for Geologic Disposal of Radioactive Waste: Interim Report", Report No. SAND78-0029, Sandia Laboratories, Albuquerque, New Mexico (1978).

2. Campbell, J. E. et al.: "Risk Methodology for Geologic Disposal of Radioactive Waste: The Network Flow and Transport (NWFT) Model", Report No. SAND79-1920, Sandia National Laboratories, Albuquerque, New Mexico (1980).

3. Dillon, R. T. et al.: "Risk Methodology for Geologic Disposal of Radioactive Waste: The Sandia Waste Isolation Flow and Transport (SWIFT) Model", Report No. SAND78-1267, Sandia Laboratories, Albuquerque, New Mexico (1978).

4. Campbell, J. E. et at.: "Risk Methodology for Geologic Disposal of Radioactive Waste: DNET: A Model for Incorporating Feedback Effects in Salt Dissolution Processes", Report No. SAND80-0067, Sandia National Laboratories, Albuquerque, New Mexico (1981).

5. Cranwell, R. M. et al.: "Risk Methodology for Geologic Disposal of Radioactive Waste: The DNET Computer Code User's Manual", Report No. SAND81-1663, Sandia National Laboratories, Albuquerque, New Mexico (1981).

DISCUSSION

H.C. BURKHOLDER, United States

Do you have any results from applying the DNET model that you can share with us ?

R.M. CRANWELL, United States

Results have been generated for a couple of release scenarios involving solution channels at the middle shale/salt interface and boreholes connecting underlying and overlying aquifers. Overrun times (i.e. times for the solution cavities to reach the waste drifts) have been observed ranging from a few thousand years to a few hundred thousand years.

H.C. BURKHOLDER, United States

What sets of conditions or assumed parameter values cause the model to predict that salt will be dissolved away and the waste exposed to flowing groundwater in less than 10,000 years ?

R.M. CRANWELL, United States

A very low exponent on the stress differential $(\sigma_1-\sigma_3)$ in the creep law and high thermal expansivity of the salt appear to be combinations that tend to enhance the relative rate of salt dissolutions. Furthermore, salt dissolutions rather appear to be quite sensitive to the rate at which the stress differential approaches zero as solution cavities or solutions channels creep close.

W.C. PATRICK, United States

How do you control the time step in your model since you use a "once through" calculational technique rather than an iterative technique ?

R.M. CRANWELL, United States

The model has an "automatic" time step capability that will either increase or decrease the time step based on the percentage of change in volumetric flow in any of the legs of the network from the previous time step. This percentage change in flow is an input parameter and so the user has the flexibility of making this decision. The model also has a built-in time step, independent of the automatic time step size, that is used in the salt creep submodel. This time step is presently set very small.

F. GERA, Italy

In many cases salt is undergoing dissolution due to either natural or man-made conditions. Do you think that comparison between real conditions and the results obtained with your model might provide a way to validate the model ?

R.M. CRANWELL, United States

Yes ! This would be an ideal method for validating such a model. We have been attempting to collect data from the Kaniai salt

mines to see if dissolution rates predicted by our model compare
favorably with this data.

P. UERPMANN, Federal Republic of Germany

Did you ever consider that the aquifer underlying your
bedded salt formation might not be saturated with sodium chloride ?
If in this case the water would come in contact with the salt formation, leaching would occur from the bottom.

R.M. CRANWELL, United States

We have been aware of this possibility. However, we have
not yet attempted to simulate the salt dissolution rates using DNET
for this setup.

M. MAKINO, Japan

You showed us the schematic diagram titled "Structure of
Methodology". Where do you locate the dynamic model in the diagram ?

R.M. CRANWELL, United States

The DNET model is included in the block entitled "Potential
Waste Release Mechanism".

TRANSPORT PHENOMENA OF WATER AND GAS COMPONENTS WITHIN
ROCK SALT IN THE TEMPERATURE FIELD OF DISPOSED HIGH LEVEL WASTE

Norbert Jockwer
Gesellschaft für Strahlen- und Umweltforschung mbH
Institut für Tieflagerung/Wissenschaftliche Abteilung
Theodor-Heuss-Str. 4
D-3300 Braunschweig

ZUSAMMENFASSUNG

Steinsalzformationen, in denen radioaktive Abfälle endgelagert werden
sollen, enthalten geringe Mengen Wasser und Gas (H_2S, CO_2 und Kohlenwasserstoffe),
die mit den Abfallbehältern bzw. den Verfestigungsprodukten reagieren und Trans-
portmedium für Radionuklide sein können. Durch die erhöhte Temperatur der
radioaktiven Abfälle werden diese Komponenten freigesetzt und können infolge
eines Gesamtdruck-, Partialdruck-, Temperatur- und Konzentrationsgradienten auf
den Kristallgrenzflächen und intergranularen Poren migrieren. Zur Erfassung der
Transportvorgänge wurde ein Diffusions- und Migrationsmodell entwickelt.

ABSTRACT

Rock-salt formations, which will be used for the disposing of high
level radioactive waste contain small amounts of water and gas components
(H_2S, CO_2 and hydrocarbons) which may interact with the waste containers or
the solidified waste products and may be the transport media for radionuclides.
These water- and gas components will be liberated in the elevated temperature
field of the disposed high level waste and may migrate on the crystal boundary
surfaces or intergranular pores as a result of the gradient of total pressure,
partial pressure, concentration and temperature. In order to calculate the
transport phenomena of these liberation products a diffusion and migration
model has been developed.

In the Federal Republic of Germany it is planned to dispose of high level radioactive waste in rock salt formations. The investigation of samples from different North German salt mines has shown that rock salt contains small amounts of water and gas components. The majority of the water derives from the hydrated minor minerals such as polyhalite and kieserite and from intergranular water adsorbed on the crystal boundaries. The amount contained in fluid inclusions in North German salt domes is comparatively low.

In order to determine the total water content in the rock salt, 202 samples from different depths and different stratigraphic layers in Asse salt mine have been investigated. Figure 1 shows the frequency distribution of the water content determined in this investigation. The distribution maximum is at 0.04 % by weight showing that the water content is not constant but depends on the amount of minor minerals present containing water of hydration. 55 % of these salt samples had a water content less than 0.1 % and 75 % less than 0.2 %. Salt samples from other North German salt domes have similar water contents.

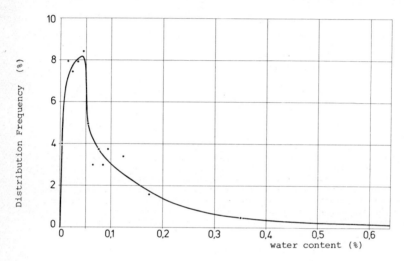

Fig. 1: Distribution frequency of the total water content of 202 rock salt samples from the Asse salt mine

In order to determine the water content within a stratigraphic layer the rock salt of a 300 m deep vertical borehole has been analysed. The result is shown in figure 2. It indicates that even in one stratigraphic layer the water content is not constant and may vary by the factor of 10 within 0.1 to 1.0 m.

The thermal behavior of the different water components within the rock salt has already been reported (7).

The gas components which have been analysed within the rock salt are:
1. H_2S within the range of 0 to 5 ppm
2. CO_2 within the range of 0 to 200 ppm
3. thermal generated HCl by decomposition of rare and trace minerals within the range of 0 to 150 ppm
4. gaseous hydrocarbons within the range of 0 to 60 ppm

The relationship between these gas components and the mineralogical composition has not yet been investigated.

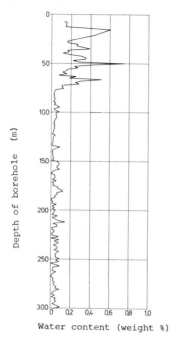

Fig. 2: Water content profile within a stratigraphic layer (Staßfurt halite Na2)

Concerning the disposal fo radioactive waste in a salt formation, the question is what happens to the water and gas components within the disposal area.

In the surrounding of a borehole with disposed high level radioactive waste the rock salt will be heated and the above mentioned components will be released into the intergranular space and may migrate on crystal boundaries and through microfissures.

The first step in calculating the transport phenomena within the disposal area is based on consideration of a porous media in which the pores and the intergranular spaces are only filled with the two components air and water in a gaseous phase. Figure 3 shows the physical situation of a pore in the rock salt.

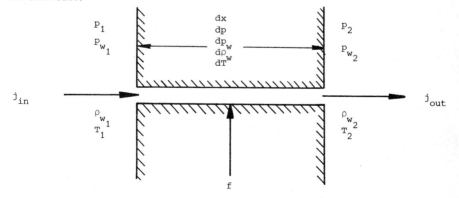

Fig. 3: Physical situation in a pore of a differential volume element

$$p \quad = \quad \text{total gas pressure}$$

$p_w \quad = \quad$ partial pressure of the gaseous water

$\rho_w \quad = \quad$ density or concentration of the water in the gas mixture in a pore

$T \quad = \quad$ absolute temperature

$f \quad = \quad$ release rate of the water in the differential volume element into the pore

$dx \quad = \quad$ length of the differential volume element

$j_{in} \,/\, j_{out} \quad = \quad$ water flux in and out of the differential volume element

Within the pore the equation of continuity can be written:

$$j_{in} - j_{out} + f = 0$$

This equation signifies that in a pore of a differential volume element the existing water flux is the sum of the flux entering the pore plus the release rate of water within that volume.

The migration of the water is caused by a total-pressure gradient, which involves viscous flow, and by a gradient of concentration, partial pressure and temperature, which involves diffusion.

Instead of the mass flux of the water within the differential volume the equation of continuity can be written as the change with time of the gas concentration in the pore

$$\frac{\partial \rho_w}{\partial t} = - \frac{d(\rho_w \cdot v)}{dx} - \frac{di}{dx} + g \, \rho_{salt}$$

$\dfrac{\partial \rho_w}{\partial t} \quad = \quad$ change with time of the water concentration within the gas in the pore

$\rho_w \cdot v \quad = \quad$ flux as a result of viscous flow

$v \quad = \quad$ velocity of viscous flow

$i \quad = \quad$ diffusion flux

$g \, \rho_{salt} \quad = \quad f \quad = \quad$ release rate of the water into the pore

$g \quad = \quad$ fractional release rate of water from the rock salt

$\rho_{salt} \quad = \quad$ density of the rock salt

According to the Darcy law:

$$\rho_w \cdot v = \rho_w \cdot \frac{k}{\eta \phi} \cdot \frac{dp}{dx}$$

$k \quad = \quad$ permeability of the rock salt

$\eta \quad = \quad$ viscosity of the gas composition in the pore

$\phi \quad = \quad$ porosity

$\dfrac{dp}{dx} \quad = \quad$ total gas-pressure gradient

The diffusion flux as a result of concentration-, partial pressure and temperature gradient is:

$$i = -D \frac{d\rho_w}{dx} - D\rho_w \frac{k_p}{P_w} \frac{dp_w}{dx} + D\rho_w \frac{k_T}{T} \frac{dT}{dx}$$

D = diffusivity

$\frac{d\rho_w}{dx}$ = gradient of water density in the pore

P_w = partial pressure of water

Dk_p = partial pressure diffusivity

k_p = coefficient of partial-pressure diffusivity

$\frac{dp_w}{dx}$ = gradient of partial pressure

DK_T = thermal diffusivity

k_T = coefficient of thermal diffusivity

T = temperature

$\frac{dT}{dx}$ = temperature gradient

As the density of the gaseous water within the gas in the pores is comparatively small and the temperature within the disposal area is greater than 100 °C, the gas law can be written as

$$p \cdot V = RT$$

$$p = \frac{R}{V} T$$

$$\frac{dp}{dx} = \frac{R}{V} \frac{dT}{dx}$$

V = molecular volume of water

R = gas constant

Then the diffusion flux can be written

$$i = -D \left[\frac{d\rho_w}{dx} + \rho_w (k_p - k_T) \frac{1}{T} \frac{dT}{dx} \right]$$

The equation for the change with time of water concentration within the gas in the pore becomes:

$$\frac{\partial \rho_w}{\partial t} = -\frac{d}{dx} \rho_w \frac{k}{\eta \phi} \frac{dp}{dx}$$

$$- \frac{d}{dx} D \left[\frac{d\rho_w}{dx} + \rho_w (k_p - k_T) \frac{1}{T} \frac{dT}{dx} \right]$$

$$+ g \rho_{salt}$$

The equation in three dimensional rectangular coordinates with the analytical abreviations div and grad in

differential form:

$$\frac{\partial \rho_w}{\partial t} = - \operatorname{div} (\rho_w \frac{k}{\eta \Phi} \operatorname{grad} p)$$

$$- \operatorname{div} \left\{ D \left[\operatorname{grad} \rho_w + \rho_w (k_p - k_T) \right. \right.$$

$$\left. \left. \cdot \frac{1}{T} \operatorname{grad} T \right] \right\}$$

$$+ g \, \rho_{salt}$$

with the Gauss-theorem: $\int_V \operatorname{div} A \, dV = \oint_F A \, dF$

the equation can be written in the

integral form:

$$\frac{\partial}{\partial t} \int_V \rho_w \, dV = - \oint_F \rho_w \frac{k}{\eta \Phi} \operatorname{grad} p \, dF$$

$$- \oint_F D \left[\operatorname{grad} \rho_w + \rho_w (k_p - k_T) \right.$$

$$\left. \cdot \frac{1}{T} \operatorname{grad} T \right] dF$$

$$+ \int_V g \, \rho_{salt} \, dV$$

This equation signifies that the change with time in the amount of water in the pore of a defined volume is equal to the amount of the water which enters through the surface as a result of viscous flow or diffusion plus the amount of water that has been released out of the rock salt into the pore.

In order to know how much water has been released to and accumulated in the borehole with the disposed high-level radioactive waste this equation has to be modified since within the borehole-volume no water will be released but there will be a loss of water by corrosion, radiolysis and escape from the incompletely sealed borehole. Therefore the change with time in amount of water within the borehole becomes

in differential form:

$$\frac{\partial \rho_w}{\partial t} = - \operatorname{div} \left[\rho_w \frac{k}{\eta \Phi} \operatorname{grad} p \right]$$

$$- \operatorname{div} \left[\operatorname{grad} \rho_w + \rho_w (k_p - k_T) \frac{1}{T} \operatorname{grad} T \right]$$

$$- 1$$

in integral form:

$$\frac{\partial}{\partial t} \int_V \rho_w \, dV = - \oint_F \rho_w \frac{k}{\eta \Phi} \text{ grad } p \, dF$$

$$- \oint_F D \left[\text{grad } \rho_w + \rho_w \, (k_p - k_T) \frac{1}{T} \text{ grad } T \right] dF$$

$$- \int_V l \, dV$$

$$l = \text{loss of water per unit of time and volume}$$

By integration over the time and total borehole-volume the amount of water within the borehole can be calculated:

$$I = \rho_w \cdot V = \int_t \left[\oint_F \left\{ - \rho_w \frac{k}{\eta \Phi} \text{ grad } p \right. \right.$$

$$- D \left[\text{grad } \rho_w + \rho_w \, (k_p - k_T) \frac{1}{T} \text{ grad } T \right] \left. \right\} dF$$

$$\left. - \int_V l \, dV \right] dt$$

The flux of water being released at the borehole-surface is:

$$j = \oint_F \left\{ - \rho_w \frac{k}{\eta \Phi} \text{ grad } p \right.$$

$$- D \left[\text{grad } \rho_w + \rho_w \, (k_p - k_T) \frac{1}{T} \text{ grad } T \right] \left. \right\} dF$$

The amount of water which has been released at the borehole-surface is:

$$J = \int_t \oint_F \left\{ - \rho_w \frac{k}{\eta \Phi} \text{ grad } p \right.$$

$$- D \left[\text{grad } \rho_w + \rho_w \, (k_p - k_T) \frac{1}{T} \text{ grad } T \right] \left. \right\} dFdt$$

These equations show that the flux and amount of water which entering the borehole depends on the gradient of concentration and temperature between borehole and pores in the rock salt of the disposal-area.

The parameters in this equation have been measured on rock salt samples in the laboratory; the results are:

1. permeability within the range of less than 0.05 to 120 μD

2. porosity   within the range of   0.01 to 0.74 volume %

3. diffusivity within the range of 0.1 $\cdot$ 10^{-4} to 2.1 $\cdot$ 10^{-4} cm^2/sec

4. diffusivity of temperature and partial pressure gradient $(k_p - k_T)$
   within the range of   - 5   to - 225

As less is known about the effects of radiolysis, corrosion and the tightness of the borehole, the concentration-gradient of water between the borehole at the pores of the rock salt in the disposal area can not be determined since conditions in the borehole are not known. Therefore it is not yet possible to calculate the flux of water being released into the borehole with disposed high-level waste.

ACKNOWLEDGEMENT

This investigation is being made within the research contracts 058-78-1 WAS D and 130-80-7 WAS D of the European Atomic Energy Community "Management and storage of radioactive waste" Action no. 7: "Disposal of radioactive waste in geological formations."

REFERENCES

[1]   Anthony, T.R. and Cline, H.E.: The thermomigration of biophase vapour-liquid droplets in solids. Acta Metallurgica Vol. 20 (1972).

[2]   Bird, R.B.; Stewart, W.E.; Lightfoot, E.N. (1964): Transport phenomena. - John Wiley + Sons, New York, London.

[3]   Förster, S. (1974): Durchlässigkeits- und Rißbildungsuntersuchungen zum Nachweis der Dichtheit von Salzkavernen. - Neue Bergbautechnik, 4. Jahrgang, Heft 4, S. 278 - 283.

[4]   Glaser, H. (1958): Wärmeleitung und Feuchtigkeitsdurchgang durch Kühlraumisolierungen. - Kältetechnik, 10. Jahrgang, Heft 3, S. 86 - 91.

[5]   Hofrichter, E. (1976): Zur Frage der Porosität und Permeabilität von Salzgesteinen. - Erdöl-Erdgas-Zeitschrift, 92. Jahrgang, Heft 3, S. 77 - 80.

[6]   Jockwer, N. (1980): Die thermische Kristallwasserfreisetzung des Carnallits in Abhängigkeit von der absoluten Luftfeuchtigkeit. - Kali und Steinsalz, Band 8, Heft 2, S. 55 - 58.

[7]   Jockwer, N. (1980): Laboratory investigation on the water content within the rock salt and its behavior in a temperature field of disposed high level waste. - Scientific basis for nuclear waste management (1980).

[8]   Jockwer, N. (1981): Die thermische Kristallwasserfreisetzung des Polyhalits und Kieserits in Abhängigkeit von der absoluten Luftfeuchtigkeit. - Kali und Steinsalz, Band 8, Heft 4, S. 126 - 128.

[9]   Jockwer, N. (1981): Untersuchungen zu Art und Menge des im Steinsalz des Zechsteins enthaltenen Wassers sowie dessen Freisetzung und Migration im Temperaturfeld endgelagerter radioaktiver Abfälle. - Dissertation, Technische Universität Clausthal.

[10]   Tollert, H. (1964): Beiträge zur Porosität im Steinsalz. - Kali und Steinsalz, Band 4, Heft 2, S. 55 - 60.

DISCUSSION

<u>R.H. KOSTER</u>, Federal Republic of Germany

What water content distribution curve could one reasonably expect for the Gorleben site salt type ?

<u>N. JOCKWER</u>, Federal Republic of Germany

I have analysed several rock salt samples from the Gorleben salt dome. This rock salt has a similar water content to that in rock salt from the Asse mine.

<u>R.H. KOSTER</u>, Federal Republic of Germany

You mentioned that the consumption of waste in the borehole cannot be calculated. I cannot agree with this ; you can very well calculate the waste consumption due to corrosion by using known area corrosion rates. Furthermore, the mean dose rates of HLW containers are known (in the order of 10^4-10^5 rads/hr), consequently, when you also use the known G-value for waste radiolysis, you can easily compute waste consumption due to radiolysis effects.

<u>N. JOCKWER</u>, Federal Republic of Germany

I have started calculating the effects of corrosion and radiolysis in order to estimate the climate within the borehole, but results are not available yet.

INVESTIGATIONS ON MODEL CAVERNS CONCERNING THE INTERACTION BETWEEN SIMULATED MLW/LLW AND SALT ROCK

W. R. Fischle
Gesellschaft für Strahlen- und Umweltforschung mbH München
Institut für Tieflagerung, Wissenschaftliche Abteilung
Schachtanlage Asse, 3341 Remlingen
Federal Republic of Germany

Zusammenfassung

In einen, von mehreren Institutionen der BRD durchgeführten Forschungsprojekt soll die behälterlose Einlagerung und Verfestigung von LAW/MAW in untertägigen Hohlräumen als Konzept für die Endlagerung untersucht werden.

Ein Teilvorhaben befaßt sich mit den Wechselwirkungen zwischen Produkt und Salz. Dabei werden vorkonditionierte Pellets mit einer Zwickelsuspension aus Zement, Wasser und Zusätzen in ausgesolte Kavernen gefüllt und relevante Parameter beim Reaktions- und Abbindeprozeß gemessen. Z.B. dient die Erfassung von Temperaturen bei unterschiedlichen Befüllraten zur Überprüfung eines mathematischen Modells und die Untersuchung von Proben aus den Kavernen dazu, Eigenschaften und Änderungen des Produktes zu erkennen.

Abstract

A research project involving a number of institutions in FRG was undertaken to investigate a concept for the containerless emplacement and solidification of LLW/MLW in underground caverns.

One task was concerned with the interaction between the waste product and salt. For this purpose preconditioned pellets and a filler suspension made of cement, water and additives were filled into a solution mined cavern and relevant experimental parameters during reaction and curing processes were measured. For instance the determination of temperature changes accompanying various fill rates served to test a mathematical model and the investigation of samples from the cavern provided information concerning product properties and changes with time.

1. INTRODUCTION AND OBJECTIVE

1.1 System Description

The previous emplacement technique for radioactive wastes in the FRG consists of transporting solidified low and medium level wastes to a storage site in drums.

The transport to the 500 - 750 m underground level is accomplished with the mine hoist through the shaft. There they are unloaded and delivered to the storage chamber. Depending upon the drum contents these are either stacked, tipped over or brought into the chamber through a special lock. In this connection it is necessary to consider the required shielding for MLW.

Alternatively and as a possible repository concept, the following project was undertaken:

"The containerless emplacement and solidification of LLW/MLW in underground caverns."

1. A waste-binder-mixture is produced at the repository site.

2. The underground delivery of the suspension is accomplished through a 1000 m long gravity-feed pipe.

3. The storage occurs in a cavern, which has a favourable rock mechanical form and can accept wastes over a number of years.

The essential advantages of this system are the following:

- fast and continuous deposition of large amounts of waste;
- saving of underground manpower;
- decreased radiation doses;
- complete filling of the cavern;
- immediate closure and binding with the rock;
- production of a compact and quasi-monolithic waste block; the resultant small volume-to-surface ratio as compared with that of waste in drums reduces potential leaching in the event of a water or brine inflow.
- the efficiency with which space is used to receive waste is substantially higher than with the stacking techniques by a factor of 6 or 7 times. Further no additional space is required for tritium liquid waste as this can be used as a make-up water in the fill material;
- the costs are substantially lower than by currently used processes;
- safety in case of an accident is greater since the direct access from above ground permits measures to be taken to restrict release of contamination.

1.2 Project Components

The basic proof of technological and licensing feasibility was established in the first phase of the project. At that time various cavity sizes and forms were investigated.

The objective of the second project phase is to develop the technical capability for the process. Necessary documentation to permit a safety evaluation of the concept shall be established. This includes material studies on inorganic binders concerning their suitability as a matrix for MLW and LLW.

This work is supported by the Ministry of Research and Development with the following firms participating under the project leadership of Kernforschungszentrum Karlsruhe/Project Reprocessing (KfK/PWA); Amtliche Materialprüfanstalt - Clausthal-Zellerfeld (AMPA); F.J. Gattys Verfahrenstechnik; Gesellschaft für Strahlen- und Umweltforschung mbH München, Institut für Tieflagerung (GSF/IfT); Kernforschungszentrum Karlsruhe/Institut für Nukleare Entsorgung (INE); NUKEM.

The emphasis in the program was directed at the following points:

1. Producing a transportable and storable product that contains a maximum of low and medium level wastes.

2. Delivery of the product to the storage site.

3. Preparation of a cavity in salt for the emplacement of radioactive wastes over a number of years and proof of its stability.

4. Studies of the emplacability of simulated MLW/LLW in salt.

These points of emphases led to the following tasks.

2.   INVESTIGATION ON MODEL CAVERNS

2.1   Test Objective

Calculations made in the first phase of the project indicated maximum temperatures in the waste product of up to 300 °C. The primary reason for this temperature is the heat of reaction due to curing of the cement. Since this has an unfavourable effect on the waste product, the rock and the control of gases in the cavity space, the temperature in the second phase of the project was limited to 90 °C. In order to reduce the maximum value it was necessary to add nonreactive materials in measured quantities.

It is desirable to precondition the waste solution with cement in order to release a part of the heat outside the cavity. This is accomplished by means of pellets in which a large part of the radioactive solution is bound. The transportability and the temperature development in pellets with the filler material and the quality of the solidified products were investigated in technological quantities and their optimization was initiated.

In this test phase the emplacement products were intended to be produced in quantities for in-situ testing. At the same time a change of about 80 metric tons of pellets had to be produced. In this connection the eventual inclusion of radioactive material had to be considered as well as the need to guarantee the constant quality of the starting product. This was also true for the transported endproduct.

It was necessary to prove that the calculational program could be adapted to a full scale model. It was required that the material properties during production, bending, hardening and post-hardening phases be determined.

Questions for potential further tests and calculations were to be determined in so far as possible. This includes the areas of transport, thermal stresses and the properties of the cement which serve as a basis for converting to the geomechanical characteristics of the rock.

2.2   Preparation of Caverns

2.2.1   Pretesting and Assembly of Brine Equipment

In order to vary test parameters, such as the filling sequences and the mixture composition a total 5 caverns, each with a volume of 10 m³ were prepared. In order to permit a uniform and symmetric heat release and flow a suitable location in the rock had to be selected.

Distances of 20 m were established, so that during preparation and later measurements no interaction between cavern and no hindrance to their filling would occur.

The only possibility for preparing the caverns was by solution mining. The brine product was to be used for later tests. For storage without damage a saturation corresponding to a density of $> 1.19 \times 10^3$ Kg/m³ was derived. Experience in the solution mining of caverns of this general size was not available. The usual methods for the preparation of large caverns such as direct and indirect solution mining appeared to be too complicated and the use of large equipment too costly.

By means of pretests at the IfT on rock salt samples solution mining times of ·5 - 2 days were established in dependence upon the volume of the water and the surface area of the sample with a saturation level of ∿1.2x10³ Kg/m³. The dissolved volume of the rock salt was approximately 10 %. This provided the means for calculating the number of solution steps required and served as a basis for requesting and obtaining bids for the solutioning work.

Preparatory work by the IfT included the following:

1. Mining a horizontal drift with a length of 123 m and a cross section of 3 by 4 m, width to height.

2. Mining of a sloping drift of 60 m length and the same cross section as above to accept the brine.

3. Preparation of 5 vertical boreholes, each with a length of 19 m for connection to the caverns. In order to permit the later instrumentation of the caverns it was necessary first to drill the holes very exactly and then enlarge them to 180 mm. As the rock had already been geologically characterized only special studies of the components of the drilling powder that were difficult to dissolve or insoluble were made and no attempt was made to investigate cores.

4. Assembly of transport tubing and storage containers for water and brine as well as making available the required quantity of fresh water.

5. In order to accelerate the saturation process during the initial phase and to influence the shape of the cavern, air was forced into the bottom of the cavern.

Corresponding tubes and equipment were assembled.

2.2.2 Solutioning Procedure and Results

The caverns were to be solution mined to a depth of 15 - 18 m under the drift floor. The lower 1 m of the borehole was to serve as a sump for the collection of insoluble and poorly soluble components. The filling tube for fresh water, which also served for the introduction of air, consequently extended to a depth of 17.50 m. The pump to suck out the brine was placed at 18.50 m, so that the suction hose, which is situated over the motor, lay at 18.00 m.

Design Conditions

Borehole: Ø 0.18 m   depth 19 m

Individual cavern volume: approx. 10 m³

Cavern shape: cylindrical

Depth of cavern: 15 - 18 m below 800 m level

Cavern height: approx. 3 m

Cavern diameter: approx. 2 m

Brine density: $\rho_s > 1.19 \times 10^3$ Kg/m³

Planning bases

Enlarging the cavern volume for complete saturation approx. 15 %; for saturation to 1.19×10^3 Kg/m³ approx. 14 %

required solution changes approx. 40

Solution times

1st - 10th change   approx. 12 h/change

11th - 25th change   approx. 24 h/change

26th - 40th change   approx. 48 h/change

until reaching the required density

required solution mining time approx. 53 days

Fresh water requirements

for 1 m³ cavern size with a density of 1.19 x 10³ kg/m³

7 - 8 m³ fresh water

for 5 x 10 m³ cavern size

350 - 400 m³ fresh water

Configuration

It is expected that the cavern will initially assume a 'flower pot shape', since the fresh water as well as the partially saturated brine will rise as a result of density differences and therefore preferentially dissolve the upper part of the cavern.

In order to maintain a cylindrical form when the upper part of the cavern reaches the required diameter, solutioning is continued using partial charges of water.

Actual Solutioning Conditions

The first filling of the caverns occurred on Jan. 9th, 1981. After 52 days including weekends the intended volume and desired configuration were reached with exceptional accuracy.

It was possible to make 2 changes per day through the 27th change; planning was based on 10 days. Looking back it is considered possible to make three changes per day in the starting phase. Finally the cavern volume was enlarged with one change per day. After reaching about 3/4 of the desired individual cavern volume it was necessary to proceed with partial charges of water. For this purpose the cavern was filled only to that depth at which the desired diameter had already been reached. During the partial filling the solutioning time was increased, because of the small wall height, to two days.

Fresh water requirements for 5 model caverns 342 m³
Recovered brine quantity approx. 380 m³

Measurements

During the solutioning process the liquid level was measured when fresh water was brought in and before pumping out the brine. Thereby the filling level could be maintained constant.

Configuration control during the solution mining process was accomplished through measurement of the cavern volume. For this purpose a defined volume of fresh water was introduced into the cavern and the resulting water level change measured. Using these two volumes the specific volume could be determined and from that the diameter calculated. After completing the solution mining process the caverns were measured as described previously. These measurements were made with saturated brine instead of water. From these, final cavern volumes and configurations were derived. In calculating the cavern diameter from the specific volume it was assumed on the basis of uniform volume distribution and the experience from other solution mining projects that the cavern form was circular and centered about the borehole. After pumping the caverns dry and disassembling the solutioning equipment this assumption was checked and confirmed using a feeler gage probe. A hinged arm was lowered to various depths by means of a square pole and there opened to specific diameters and then turned about the vertical axis for this purpose.

The specific final cavern volumes were:

cavern 1	9.475	m³
c 2	9.440	m³
c 3	10.500	m³
c 4	9.707	m³
c 5	9.698	m³

The cavern configurations were essentially cylindrical. The stepped form of the cavern roof is traceable to the changing filling level caused by the changing solid content. Through the deposition of insoluble and poorly soluble components the cavern sumps were raised by 1 - 1.5 m.

The constriction in the sump area is the lower part of the expected "flower pot" configuration. It serves as the seat for the thermocouples in the salt and therefore was not expanded to cylindrical form.

2.3   Measuring Instrumentation

Temperature Measurements

In each cavern 7 temperature measurement points were provided. These were so arranged, that the temperature profile could be measured at the product/ salt transition point as well as along the axis of the cavern. One thermocouple was placed in the drift, so that a total of 36 elements were involved. The maximum expected temperature was 90 °C. Consequently a measuring range of 0 °C to 150 °C was selected. NiCr-Ni-elements were chosen. These show a linear temperature dependence in the above range. On the average they exhibit a deviation of 3 °C. They have shown good performance in salt under various conditions. The elements are fastened on a pole with hinged arms. The measuring points were positioned exactly and the pole fastened with a wire.

In the test drift the thermocouples were attached to three potentiometers. The received signals were converted and amplified. The data is called up hourly by the central collection system and was stored on a magnetic tape above ground. It is possible at any time to call up measuring point analogues.

Climate in the Cavern Space

The climate in the void space of partially filled caverns was determined. To do this two humidity probes were hung in the cavern. Calibration was accomplished above ground over various steam saturated solutions. Each humidity probe was tied in with a separate temperature measuring point. These signals were also stored on the magnetic tape.

Stress in the Product

In order to measure loads in the product e.g. as a result of rock loading, two pressure pillows were inserted in each of two caverns. They were placed in the middle of the cavern, one in the horizontal and one in the vertical direction. The orientation was such that the vertical element was subject to loads from the flanks of the mountain range.

2.4   Proportioning, Mixing and Filling

2.4.1   Starting Products for Filling

Pellets

To fill the caverns 80 metric tons of pellets had to be produced. The main component of a simulated waste solution was $NaNO_3$. It was decided not to add Cs and Sr. As a binding material Portland cement PZ 35 f was chosen instead of HOZ 35 L-NW/HS/NA for the following reasons: The heat of hydration plays no role in the case of the pellets. The hardening kinetics, starting strength and fracture tendence is more favourable with PZ 35 f than with HOZ 35 L-NW/HS/NA. The quinary brine corrosion resistance is nearly the same.

Proportioning during the pellet production makes it possible to establish an exact W/C-ratio, so that the separation does not occur. The goal was to prepare pellets with a high salt loading (495.5 g/l $NaNO_3$). At the same time the pellet size distribution was to be maintained between O and 5 mm diameter.

Filler Suspension

For this purpose HOZ 35 L-NW/HS/NA was chosen. Besides a low heat evolution this cement shows good Cs and Sr-retention. Additions of liquifier and retarder were made in order to reduce the water requirements and to increase the delivery time. Excess water has a detrimental effect on cement properties and can dissolve the salt surface. In the case of accidents or delays in delivery the product should remain flowable for about 18 hours.

2.4.2   Composition of the Mixture

The pellet/filler mixture was optimized following delivery and cement technology studies. The ratio of

Pellet : Filler was 60 : 40 vol.-%

Delivery tests produced a W/C-ratio of 0.45 for the filler suspension. If required this can be increased in a stepwise fashion. The filler suspension was calculated on the basis of a material balance. The following amounts were determined.

		weight-%
Pellets		71
Cement		68.7
Water		30.9
Liquifier	29	0.13
Retarder		0.2

Theoretical density 1.8×10^3 Kg/m^3

Pore Space   12 %

For mixing, a free-fall mixer with a fresh cement volume of 200 liter was available. The mixing was based on 100 kg cement corresponding to two sacks of cement.

The pellets were delivered in big bags of about 1.2 metric tons each. They could be weight on a scale on the charging equipment and added in the required quantity. The cement additives were prepackaged in the required amount and were distributed over the dry material.

About 2/3 of the necessary water was first added into the mixing drum. After subsequent filling with the dry materials the remaining water was added. The mixing time was three minutes. This guaranteed the internal mixing of the liquifier and retarder with the other components. At the end of the mixing period, tests were made to determine the deliverability. If the consistency of the mixture was under the desired value, water was added and mixing was continued for another minute.

For the first filling tests in three caverns the fill rate was varied keeping the mixture composition constant. Thereby it was possible to establish various boundary conditions for the calculation. The other two caverns will be filled with mixtures of varying compositions. These are intended to answer questions concerning the final product and the delivery relationships.

2.5   Filling Status

Currently two caverns are filled and one is partially filled.

contents	cavern no. 1	cavern no. 2	cavern no. 5
V_{cavern}	9.475 m³	9.440 m³	9.698 m³
$V_{borehole}$	0.327 m³	0.331 m³	0.343 m³
V_{total}	9.802 m³	9.771 m³	10.041 m³
Fill volume/charge	1/6	1/1	1/2
Filling date and time	14.05. 3 h 40' 10.06. 2 h 12' 10.07. 1 h 04.08. 1 h 15'	23.06. 7 h 25'	06.07. 3 h 35' 08.07. 2 h 25'
Loads/charge	8 9 8 8	50	22 22
Fresh concrete/load		195 l	228 l
Fill opening	180 mm-borehole	180 mm-borehole	3"-tube
W/C	0.45 0.5/0.45/0.45... 0.5/0.48/0.48... 0.5/0.5/0.45/0.45...	0.45/0.48/0.45/0.45...	0.48/0.50/0.49/ 0.48 varied

3. MEASUREMENT RESULTS

3.1 Temperature Changes

The thermocouples were emplaced several days before the cavern filling permitting the starting temperature to be determined. The days of measurement were counted only from the day before filling. The indicated hours relate to the beginning of filling if nothing else is said. Because of a mechanical defect it was not possible to make measurements when the first 1/6 filling of cavern 1 was started. All curves show a short cooling period on the filling day. Subsequently the maximum temperature occurs between 1 and 3 days. The falling part of the temperature curve is relatively flat. 4 weeks after filling the temperature is never more than 10 °C above the initial cavern temperature. After 45 days the temperature in cavern 2, which was filled in 1 step, was only 1 - 3 °C above the initial cavern temperature.

Cavern 1 was to be filled in charges of 1/6 volume each. Therefore the 5 thermocouples situated one above the other (Th 1 and 4 - 7) are not always in the centre of a charge. So they do not always show the maximum temperature reached in the concrete. The temperature decrease is most noticeable in the cavern space. It is not as evident in the solidified concrete. The highest temperature occurs 30 - 60 h after filling with a given charge. The highest values are measured by the thermocouple closest to that charge.

Temperature in the Caverns

Cavern 1

Observations	Measurement time day h	Thermocouples						
		Th 1	Th 2	Th 3	Th 4	Th 5	Th 6	Th 7
no measurements	1							
1/6 filling	2		concrete					
starting time								
7.20 h	15	36.1						
	28	35.3	35.3	35.2	34.9	35.0	34.6	34.5
2/6 filling	29 10.30	35.2	35.2	35.1	34.1	34.4	33.5	33.0
starting time								
7.05 h	30 13.45				59.2			
	16.15				58.2			
	37				38.9			
	44				36.6			
	51	36.0			35.8			
	58 8.00	35.7			35.5	35.1		
3/6 filling	59 7.50	35.6	35.0	35.4	35.4	35.3	34.7	27.1*
starting time	13.20	35.6	35.4	35.3	35.2	34.4	33.1	26.0*
8.30 h	22.15				38.0	37.7	36.0	
	60 14.00				45.4			
	61 19.00				46.8			
	62				46.2			
	69				39.0			
	76				37.0			
	83				36.3	35.4		
4/6 filling	84 5.55	35.7	35.6	35.4	36.2	35.4	34.9	27.2*
starting time								
6.45	85 10.15				64.5			

*: loss of measurement

On the 44th day of measurement thermocouple No. 7 showed a significant jump in temperature of about 4 °C. This was traced to a thermocouple defect. Following measurements were utilized to indicate subsequent relative temperature changes.

Cavern 2

Observations	Measurement time day h	Thermocouples						
		Th 1	Th 2	Th 3	Th 4	Th 5	Th 6	Th 7
	1	33.8	34.2	34.0	33.6	34.2	34.1	34.1
1/1 filling	2 13.00	35.9	34.5	34.8	33.6	31.7	31.4	31.9
starting time	3	48.7	42.0	39.0	64.9	64.3	55.3	41.9
7.05 h	4 8.00			40.0	68.8	71.6		
	13.00			40.0	68.6	72.1		
	16.00			40.0	68.5	72.3		
	5 8.15	47.2	42.0	40.0	67.6	72.6	63.6	45.4
	6 21.15			39.7	64.1	70.4		
	7 19.30			39.4		68.2		
	8			39.3		66.0		
	15			37.6		50.8		
	22			36.4		42.9		
	29	36.0	35.9	35.8	37.7	39.2	38.0	35.8

The hydration of the cement lime begins as soon as water comes in contact with the cement powder. A temperature increase can be observed on the bottom of the cavern with Th 1 already 6 1/2 h after filling of the cavern has begun, while the temperature of the top has decreased relative to the temperature in the cavern space as seen by Th 5 - Th 7.

The maximum temperature occurs in the centre of the block 66 h after filling began. The temperature increase is 38.4 °C. The maximum temperature difference between two neighbouring couples occurs after 49 h. At this point the temperature difference between Th 3 and Th 4 is 28.8 °C. On the 29th measurement day temperatures decreased to between 35.8 °C (Th 3 and 7) and 39.2 °C (Th 5).

Cavern 5

Observations	Measurement time day h		Thermocouples						
			Th 1	Th 2	Th 3	Th 4	Th 5	Th 6	Th 7
	1		31.9	31.7	31.8	31.6	31.9	32.1	31.9
					concrete				
1/2 filling	2	7.35	31.7	31.7	31.9	30.3	31.0	30.4	30.1
starting time		16.00	37.4				31.1		
7.50 h	3	8.00	66.5		38.8	67.5			33.6
		13.30			39.3	69.8			33.7
		16.15			39.5	70.5			33.8
1/2 filling	4	6.00	69.5	42.2	39.5	71.0	36.4	36.2	34.0
starting time		8.00			39.8	70.9	30.9	30.8	31.1
7.20 h		13.30			39.6	70.1			33.3
		16.20			39.5	69.7			33.9
	5				39.2	68.3	66.9		38.2
					39.2		69.6		38.3
							70.6		38.1
	6	7.50					73.8		37.6
		13.20					74.3_5		37.4
		22.15					74.8_5		37.0
	7	14.00					74.6_5		36.7
	13						58.6		35.1
	20						45.0		34.0
	26		36.8	34.1	34.1	39.1	39.5	35.8	33.4

Here also the temperature shows a decrease during filling. With charge 1 the maximum decrease near the cavern top is 1.8 °C (Th 7). Starting with an increase in surrounding temperature in the cavern space during the filling of charge 2 the temperature decreases by 5.5 °C (Th 5). The thermocouples 4 and 5 are centred respectively within charges 1 and 2. The maximum temperatures are 71.0 °C and 74.8 °C respectively. They were achieved 46 and 63 h respectively after filling of each charge was begun. The maximum temperature difference between two neighbouring thermocouples was 31.5 °C. It occurred between Th 3 and Th 4 47 h after beginning filling with the first charge. On the 26th measurement day temperatures are between 33.4 °C (Th 7) and 39.5 °C (Th 5).

3.2   Pressure Measurements

The pressure and expected stress increase in the filled cavern is dependent on the convergence rate of the rock and the resultant application of stress to the concrete. Long time observations in salt caverns indicate low convergence rates. This means that the investigation of load changes is a long term affair. After zero load measurements on the installed pillows pressure measurements were made in weekly intervals starting 7 days after filling. An increase in pressure after 7 days, above the zero load measurement, disappeared within 3 weeks.

3.3 Humidity Measurements

In the space of each partially filled cavern the relative humidity as well as the temperature was measured. Questions concerning humidity exchange with the drift, precipitation on the cavern and borehole walls as well as the dependence of product strength on the cavern climate were to be answered.

The emplaced humidity probes were very sensitive to mechanical loading and corrosive attack. In addition their calibration in the 92 % humidity range is not very simple since the sensors show a nonlinear dependence. Both probes failed, although one was repaired and replaced in service. The relative humidity in cavern 1 was determined in the middle of the remaining cavern space immediately following each filling charge. The measured values lay between 43 and 57 %.

4. EVALUATION

4.1 Current Results

The evaluation of temperature measurements shows that the goal of keeping the product temperature below 90 °C was met. The curves show an observable temperature increase indicating that the curing process took place. The filling data provide input for the calculation program. Comparative calculations can be made.

Temperatures in the cavity space and at the edge of the cavern were observed to decrease before the maximum temperature within the product was reached. This can provide an indicator for charge sizes and timing in a large full-size cavern. The temperature increase due to curing of the fresh concrete charge has little effect on other already emplaced layers. The initial drop in temperature within the fresh product and in the overlying cavity space has several explanations.

1. During the curing process heat is required.
2. The temperatures of the freshly introduced product is below that of the cavern temperature.
3. During filling an air exchange occurs with the cooler air of the drift and that of the cavern.

Which of these has the greatest influence and to what extent the temperature of a large cavern would be determined thereby will have to be investigated by further tests in the laboratory and with similar product and temperature environment.

The measured curves provide the basis for calculating thermal stresses in the concrete as well as resulting loads. Investigation of cores will permit determination of stress overloads and fracture development.

After three months the emplaced thermocouples in two unfilled caverns had corroded. The measurement assembly was sent to the fabricator for repair. The resulting delay will be used for laboratory studies on product improvement and to investigate the influence of additives.

The planned W/C-ratio of 0.45 could not be maintained. In order to obtain consistent fresh concrete properties it was necessary to vary the ratio between 0.48 and 0.5. The flowability and deliverability as well as the pore content and later product strength set relatively narrow limits.

Further studies in the laboratory should permit the reasons for the variable W/C-ratio to be determined and improve the constancy of this ratio.

The following influences could play a role:

1. Variable pellet size distribution from charge to charge.
2. Variable and insufficient proportion of pellet fines.
3. Increased surrounding temperature during production of the fill product.
4. Changing effect of the liquifier and retarder additives.

Setting tests with fresh concrete showed no separation in the charge by sinking of the pellets or by bleeding. Consequently one can expect that no separation will occur in fresh concrete charges.

After mixing, the pore space was between 10 and 15 %. If changes in this value occur following delivery and curing, it can only be determined after test cores have been obtained from the caverns.

During the filling, test samples were produced. The 7 day strength was in the range of 20 N/mm².

A number of fill-level measurements were made while filling each cavern. From the number of loads and the total cavern volume the fresh concrete amount per load and resulting fill height after each loading can be calculated.

From the difference from the calculated and measured fill height the slumping angle can be determined. It was found that the angle was directly dependent upon the flowability of the load and in general was very low. By this means an optimum filling of the cavern can be achieved without leaving a substantial void in the roof area. It is possible during filling that the underlying layers of concrete may densify. It is also possible that air pores in the fresh concrete will rise. The extent to which the concrete density varies and an air pocket has formed at the roof of the cavern will be established by means of later coring.

4.2   Further Investigations

56 days after filling each cavern it will be bored and samples taken. This will permit a comparison with test samples that were stored in a normal atmosphere and also with samples that were emplaced in 30 cm deep boreholes in the drift floor. To be investigated as a function of cavern depth are the uniaxial and triaxial strength, the density, the pore volume, the dynamic and the static modulus of elasticity, the depth of water penetration, the humidity content, particularly in the product-salt interface region as well as chemical and physical changes in the product under the influence of the salt and the special climate.

The stress distribution in the concrete, its change with time as well as its transfer to salt characteristics is to be investigated by overboring techniques. Load increases to the product as a result of the converging rock will be monitored by means of pressure pillows.

Prior experience has shown that additional studies are necessary. Thus from the two caverns in which thermocouples corroded, gas samples have been drawn to investigate the presence of corrosive gases. The results should permit conclusions concerning the selection of materials for installations in the large caverns.

The relative humidity measurements did not answer questions concerning the cavern climate. It is not clear if a saturation of the air equivalent to that either over a NaCl-solution or of a steam atmosphere can be reached. For this purpose measurements of absolute humidity are needed. The filler/pellet product composition must be checked by means of delivery tests and its flow properties improved before filling of the last caverns begins. The entry of brine into the pellet concrete from a pressurized borehole will be investigated in the caverns.

5.   CONCLUSIONS

Although only half of the caverns are filled, it can be concluded that the program 'Investigation of model caverns' has been successful. It shows how necessary such investigations are for the transfer from calculational models and laboratory data to the in situ full scale.

In this case other more realistic boundary conditions and parameters develop e.g. the uniformity of the starting product and the continuity of mixture production such as does not occur in the laboratory. Also the climatic conditions for hardening of the product are produced as they later will be for the containerless emplacement.

The potential for measuring loads and stresses is an additional possibility whose sucessful application remains to be proven.

Acknowledgement

Looking at the results one should not underestimate the level and comprehensiveness of the effort involved in this work.

My thanks are expressed to the technical department of the institute in particular to Mr. Hartwig, who by opening the test drift made this work possible and made sufficient personal available for filling of the caverns who showed the required understanding of the need to interrupt work in order to make exact measurements.

I also thank the whole Geotechnics group with their leader Mr. Dipl. Geol. Schmidt and particularly Mr. Dipl. Geol. Raab who during the planning and the assembly of the solution mining equipment were heavily involved.

All coworkers of the institute are thanked for their helpfulness, support and suggestions which helped the successful performance of the program.

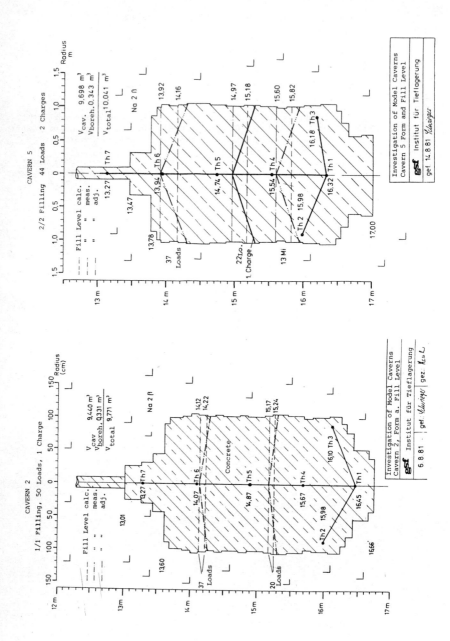

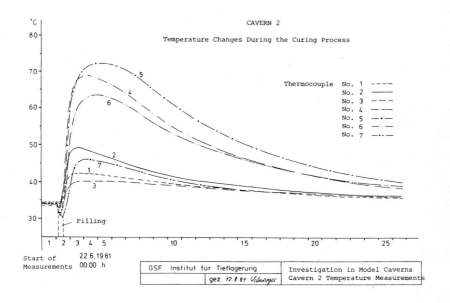

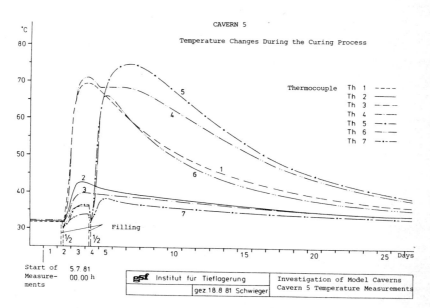

DISCUSSION

R.H. KOSTER, Federal Republic of Germany

The German concept of an integrated fuel "back-end" centre - at *one* site - was the basis for the so-called in-situ stabilisation concept.

The main features of this containerless emplacement concept are :

- prefabrication of granules with aqueous MLW/LLW and inorganic binder above ground ;

- transportation of the granules in cement based grout via a vertical pipeline into a 75,000 m^3 cavity located below ground ;

- in-situ solidification of the concrete like waste product forming layers which represent subsequent batches.

W.C. PATRICK, United States

You mentioned problems with thermocouple corrosion. What type of shielding was used on the Ni-NiCr thermocouple wires ?

W.R. FISCHLE, Federal Republic of Germany

I do not know the alloying. If you are interested, please contact me later.

THERMOCREEP ANALYSIS OF SALT-LAYERED MEDIA FOR A NUCLEAR WASTE REPOSITORY

R. A. A. Todeschini[1], I. Kadar[2], and C. L. Wu[3]
1. Research and Engineering, Bechtel Group, Inc.
2. Hydro and Community Facilities, Bechtel Civil and Minerals, Inc.
3. Nuclear Fuels Operation, Bechtel National, Inc.
   San Francisco, California (USA)

ABSTRACT

This study addressed the relative merits of different underground room configurations using the criterion that nuclear wastes be retrievable for a period of several years following emplacement. Retrievability is related to the problem of room closure, where geologic creep causes inward displacements of the surfaces of storage rooms. The storage facility is being considered for construction within salt-layered media interspersed with clay seams and anhydrite layers. Both single and multiple storage rooms were analyzed. Creep analyses were performed for raised and ambient temperature conditions but only partial results are reported. The MARC [1] computer program was used in determining the solution. The thermo-mechanical response of salt was modeled by a secondary creep relationship which included nonlinear temperature-dependent material behavior.

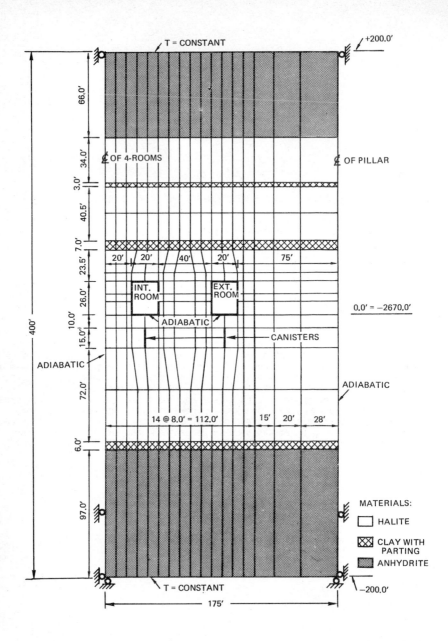

Figure 1 FINITE ELEMENT MODEL FOR THERMAL ANALYSIS, 4-ROOM 20 x 26 FT.

1.0   INTRODUCTION

The design of openings for an underground repository to manage nuclear wastes relies heavily on rock mechanics analyses which determine the trade-offs between
various design configurations. This study addresses the relative merits of different underground room configurations using the criterion that stored wastes be
retrievable for a period of several years following emplacement. Retrievability
of wastes is related to the problem of room deformation or closure, where geologic
creep causes inward displacements of the ceiling, floor, and walls of storage
rooms.

Creep analyses were performed for single and multiple storage rooms excavated in a
salt-layered medium with clay seams and anhydrite layers, subjected to thermal and
mechanical loading. Both raised and ambient temperature conditions were considered. The nature of bedded salt with horizontal clay-filled discontinuities and
elastic anhydrite layers, lends itself to the finite element technique of analysis. In this study the MARC [1] program was employed.

Because the storage rooms are relatively long, a two-dimensional plane-strain
model was used. In addition, the rooms were assumed to be centrally located within the repository, hence the results of the analyses are not applicable to room
ends or intersections or to storage rooms at the perimeter of the repository. The
thermomechanical response of salt was modeled using a secondary creep constitutive
relationship which incorporates nonlinear temperature-dependent material behavior.
This relationship neglects the primary creep response because it is assumed that
the deformations resulting from secondary creep are dominant.

The results presented herein were obtained by the authors in calculations performed during the last three years [2,3]. Due to space limitations, only a
portion of the results are reported.

Section 2, provides results of the heat transfer solution for a four-room system
with heat sources along the floor centerlines. In Section 3, the displacments and
stress distributions for single- and four-room configuration subjected to mechanical loads under isothermal conditions are given.

2.   THERMAL ANALYSIS

2.1   Finite Element Model

2.1.1   Geometry

The model analyzed is a four-room arrangement with rooms 20 ft wide by 26 ft high
and with the floor horizon located at 2670 ft below the surface. The selected
geometry for the finite element model considers a width of 175 ft and a total
height of 400 ft as shown in Figure 1.

Due to symmetry considerations, the model can be considered extending from the
centerline of the four-room arrangement to the centerline of the abutment pillar.
The layer thicknesses selected approximate the varying geologic conditions. Refined element sizes were chosen for elements closer to the openings where stress
and stress concentrations are expected.

2.1.2   Boundary Conditions

For heat transfer analysis, the vertical planes bounding the model and the room
surfaces were taken as adiabatic boundaries. This assumption is supported by
considerations of symmetry.

At the top and bottom horizontal edges of the model a constant fixed temperature
of 80.6°F was assumed. This boundary condition is reasonable since the boundary
is far away from the heat source. To this end, the top and bottom model boundaries were taken about 200 ft from the heat source.

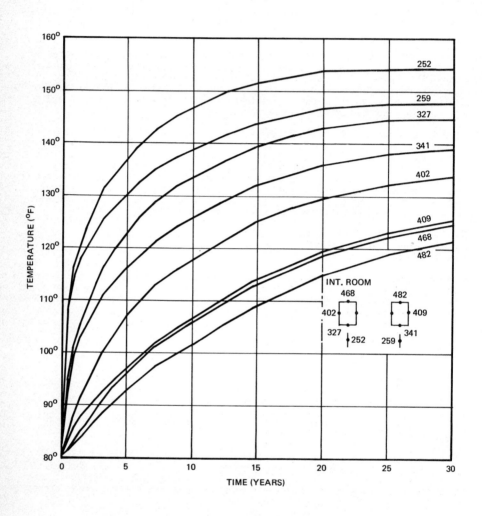

Figure 2 NODAL TEMPERATURE TIME HISTORIES, 4-ROOM 20 x 26 FT.

2.1.3 Material Properties

Three different materials were used in the analysis, salt (halite), salt with clay, and anhydrite. The material properties and their sources are provided in Reference [2].

2.2 Heat Transfer Analysis

The heat sources were assumed to be spent fuel canisters emplaced along the centerline of the opening 10 ft below its floor. The canister spacing in the room's longitudinal direction is 10 ft corresponding to a heat load of 30 kW/acre with each canister having 600 watts of power output. This power source is assumed to decay exponentially and to have a half-life of 30 years. In modeling the repository, the individual heat sources were assumed to represent a planar heat source 15 ft high by 1 ft thick, with uniform heat generation along the length and height of the plane. Due to this assumption, the accuracy of the computed temperature distribution in the vicinity of the heat sources is reduced. However, this approximation is acceptable since the purpose of this calculation is the determination of temperatures in the pillars and in regions away from the heat sources. As stated earlier, an initial temperature of 80.6°F was assumed throughout the model.

2.3 Temperature Distribution Results

The temperature-time history plot of selected nodal points around each room cross-section is given in Figure 2. It could be observed that the inner room is exposed to high temperatures than the outer room. However, the maximum temperature difference between the rooms at similar locations is less than 10°F. Temperature contours at 30 years are shown in Figure 3. They indicate that the approximate maximum temperature rise after 30 years is 70°F at room floors, 50°F at interior pillars, and 40°F at exterior pillars.

3.0 ISOTHERMAL CREEP ANALYSES

3.1 Creep Constitutive Relationship

The creep constitutive relationship used for the long-term room closure calculations was developed by S. Serata [4]. In adapting this generalized model to the analysis of salt openings, it was simplified to consider viscoplastic deformations only, since the latter are expected to be much larger than the viscoelastic deformations for salt openings.

3.1.1 Salt Creep Constitutive Relationship

It is assumed that viscoplastic flow takes place only when the octahedral shearing stress exceeds the value of the octahedral shearing strength, K_0 as follows:

$$\dot{\gamma}^P_0 = \frac{K_0}{V_4} \left(\frac{\tau_0 - K_0}{K_0} \right)^{2.62}$$

where

$\quad K_0$ = octahedral shearing strength (650 psi at 20°C)

$\quad V_4$ = viscoplastic viscosity (1.5 x 10^6 psi-day at 20°C)

$\quad \tau_0$ = octahedral shearing stress

$\quad \dot{\gamma}^P_0$ = octahedral shearing strain rate

The octahedral shearing strength, K_0 is function of the mean stress σ_m, the temperature T, and the cumulative octahedral shearing strain γ_0, according to:

$$K_0 = K'_A + (K_B - K'_A)(1 - e^{-\alpha\sigma_m})$$

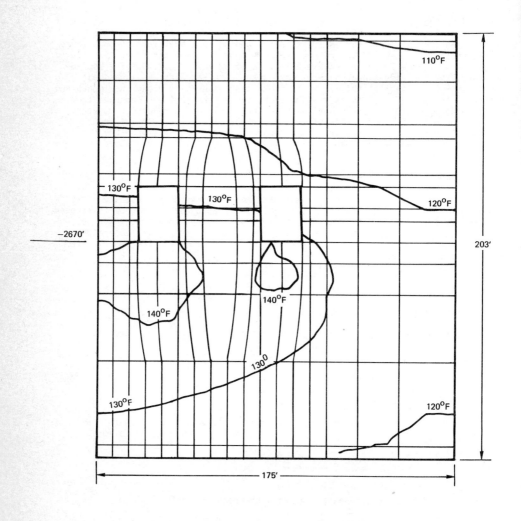

Figure 3 TEMPERATURE CONTOURS AT 30.0 YEARS, 4-ROOM 20 x 26 FT.

where
$$K'_A = \frac{K_B}{3} \left(\frac{\gamma_c - \gamma_o}{\gamma_c} \right), \quad \gamma_c = 0.02$$

$$\alpha = 0.0015 \text{ psi}^{-1}$$

$$K_B = 650 \, e^{-0.00583(T-20)}, \quad T \text{ in } °C \text{ and } K_B \text{ in psi}$$

The effect of temperature on viscoplastic flow rate $\dot{\gamma}^P_o$ is approximated by the expression

$$\dot{\gamma}^P_o = \frac{K_o}{V_4} \left(\frac{T + 273}{293} \right)^{9.7} \left(\frac{\tau_o - K_o}{K_o} \right)^{2.62}, \quad T \text{ in } °C$$

3.1.2   Clay Seam Constitutive Relationship

The creep strain rate of the clay seams was approximated according to the following expression

$$\dot{\gamma}^P_o = \frac{K_o}{V_4} \left(\frac{\tau_o - K_o}{K_o} \right)^{2.62}$$

where
$K_o = 60 + 0.4\sigma_m$, in psi
$V_4 = 1.5 \times 10^5$ psi-day

3.2   Finite Element Models

3.2.1   Geometry

Three finite element models were used to study the effect of pillar width. The geological layers, room sizes, and extraction ratios of all models were the same. Breaking up the continuity of the salt formation are six thin clay seams and four anhydrite layers in each model as shown in figures 4 and 5. All rooms are 13 ft high by 33 ft wide. Model a (not shown) is a four-room arrangement with 25 ft interior pillar width and 300 ft exterior pillar. Model b (Figure 4) is a four-room system with 50 ft interior pillar and 225 ft exterior pillar. Model c (Figure 5) is a single room configuration with both interior and exterior pillar of 94 ft width.

3.2.2   Boundary Conditions

Roller supports were assigned to the nodes along the vertical edges of each model to restrain horizontal movement and along the bottom edge to restrain vertical movement.

3.2.3   Loading Conditions

The applied loading consisted of overburden pressure applied at the top of the models and a gravity body force throughout. The models were initially prestressed with a hydrostatic stress state ($\sigma_x = \sigma_y = \sigma_z$) with the element stresses determined by the weight density multiplied by the element depth. These represent the original undisturbed state of stress before excavation. The use of a pre-stress loading insures that after excavation the elements located far from the room openings will not creep, since their deviatoric (Von Mises) stresses are near zero. At the room surfaces due to the elimination of supporting forces, the deviatoric stresses are large. This means that initially the loadings on the structure are concentrated on the free room surfaces. These internal resisting forces are the forces that produce the closure of the rooms. With the resulting creep these forces redistribute themselves through the elements located closest to the rooms.

3.3   Analysis Results

For Model a, creep analysis reached 5.043 years in 408 time increments. Model b, attained 5.123 years in 192 time increments. Model c, reached 25.39 years in 102

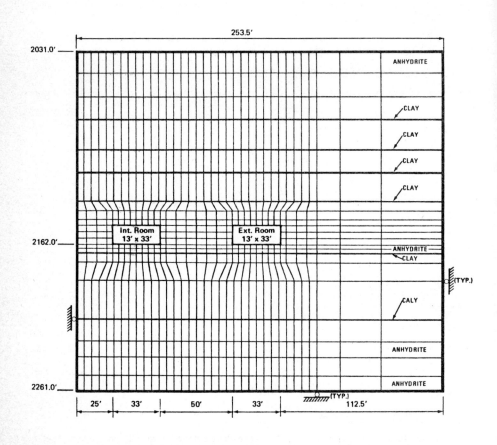

Figure 4 FINITE ELEMENT MODEL B, 4-ROOM, 50-FT PILLAR (897 ELEMENTS AND 980 NODES)

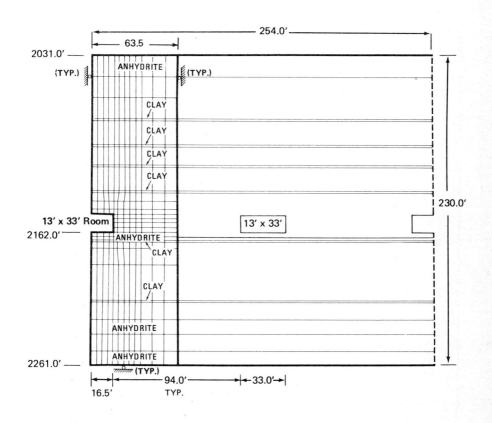

Figure 5 FINITE ELEMENT MODEL C, SINGLE ROOM (320 ELEMENTS AND 365 NODES)

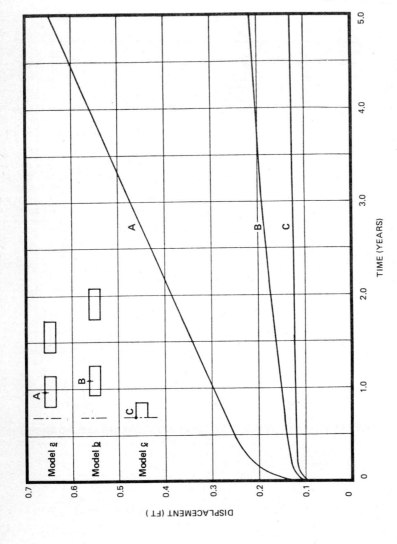

Figure 6 COMPARISON OF VERTICAL ROOF DISPLACEMENTS AMONG MODELS A, B, AND C.

time increments. Room closure results for the three models at several time intervals are presented in Table I through III. Since Models a and b reached only 5 years of creep time, the 10-year closure values were estimated by linear extrapolation based on the closure rates at the end of their respective times. This approximation is acceptable since the closure rates are almost constant at that creep stage. A comparison of vertical roof displacements versus time among the three models is shown in Figure 6. The development of creep modifies the stress concentrations observed near the room surfaces immediately after excavation. As time progresses, the high peak stresses move inside the pillar and away from the openings. Figure 7 shows the variation of vertical stresses in a plane located at room midheight at different creep times for Model b. The plots clearly shows the relaxation of vertical stresses in the proximity of the room surfaces. The peak vertical stress moves about 25 ft inside the abutment (exterior) pillar in 0.9 year. Similarly, Figure 8 shows the relaxation of vertical stresses that occur in the 94 ft pillar of Model c. In the horizontal direction the normal stresses reveal a relaxation pattern around the openings similar to the one observed in the vertical direction. The only exception to this is the high horizontal normal stresses developed in the very stiff anhydrite layer below the room. This layer attract very high stresses (3800 psi) and does not relax appreciably preventing the development of high deviatoric stresses in the surrounding salt. The anhydrite layer thus is beneficial in decreasing the creep closure from the room floor. However there is some uncertainty from the unknown load carrying capacity and initial stress condition of this anhydrite layer. If the anhydrite is weaker than presently modeled it may result in a sudden brittle failure with increased creep from the floor and walls of the room resulting from the transfer of forces.

4.0   CONCLUSIONS

The 13 ft rooms on the repository level were analyzed considering three pillar widths. The results presented herein provide information of vertical roof displacements, closure of rooms and vertical normal stress distributions for the three models analyzed. The most signficant effect of pillar width on the room closure was summarized in Table I through III. At 5 years, the horizontal room closure reached 1.31 ft, 0.36 ft and 0.12 ft for Models a, b and c, respectively. Similar trend was observed in the room closure in the vertical direction.

Smaller pillar width resulted in greater stress relaxation near the rooms through creep deformations and in a corresponding transfer of loads to the exterior (abutment) pillar.

Utilizing principal stress plots (not reported herein) at different time stages, stress envelopes were drawn, and the effect of excavation and creep on principal stresses was clearly noticed. Comparison of the three models after 5 years of creep indicate that the four-room case with 25 ft interior pillar had the highest stress relief in the area surrounding the rooms described by secondary stress envelopes [5]. The fact that stress was relieved due to the rapid deformation of narrow pillars seemed to agree with the yield pillar concept [6]. However, results showed that such relief of stresses did not prevent large room closures and closure rates.

The clay seams contributed to the increase of the closure rates of the rooms but they did not seem to pose any threat of roof structural instabilities. No reliable field data currently exist to verify steady state closure rates in salt. Minimal closure information is available in potash mines. Steady state closures for single isolated room of the order of 0.5 to 1.0 inch per year have been reported from Canadian potash mines [7,8].

The total closures calculated in this analysis indicate approximately 2.33 ft horizontal convergence in 10 years for the 25 ft pillar and 0.14 ft for the single room case. To determine the most suitable pillar width or room layout, the effects of stress relief and room closure must be carefully considered. However, a failure criterion has yet to be established from field measurements and observations from a proposed field test program.

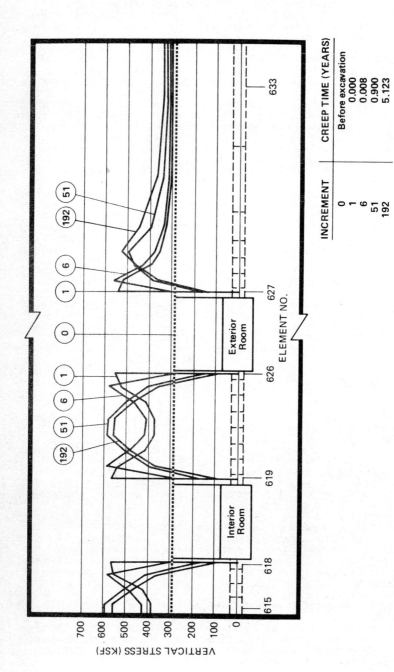

Figure 7 VARIATION OF VERTICAL STRESS, 4-ROOM 50-FT PILLAR, MODEL B

INCREMENT	CREEP TIME (YEARS)
0	Before excavation
1	0.000
6	0.008
51	0.900
192	5.123

G1002891-21

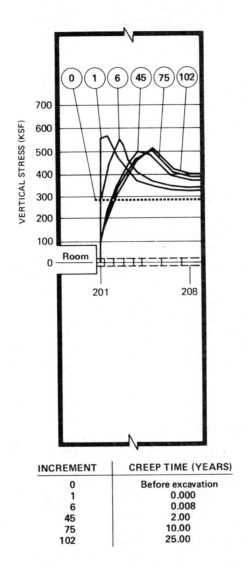

INCREMENT	CREEP TIME (YEARS)
0	Before excavation
1	0.000
6	0.008
45	2.00
75	10.00
102	25.00

Figure 8 VARIATION OF VERTICAL STRESS, SINGLE ROOM, MODEL C

5.0 REFERENCES

1. MARC, General Purpose Finite Element Analysis Program, Rev. J-1, MARC Analysis Research Corp, Palo Alto, California, 1980.

2. Bechtel, Stability Analysis of Underground Openings - RMC-1, Document No. 51-R-510-02, Rev. 1, Bechtel Corp, San Francisco, California, 1979.

3. Bechtel, Stability Analysis of Underground Openings-RMC-2, Document No. 51-R-510-04, Rev. 0, Bechtel National, Inc., San Francisco, California (to be issued).

4. Serata, S. "Design Basis of Underground Salt Openings in the WIPP Project", Serata Geomechanics, Inc., Berkeley, California, July 1979.

5. Serata, S., Empirical Design Basis of Underground Salt Openings in the WIPP Project, Document No. DR-51-R-1, Rev. 0.

6. Serata, S, "Stress Control Technique - An Alternative to Roof Bolting?", Mining Engineering, pp. 51-56, May, 1976.

7. Lopez, G. B., The Rock Mechanics Program at the Rocanville Division of P.C.S., Annual Western Meeting of CIM, Saskatchewan, Canada.

8. Mraz, D., Theoretical Predictions Confirmed by In-Situ Rock Behavior in Deep Potash Mines, Nineteenth Symposium on Rock Mechanics, pp. 468-475, 1978.

Table I
CLOSURE OF ROOMS, MODEL a

| ROOMS | DIRECTIONS | CLOSURE-COMPUTER RESULT (ft) | | | EXTRAPOLATED CLOSURE (ft) |
		0.98 yr	2.37 yr	5.04 yr	10 yr
Interior	Vertical	0.31	0.43	0.66	1.10
	Horizontal	0.49	0.78	1.31	2.33
Exterior	Vertical	0.28	0.40	0.63	1.07
	Horizontal	0.36	0.55	0.91	1.58

Table II
CLOSURE OF ROOMS, Model b

| ROOMS | DIRECTIONS | CLOSURE-COMPUTER RESULT (ft) | | | EXTRAPOLATED CLOSURE (ft) |
		0.90 yr	2.33 yr	5.123 yr	10.0 yr
Interior	Vertical	0.155	0.186	0.224	0.290
	Horizontal	0.162	0.247	0.360	0.558
Exterior	Vertical	0.141	0.161	0.190	0.239
	Horizontal	0.127	0.175	0.240	0.355

Table III
CLOSURE OF ROOMS, Model c

| DIRECTIONS | CLOSURE-COMPUTER RESULT (ft) | | | | | |
	1.06 yr	2.45 yr	5.075 yr	9.99 yr	19.2 yr	25.39 yr
Vertical	0.132	0.136	0.140	0.143	0.145	0.146
Horizontal	0.104	0.123	0.130	0.130	0.135	0.138

DISCUSSION

M. WALLNER, Federal Republic of Germany

You mentioned the strong effect of the stiff but very small anhydrite layer beneath the storage room which kept the floor upheave very small. On the other hand high stresses accumulated in this layer. Would this behavior not change if you take failure into account which probably could occur in the anhydrite layer ?

R.A. TODESCHINI, United States

I mentioned this effect in the paper. Although the computed stresses (3800 psi) are well below the anhydrite compressive strength (13,200 psi) we do not know the initial state of stresses in that layer which could lead to a sudden brittle failure. That effect is going to be included in the future, but we need to develop a failure model to incorporate this effect.

W.C. PATRICK, United States

Although we have discussed this at some length, I feel that it is worth mentioning the importance of boundary conditions. The mined openings should *not* be treated as adiabatic but *should* have either convection or/and radiation transport mechanisms included.

R.A. TODESCHINI, United States

Although the adiabatic boundary conditions to model the room surfaces might be questioned, we felt that for our engineering purposes was good enough. The temperature differences between the roof and floor locations did not exceed 25°F which does not have a big impact in the thermoelastic stresses due to the creep relaxation properties of salt.

THE ROLE OF BENCHMARKING IN ASSESSING THE
CAPABILITY TO PREDICT ROOM RESPONSE IN
BEDDED SALT REPOSITORIES*

R. V. Matalucci, H. S. Morgan, and R. D. Krieg
Sandia National Laboratories
Albuquerque, New Mexico 87185
U.S.A.

ABSTRACT

    An overview is presented of the Waste Isolation Pilot
Plant (WIPP) Benchmark II study which was used to assess
the accuracy and reliability of computer codes used to
predict drift response. Emphasis is placed on the objectives and philosophy of Benchmark II and on procedures used
to ensure that valid comparisons of results could be made.
The problem definition, the selection of participants, and
review procedures are all discussed. A sampling of results
is also presented to illustrate the degree of variability
in the results obtained with various computer codes.

─────────────
*This work supported by the U.S. Department of Energy.

Introduction

Sandia National Laboratories (SNL) is developing performance assessment techniques in support of the U.S. Department of Energy's (DOE) Waste Isolation Pilot Plant (WIPP) which is sited in south-eastern New Mexico (SENM). The WIPP project provides for the storage of defense transuranic waste within excavated rooms situated in the bedded salt layers of the Salado Formation and includes an under-ground research and development facility to test the response of the rock to the emplacement of heat- and nonheat-producing wastes. The design of both the facility for storage of defense transuranic waste and the facility for the experimental program of the WIPP project require analytical methods to predict the response of the surrounding rock mass and more specifically to predict the closure rates of the storage rooms following excavation and waste emplacement.

A part of the WIPP experimental program includes an effort to ensure that computer codes used for room stability calculations are accurate, reliable, and useful for evaluating the response of open-ings in the bedded salt formation. Numerous computer structural analysis methods have been applied to problems of this type for various repository configurations in different lithologies [1,2]. However, in order to evaluate the capability of different computer codes to solve more specific thermal/structural interaction problems associated with a bedded salt repository, a benchmark series was developed at SNL which would allow a direct code-to-code comparison of the results obtained with the different codes.

The benchmark series was initiated at SNL in 1979 with the Benchmark I problem [1]. The emphasis of this study, similar to that of the Office of Nuclear Waste Isolation (ONWI) benchmark [2], was on modeling methods and obtaining a solution to a physical problem. The Benchmark I problem consisted of two parts: an isothermal room configuration and a heated room configuration. Both rooms were excavated in a hypothetical uniform rock salt lithology and were situated 825 meters below the earth's surface. The salt was modeled as a creeping medium which followed a prescribed secondary creep law. Elastic terms in the constitutive equation were optional. Other mechanical and thermal properties, the drift configuration, and depth were also specified, and the 25-year response of the isothermal room and the 10-year response of the heated room were to be pre-dicted. Five codes, COUPLEFLO, JAC, SANCHO, SPECTROM, and STEALTH, were evaluated in Benchmark I for applicability to the WIPP pro-ject. The results were somewhat difficult to evaluate in that most participants predicted the response for periods much shorter than the prescribed periods of 25 and 10 years. When comparisons could be made, the variations were considered acceptable.

Benchmark I was a valuable exercise. Computational experiences, difficulties, drift modeling techniques, and code capabilities and features were examined at great length by the participants. At that time detailed drift modeling in a creeping medium was relatively new so Benchmark I represented a state-of-the-art exercise. However, actual code comparisons were somewhat difficult to make due to the Benchmark I problem definition and philosophy. Although appropriate at that time, the Benchmark I problem was not formulated to criti-cally examine codes but rather to examine modeling techniques. The drift dimensions, spacing, and depth were all specified along with the suggested creep model for salt, but the boundary conditions, the vertical extent to be modeled, and even the option to include elastic terms in the rock salt constitutive model were all left to the discretion of the participants. The emphasis was on the solution to a physical problem. This emphasis was also evident in the type of answers requested from the participants. Only parameters of importance to a waste disposal repository design were requested.

Benchmark II, the second WIPP code-comparative study, had a different objective and hence different philosophy and problem definition. At the time Benchmark II was formulated, the state of the art for room response calculations had progressed considerably so several factors which were ignored by Benchmark I participants were included in Benchmark II. As a result, Benchmark II was a more challenging technical problem. The new benchmark was intended to look not at modeling but rather at the codes which had been developed. For this reason, the problem was defined in minute detail [3]. The participants were allowed virtually no freedom in problem definition, only in code specific details such as mesh refinement, error bounds on iterative processes, and time step size. The type of output requested was also much different than Benchmark I. Some of the data requested had virtually no impact on drift design but illuminated particular numerical aspects of the problem such as the accuracy of a slip line algorithm.

In this paper, Benchmark II is used to illustrate the role of benchmarking in assessing code capability to perform room response calculations. Procedures used to provide meaningful evaluations of the participating codes are described in detail. First, the importance of code-code comparisons in relation to code-field data comparisons is discussed. Then, the objectives and philosophy of the Benchmark II study are presented. Next, the formulation and specification of the problem are described. The subsequent selection of participants is reviewed in the next section. This paper then presents the procedures used to ensure adherence to issued instructions and explains the method used to review results during selected stages of the study and describes how this information was communicated to the participants. Following this section, some characteristic results are presented to illustrate the extent of variability which existed among the participating codes. A cost comparison technique is also described which was used to evaluate operating efficiency of the codes. The paper concludes with the proposed next step to complete the code qualification process and establish confidence in the ability of codes to evaluate room stability for the WIPP project.

Code-Experiment Versus Code-Code Comparisons

Although agreement with field data is ultimately necessary for final code qualification, this agreement alone does not provide sufficient confidence that all aspects of a problem are being properly modeled. First, there are only a few parameters that can be measured accurately in the field such as displacement or strain which can be used to compare against code results. Secondly, experimental uncertainties can be large and strongly dependent on the lithology, material properties, and local inhomogenieties near the installed instrumentation. Thirdly, certain field measurements may coincidentally compare well with code calculations. The calculations could easily be erroneous and still compare well with measured data because one error compensates for another. As a result, in a code-experiment comparison the influence of a particular material property or nonlinearity cannot be isolated and studied independently to determine its effects on the problem.

The other aspect of using experimental data for code qualification concerns the availability and reliability of rock mechanics data from a site which is well characterized. Only at a site where the material properties and stratigraphy are well understood and where underground measurements are accurately obtained can the process of code-experiment comparison be adequately applied. An additional constraint to the use of field data in this regard is the expense of characterizing time-dependent materials such as salt. The manpower and expense required to obtain good long-term creep data at a characterized site is usually prohibitively excessive.

Furthermore, if for example a check on a 25-year creep closure calculation is necessary, it would take the full 25 years to obtain the required field data.

Laboratory and bench-scale experimental data, although very helpful and frequently used in the code qualification process, are also limited with respect to code qualification due to complications in scaling the experiment and to the small number of parameters that can be examined at one time. Subsequent integration of the scaled effects of the numerous laboratory-measured parameters that would actually interact in a field configuration is usually filled with uncertainty.

A useful and important forerunner to field proof-testing is a controlled comparison of codes with each other. This comparison was successfully achieved in the Benchmark II study by defining two problems which resembled room response calculations for the WIPP project but were in fact well-specified boundary value problems. A host of structural features such as slide lines and interbedded stiff and soft layers and time-dependent and nonlinear material properties were used and many parameters were compared individually. The description of the problems and the benchmarking process used are detailed in subsequent sections.

Benchmark II Objective and Philosophy

The overall objective of the Benchmark II study was to assess the capabilities of various thermal/structure computational codes to solve two complex boundary value problems which represent idealizations of real drifts located in the WIPP reference stratigraphy. Direct comparisons of drift closure, stresses, relative displacements, and temperatures obtained with the codes were used to identify disparities in the results. By investigating the causes of these disparities, possible code deficiencies were to be revealed and necessary effort was to be made to improve that weakness.

A beneficial aspect of a code-comparison study is the exchange of information on code capabilities, modeling techniques, and approaches used in solving a common problem. Peer reviews provide the essential ingredient in establishing the capability of different codes to perform thermal/structural interaction calculations of the type needed for the WIPP project or similar nuclear waste repository programs. These reviews also help to establish a consensus on how to perform the calculations and give confidence to the analysts that an accepted method is logical and free of deficiencies.

In order to ensure that the benchmark study was oriented in the desired direction, an overall philosophy and approach was adopted early. The emphasis of the approach was on a code comparison not an operator or analyst evaluation. Assuming that only experienced analysts were participating, no attempt was made to evaluate their capability or judgement except where necessary to ensure that results that could be validly compared were obtained. However, participants were cautioned about possible errors and certain review procedures were implemented to assist in obtaining accurate computational results. These review procedures are discussed in a subsequent section.

Formulation of the Problem

The Benchmark II problem was formulated to represent hypothetical design calculations for repositories located in the WIPP stratigraphy [3]. Two rectangular room configurations, located approximately 650 meters from the surface as shown in Figure 1, were specified in the problem. One configuration simulated a room designed to hold nonheat-producing waste and was designated as the isothermal room; the other configuration simulated a room designed to hold heat-producing waste and was designated as the heated room. The heated room had a thermal power loading of approximately 30 kW/acre (7.5 W/m^2). The stratigraphy modeled in the problem consisted of horizontally bedded layers of halite, anhydrite, polyhalite, a mixture of these three, argillaceous halite, and clay seams (Figure 1). Thermal and mechanical properties of these materials were specified in the problem description.

Plane strain (2D) and symmetry assumptions were used to model the media around each room. Boundary conditions that allowed vertical motion but prevented horizontal motion were imposed on the lines of symmetry. Vertical pressures were placed at the top and bottom of the layered configuration to replace the overburden and reaction to the gravity force, respectively. Clay seams located at four specified levels in each problem were modeled as slide lines with a zero coefficient of friction (perfect slip). All the layers had elastic properties, some with higher stiffness (anhydrite and polyhalite). Some of the layers also had a prescribed secondary creep law. A nonstructural "thermal material" was used in the drift area to simulate radiation heat transfer in a closed cavity. The conductivity of this material was chosen so that it would produce equivalent temperature histories around the room as though radiation heat transfer were included. This was done to allow codes without closed cavity radiation capabilities to participate in this study.

The format of the calculational results required for the Benchmark II problem was prescribed in detail to facilitate direct code comparisons [3] and included:

1. Computational mesh statistics,

2. Deformed shape of the configuration at 10 years,

3. Displacement history plots,

4. Plots of relative displacement across slip lines at 1, 2, and 10 years,

5. Stress profile plots at 1, 2, and 10 years,

6. Effective stress contours around the rooms at 1 and 10 years (to be used qualitatively only),

7. Temperature history plots for the heated rooms,

8. A computational cost parameter used to compare operating efficiencies of codes, and

9. A discussion of certain specific issues pertaining to the benchmark problem.

These results were specified to check that all aspects of the problem were being solved correctly.

Prior to publishing the Benchmark II definition document [3], draft copies were furnished to perspective participants for comments to determine if the problems were logically and completely defined and could be solved with their existing codes. The replies received from some of the participants included questions on the specified boundary conditions, suggestions on adding other material parameters such as failure criteria, suspected difficulties with certain problem features such as time-dependent material response and slidelines, and other comments on the form in which data were to be submitted for processing by Sandia. Each comment was evaluated and necessary clarification or modifications to the problems were made.

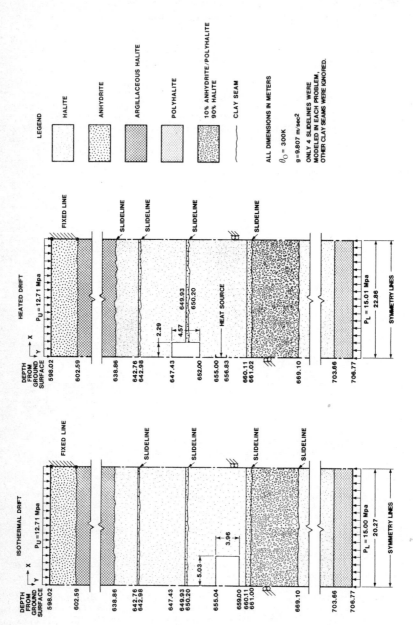

Figure 1. Repository Room Configurations in the WIPP Stratigraphy

After the agreed upon definition of the Benchmark II problem was published, participants were each carefully instructed to adhere to the descriptions and configurations of the problems. No variations or alterations were permitted which could significantly affect the comparison of results. However, each participant was permitted to develop his own mesh and use technical judgement for selecting time steps and convergence tolerances. Changes in the problem definition which were judged to be so significant that results could not be directly compared with others were not included. This occurred in one case.

Selection of Participants

In addition to performing the calculation on three codes at Sandia, invitations were made by SNL to all known prospective participants involved in calculations for nuclear waste isolation in geologic media. Inquiries were made also to government agencies, including other national laboratories, and contractors who either were actively applying computer techniques to thermal/structural interaction problems or who were developing codes to perform such calculations. Arrangments were made by SNL for those to participate either through a direct SNL contract, a DOE WIPP contract, a cooperative program, or on a voluntary basis.

Six organizations participated under SNL funding; two under direct DOE arrangements; and one under a cooperative program. The following participants performed the calculations using the codes indicated in parentheses: (1) Bundesanstalt für Geowissenschaften und Rohstoffe (BGR) of the Federal Republic of Germany (ANSALT), under the United States/Federal Republic of Germany Cooperative program (heated room calculations only); (2) D'Appolonia, Technical Support Contractor for WIPP (DAPROK), under a DOE contract (isothermal room calculations only); (3) SNL (JAC); (4) Bechtel National, Inc., architect-engineer for WIPP, under a DOE contract (MARC) (isothermal room calculations only); (5) SNL (MARC); (6) Serata Geomechanics, Inc., under a direct SNL contract (REM); (7) SNL (SANCHO); (8) RE/SPEC, Inc., under a direct SNL contract (SPECTROM); and (9) Science Applications, Inc. (STEALTH), also under a direct SNL contract. Other information about the participating codes is shown in Table I. Although large strain and large deformation capabilities were only marginally needed for Benchmark II, they are often required in drift calculations and hence are included in Table I. All the codes are finite element codes except STEALTH which uses a finite difference formulation.

Several other interested organizations, using different codes, expressed their desire to participate but could not do so due to limited funding and/or other priorities. These included Los Alamos National Laboratories (SANGRE), Lawrence Livermore National Laboratories (NIKE-2D), and Sandia National Laboratories, Livermore (GNATS).

Review Procedures and Participant Interactions

The review procedures for this program were developed early to assure that the problems calculated were those specified and that the results obtained from the different codes were not influenced by operator error. Operator errors, of course, would make comparisons difficult or invalid.

During the course of the calculations, participants were informed of a "check system" arranged by SNL for assisting in checking the preliminary results for gross errors. Values of room floor-to-ceiling closures and temperatures at a point for 1-, 2-, and 10 years were obtained with two codes at SNL. These values were then

TABLE I

Benchmark II Problem
Participating Code Characteristics

Code Name	Type	Large Deformation	Large Strain	Participant
ANSALT	FE	Yes	No	BGR of the FRG
DAPROK	FE	No	No	D'Appolonia
JAC	FE	Yes	Yes	Sandia
MARC (B)	FE	Yes	No	Bechtel National, Inc.
MARC (S)	FE	Yes	No	Sandia
REM	FE	No	No	Serata Geomechanics, Inc.
SANCHO	FE	Yes	Yes	Sandia
SPECTROM	FE	No	No	RE/SPEC, Inc.
STEALTH	FD	Yes	Yes	Science Applications, Inc.

FE--Finite Element
FD--Finite Difference

used to define a liberal range of values in which the results of other participants should fall. Each participant was informed whether his results fell within the "acceptable" ranges. This check provided the participants a means for ascertaining the accuracy of their preliminary results prior to finalizing the calculations. In-house reviews and checks were also periodically suggested by SNL.

Frequent discussions between SNL and participants were encouraged. Whenever questions arose which affected other partici- pants, clarification letters were distributed to ensure that all participants were following the same issued instructions.

Computational results were submitted on tape or on cards in a prescribed format for plotting by SNL. Each participant's results were then returned to him for review. If the participant agreed that his results were plotted correctly, they were then ready for com- parison with others. If the participant discovered errors in the SNL plots, these were corrected at SNL before comparisons were made.

A review conference was held at SNL where the results from all participants were presented. Participants were given an opportunity at the conference to compare their results with those of others for the first time. Each participant was given a copy of the SNL- compiled preliminary result summary and the reports of the other participants. After SNL presented the collective results in a com- parative manner, each participant presented his results and discussed his code capabilities. Participants were then requested to comment on the results of others and discuss the possible causes of the dis- parities in the results. The minutes of the review conference were documented and distributed for comment.

Following the review conference, each participant was offered an opportunity to perform additional checks and calculations to assist in resolving the causes of the identified differences. Checks were made on all input data; models, mesh sizes, time-steps, and tolerances used; and on all other possible sources of error or coding variations. These final results were again compiled by SNL, and the final comparative analysis was completed and documented [4]. The report presents the preliminary and revised results where they exist, discusses the causes of disparities where noted, and draws conclusions about the capability of current codes to perform calculations for bedded salt formations such as for the WIPP project.

Characteristic Results

In this section, a sampling of the original and revised results are presented to show the extent of the disparities that were found and how they changed upon revision. No attempt will be made to discuss reasons for these disparities because they are thoroughly analyzed in the Benchmark II report [4].

The preliminary isothermal room vertical closure histories for 10 years are shown in Figure 2. These results were submitted by the participants prior to the review conference. The disparity is large in Figure 2 with the REM and SPECTROM results being substantially different from the others. Figure 3 shows the results which were revised after the review conference. Although the disparities are generally not as large as in Figure 2, the REM results are still significantly different from the others, and the STEALTH results deviate farther from the others than they did originally.

The preliminary effective stress profiles across the salt pillar of the isothermal room at 10 years are illustrated in Figure 4. The plots for the results from all the codes are not labeled because most of them are very similar. Since the effective stress drives the creep process, this is a very meaningful plot for rock salt repository calculations. The REM and SPECTROM results are labeled because they are significantly different from the others. Revised effective stress profiles are shown in Figure 5. The revised SPECTROM stresses fall in line with the others, but the REM stresses still do not.

The preliminary heated room vertical closure histories for 10 years are illustrated in Figure 6. The variation of results is again large. ANSALT results do not appear in Figure 6 because the ANSALT calculations were completed after the review conference. ANSALT results and the revised results of the other participants are plotted in Figure 7. REM and STEALTH results do not appear in Figure 7 because although the preliminary calculations were found to contain errors, revised calculations were judged unnecessary considering time and funding limitations.

The computational cost parameter (CCP) was developed to form a basis for comparing the operating efficiency of the codes participating. This comparison was not straightforward because a wide variety of computers ranging from a minicomputer to a CDC STAR were used. A small computer program was written to simulate the computational characteristics of a large finite element code [3]. Participants provided SNL the ratio of costs to run the Benchmark II problems to the costs to run the given computer program. Each of these ratios were then reduced by dividing by the number of degrees of freedom used in each code computation. The range of values indicated in Table II shows that REM was the least expensive and MARC the most costly to operate. From a cost efficiency standpoint, all the other codes were in the middle of this range.

Conclusions

The benchmark series has been extremely useful in comparing the capabilities of thermomechanical codes to solve boundary value problems. This code-code comparison study has been a significant step in gaining confidence in the ability of codes to predict room response in bedded salt for the WIPP project. This study has confirmed the valuable role benchmarking plays in the code qualification

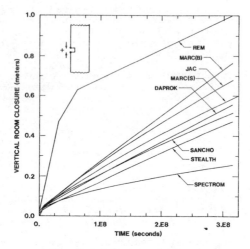

Figure 2. Preliminary Displacement Histories for the Isothermal Room

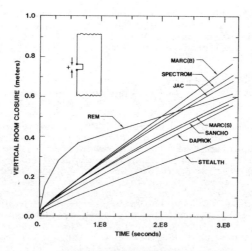

Figure 3. Revised Displacement Histories for the Isothermal Room

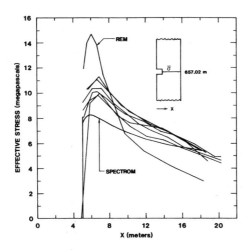

**Figure 4. Preliminary Effective Stress Profiles
for the Isothermal Room at 10-Years**

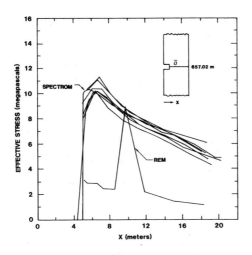

**Figure 5. Revised Effective Stress Profiles
for the Isothermal Room at 10-Years**

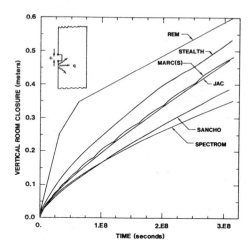

**Figure 6. Preliminary Displacement Histories
for the Heated Room**

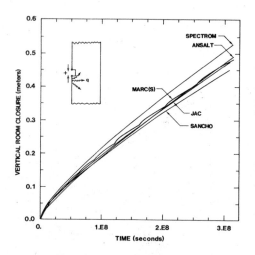

**Figure 7. Revised Displacement Histories
for the Heated Room**

Table II

Benchmark II Problem Degrees of Freedom (DOF) and Cost Comparison Parameters (CCP)

Code Name	Isothermal		Heated	
	DOF	$CCP/10^3 DOF$	DOF	$CCP/10^3 DOF$
ANSALT	--	--	1392	6.0
DAPROK	1004	5.1	--	--
JAC	1494	7.1	1475	9.9
MARC (B)	912	23	--	--
MARC (S)	1610	60	1957	76
REM	1146	.013	1208	.022
SANCHO	1311	6.4	1432	3.2
SPECTROM	2300	2.3	2400	2.0
STEALTH	1733	3.0	2109	7.3

process. Further, this benchmarking procedure has motivated improvements in many of the participating codes and has highlighted certain deficiencies in some codes to perform these particular types of calculations. Despite some minor code limitations, the results were impressively similar considering the complexity of these highly nonlinear problems.

During the course of the Benchmark II study, considerable additional information was gathered on the influence of the operator and his judgement in a complicated thermomechanical calculation. The review procedures which were followed by the participants and SNL were necessary to ensure that a valid code comparison could be achieved. Numerous discrepancies and input errors were discovered and corrected which normally would not have been possible if only one calculation was performed. The existence of a large number of input errors highlights the need to consider the use of more than one code for future critical calculations.

The final code qualification step will ultimately be the correlation of field data with calculational predictions using the material models and properties determined from laboratory experimentation. The upcoming WIPP Site and Preliminary Design Validation (SPDV) experimental program in SENM will provide the in situ measurements which will be used in the next step of field confirmation.

References

1. Wayland, J. R., and Bertholf, L. D., "A Comparison of Thermomechanical Calculations for Highly Nonlinear Quasistatic Drift Deformations," SAND80-0149, Sandia National Laboratories, Albuquerque, NM, July 1980.

2. Wigley, M. R., and Russel, J. E., "Comparison of Solutions to Benchmark Problems in Salt Using Different Numerical Methods," Proc 1980 National Waste Terminal Storage Program Information Meeting, December 9-11, 1980, Office of Nuclear Waste Isolation, Columbus, Ohio, pp 320-325.

3. Krieg, R. D., Morgan, H. S., Hunter, T. O., "Second Benchmark Problem for WIPP Structural Computations," SAND80-1331, Sandia National Laboratories, Albuquerque, NM, December 1980.

4. Morgan, H. S., Krieg, R. D., and Matalucci, R. V., "Comparative Analysis of Nine Structural Codes Used in the Second WIPP Benchmark Problem," SAND81-1389, Sandia National Laboratories, Albuquerque, NM, (in preparation).

DISCUSSION

K. SHULTZ, Canada

       First of all I would like to express to Mr. Matalucci my
appreciation for his clear description of the process and pitfalls
of benchmarking. My comment arises from his paper but it concerns
English usage. In this international group many participants use English
as a second, or even as a third language, so I urge them to be cautions
when they encounter the terms "benchmark" and "validation". In the
jargon of the nuclear industry both have been given rigorous meanings
but I noted that this community and sometimes other segments of the
industry often use these terms in a less rigorous sense. I will explain
in the hope that communication will be clarified.

       "Benchmark" has been used in this paper in its rigorous
sense. It means a well defined problem that is to be solved ; usually
using some complex mathematical model. Its utility for code comparison was described. Most often a benchmark problem is based on an
experiment with an unambiguous result but sometimes it refers to a
hypothetical or other well defined problem. The latter type of benchmark can give a relative comparison as was done in this paper but
absolute comparisons can only be done if an experimental benchmark is
used.

       "Validation" is a closely related term. A rigorous meaning
for it is set out in ANSI N.16-9. While it is given in the context of
nuclear criticality safety its usage is not restricted to this disci-
pline. In this standard, "validation" is a process of systematically
applying a computer code to benchmarks or other documented experimental
data to determine any systematic (or absolute) bias inherent to the
model being validated, and to determine relative error occurring with
the use of the model ("bias" and "error" also have specific meanings
described in the standard). Unfortunately, "validation" is frequently
used to mean any process which increases confidence that a code will
give reasonable results regardless of the amount of increase of that
confidence. Perhaps another term should be used for this looser usage.

       As to models being developed to describe nuclear waste
repository performance, they are often conglomerations of several
codes which may be simplifications or modifications of existing codes.
Individually they may or may not have been validated to varying
degrees in the disciplines that originated them. In the rigorous sense
of these terms it may be possible to benchmark or validate isolated
portions of the larger models, now or in the near future, but it is
unlikely the larger models will be benchmarked for some time to come,
perhaps years. It is even less likely they will be validated. Nevertheless both are worthwhile goals to seek and progress is being made
as we have heard in the past two days.

H.C. BURKHOLDER, United States

       It may be useful to clarify the definitions of the words
"verification" and "validation". Verification is the process by which
one becomes confident that the computer code correctly performs the
calculation it is asserted to perform. Validation is the process by
which one becomes confident that the predictions of the code adequately
describe real behavior.

Session 6

BARRIER EFFECTS

Chairman - Président

R.H. HEREMANS

(Belgium)

Séance 6

ACTION DES BARRIERES

THERMAL PROPERTIES OF BUFFER/BACKFILL MATERIALS
AND THEIR EFFECTS ON THE NEAR-FIELD THERMAL REGIME IN
A NUCLEAR FUEL WASTE DISPOSAL VAULT

H.S. Radhakrishna and K.K. Tsui

Ontario Hydro, Toronto, Canada

ABSTRACT

The near-field thermal regime in a nuclear fuel waste disposal vault will be governed by the waste loading and the thermal properties of the media surrounding the waste containers. Elevated temperatures and temperature gradients affect the geomechanical and geochemical performance of the various vault components. Therefore thermal properties must be considered in selecting backfill and buffer materials.

This paper presents the results of laboratory investigations of the thermal properties of bentonite, kaolinite, and illite based mixtures over a range of minerological composition, density, moisture content and temperature conditions. A near-field thermal analysis of a conceptual vault design is presented which demonstrates the significance of the thermal properties of the buffer and backfill on the temperature distribution in the vault as a function of time.

*This investigation was carried out at Ontario Hydro as a part of technical assistance to the High Level Nuclear Waste Management studies by Atomic Energy of Canada Limited.

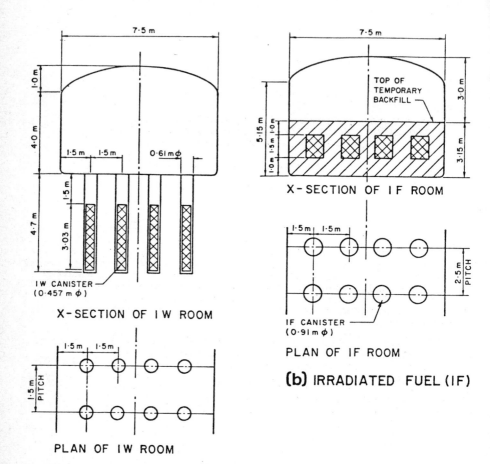

FIGURE 1

GEOMETRY OF IMMOBILIZED WASTE (IW) AND
IRRADIATED FUEL (IF) ROOM AND CANISTER

1. INTRODUCTION

In the Canadian concept of high-level nuclear reactor waste disposal in the geologic medium, two baseline designs have been developed (Burgess and Sandstorm, 1980)[1]. Immobilized reactor waste (IW) is packaged in cylindrical canisters and emplaced in holes drilled in the floor of an underground room. Immobilized CANDU reactor fuel (IF) on the otherhand is emplaced in the rooms which are partially backfilled for up to 20 years before the rooms are totally backfilled and sealed. In both these designs the canisters are surrounded by a material called Buffer, while the remaining material is termed Vault backfill (Figure 1). Buffer, backfill and various other seals to be used in the repository closure constitute an engineered barrier system.

Since the primary purpose of the engineered barriers and the host rock is to provide an effective containment system for the escaping radioactive nuclides from the waste form to the geosphere, the various processes of heat, liquid and mass transport both in the near-field and far-field have to be examined.

For simplicity the hydro-thermal regime in the backfill and buffer filled zones can be considered to have 3 significant phases:

(i)   initial thermal drying

(ii)   resaturating of the repository rooms

(iii) thermal induced water circulation and mass transport

In order to model these near-field phenomena the thermal and hydraulic conductivity properties of the materials and their associated effects on the flow regimes are to be analyzed.

In this paper an effort is made to examine the thermal properties of the buffer and backfill materials and their effect on the near-field thermal regime.

Assuming that the waste emplaced in the repository is 10 years precooled under water and its thermal power output is 269 W/canister, rates of the radiogenic heat release from IF and IW waste packages are calculated as shown in Figure 2.

2. BUFFER AND BACKFILL MATERIALS

As part of the engineered barrier system for the radionuclide transport from the waste to biosphere the buffer and backfills are required to perform the following functions:

1) Buffer the chemistry of the ground water as it reaches the canister in such a way as to enhance the life of the canister.

2) Capture and retard the migration of radionuclides released from the waste form.

3) Provide mechanical support to the canister and absorb stresses induced by possible rock movement.

4) Dissipate the heat generated by the waste, in the process of its decay to the surrounding rock mass.

The physical and chemical attributes of the buffer and backfill materials required for the proper performance of an engineered barrier for radionuclide migration are discussed in detail by Bird, 1979 [2]. Of these attributes, the thermal properties of the materials control the near-field thermal regime.

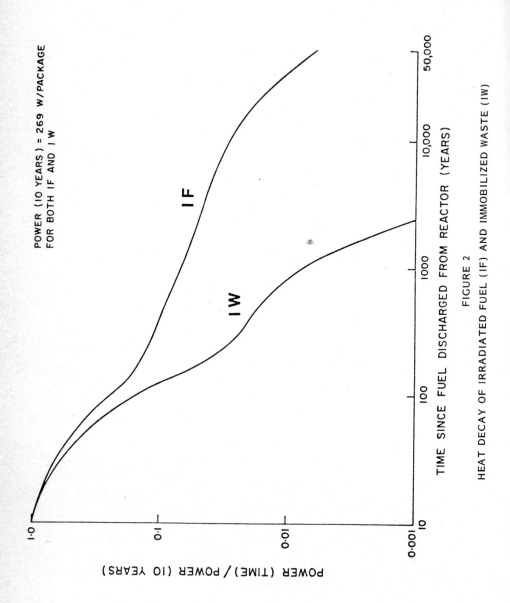

POWER (10 YEARS) = 269 W/PACKAGE FOR BOTH IF AND IW

TIME SINCE FUEL DISCHARGED FROM REACTOR (YEARS)

POWER (TIME)/POWER (10 YEARS)

FIGURE 2

HEAT DECAY OF IRRADIATED FUEL (IF) AND IMMOBILIZED WASTE (IW)

For reasons of longevity, durability and comptability with the host rock, geological products such as rock flour and clay materials are the most favoured candidates for buffer and backfill materials. Swedish researchers are investigating bentonite based buffer, backfill and other seals for their deep underground disposal vaults in crystalline rock (Pusch, 1979) [3].

Some of the disadvantages of using bentonite are its low thermal conductivity, high swelling potential and chemical instability at temperatures of 100°C and above. In the Swedish conceptual design, the maximum temperature in the buffer mass is limited to 100°C, however in the Canadian reference design, the temperature at the canister surface could be well above 100°C. The limitations on maximum permissible temperatures in the buffer mass and host rock (for reasons of thermal spalling) will obviously limit the amount of radioactive waste that can be emplaced per unit floor area.

Research programs in the United States are currently investigating a range of materials (including cement, chemical resins and geological products) for various vault seals. The chemical behaviour of clays and crushed rock under the anticipated vault environment is being evaluated. Specially formulated backfills designed to adsorb actinides are suggested by Bell and Allard [4]. Ground basalt and salt beds are also being examined for the backfill materials [5].

3. THERMAL PROPERTIES OF BUFFER MATERIALS

In the IW vault after the immobilized waste is emplaced in the boreholes and the room is backfilled, the primary mode of heat dispersion from the canisters would be by conduction through buffer, backfill and the host rock. Since the buffer and backfill materials are likely to have low permeability the heat transport by moisture convection would be negligible compared to conduction.

In the case of IF vault which will be partially backfilled in the initial period of 20 years after emplacement, the heat will conduct through the buffer mass and then transported by radiation/convection to the host rock. In both cases the thermal conductivity of the buffer-backfills materials will influence the canister skin temperature and the temperature field within the buffer-backfill zone.

It is known that the effective thermal conductivity of a 3-phase particulate material such as unsaturated soil is a function of material composition, thermal properties of the components and their geometrical arrangement [6,7]. Of the 3-phases in the media air has the lowest thermal conductivity, water has a thermal conductivity of about 25 times that of air, and solids have thermal conductivities varying from one to 15 times that of water. Because of this wide variation in the thermal conductivities of the component materials, the effective thermal conductivity of the medium is dependent on the volumetric proportions of each component. Also since the heat flows through the path of least resistance the thermal resistance at the solid to solid and solid to water contacts significantly controls the effective thermal conductivity of the media.

As a part of the material characterization phase of the buffer-backfill studies, the thermal conductivity of mixtures of coarse grained (sand or crushed rock) and fine grained materials (clays) in different proportions were measured as a function of their composition, density, moisture content and temperature. Commonly occurring clay minerals namely kaolin, illite and bentonite mixed with silica sand or crushed granite (from Lac du Bonnet batholith near

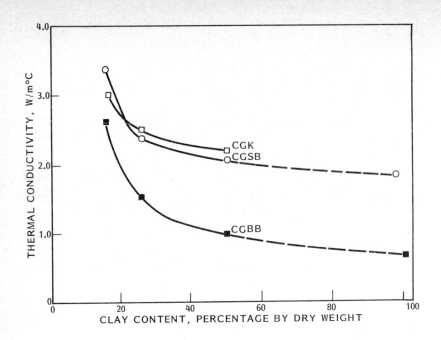

FIGURE 3

THERMAL CONDUCTIVITY AS A FUNCTION OF CLAY CONTENT
IN CRUSHED GRANITE AND CLAY MIXES

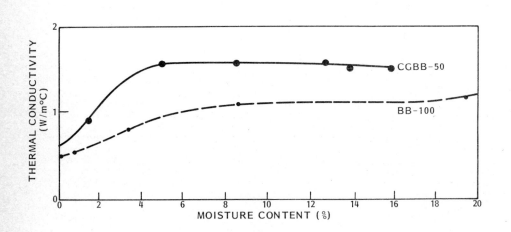

FIGURE 4

THERMAL CONDUCTIVITY VS MOISTURE CONTENT RELATIONSHIP AT
CONSTANT DENSITY FOR CRUSHED GRANITE AND BENTOITE BUFFER

Pinawa) were used in this study. The physical and chemical properties of these materials are summarized in Table I.

The measured thermal conductivity values of the compacted samples of mixes in both moist and dry states are summarized in Table II.

Among the various factors investigated the moisture content and the mix composition appeared to have the greatest effect on the thermal conductivities of clay based materials (Figures 3 to 5). The effects of ambient temperature were generally not significant up to 100°C above which the thermal conductivity values decreased but not below those obtained for oven dried conditions (Figure 6). Even though the test specimens were sealed against moisture and vapour losses and no loss in total moisture was evident, the vapour-ization of moisture within the sample would largely account for the reduction in thermal conductivity between 100°C and 120°C ambient temperature. Subsequent measurement of thermal conductivity of oven-dried samples at both room temperature and 120°C did not show significant change in their values.

Bentonite based mixes generally gave the lowest thermal con-ductivity in comparison to the others. In an oven dried state the compacted clays with no silica or crushed granite gave thermal con-ductivity values ranging from 0.5 W/cm°C for bentonite to 0.7 W/m°C for illite (seal bond) (Table III). By adding silica sand or crushed granite the thermal conductivity values improved significantly (Figure 3).

4.   NEAR-FIELD THERMAL ANALYSIS

While the selection of materials for the buffer and backfill zones will be largely determined by their hydraulic conductivity, radionuclide retardation capacity and long-term durability, the effect of their thermal properties on the near field thermal regime is of interest, particularly in estimating the maximum canister sur-face temperature and the thermal gradients within the buffer mass.

A near-field thermal analysis of IW-vault with waste emplaced in 4 rows of boreholes (Figure 1) was carried out by using the range of thermal conductivity values measured for buffer and backfill.

The analysis was carried out by using a 2-D finite element model. The following simplifying assumptions were made in the analysis:

1)   The cylindrical heat sources (canisters) were replaced by equivalent trench type heat sources running perpendicular to the cross section of room (Figure 1). This assumption is likely to predict slightly lower temperatures for the canister-buffer interface.

2)   The room is backfilled soon after the waste is emplaced in boreholes. This assumption will give somewhat higher tempera-tures in the near-field. Since the equivalent conductance of an air space in the large room is about 15 W/m°C as compared to the assumed thermal conductivity for backfill of 2 W/m°C (same as that of the host rock).

3)   Thermal conductivity of buffer, backfill and host rock are temperature and moisture independent. As discussed already this is not true. Methods of handling the non-linear thermal properties are being developed.

Figure 7 shows the temperature-time history of the canister-buffer interface for different thermal donductivity values of buffer. The thermal conductivity value of 0.5 W/m°C is typical of bentonite

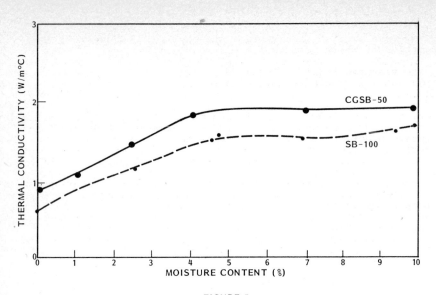

FIGURE 5

THERMAL CONDUCTIVITY VS MOISTURE CONTENT RELATIONSHIP AT
CONSTANT DENSITY FOR CRUSHED GRANITE AND SEALBOND BUFFER

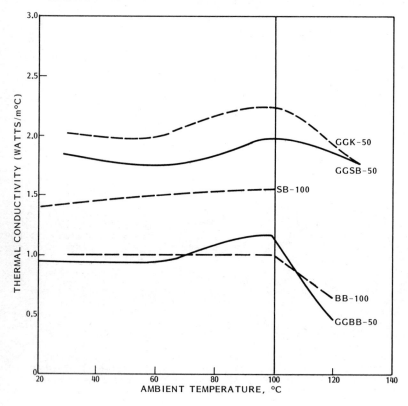

FIGURE 6

THERMAL CONDUCTIVITY AS A FUNCTION OF AMBIENT TEMPERATURE

buffer in its dried out state, and thus represents the worst conditions for canister temperature. The effect of thermal drying in the buffer zone will increase the canister temperature by about 8 to 10°C. No significant benefit is seen in using materials with thermal conductivities higher than those of the host rock.

The temperature profile across the centreline of the canisters in the IW valult (Figure 8) shows the effect of compacted bentonite buffer in its dried out condition in comparison to an ideal case of buffer with its thermal conductivity the same as that of rock (shown in solid lines). The canister, buffer and rock temperatures are increased by about 6 to 10°C within the placement zones. Beyond the extreme row of canisters the rock temperatures are generally not affected by the thermal conductance of buffer materials. Also the maximum temperature gradients in the buffer zone (76 mm wide) are about 80°C/m, as compared to the maximum thermal gradients in the host rock of 15°C/m (Figure 7).

Influence of the buffer material properties on the near-field temperatures and thermal gradients appear to continue for a period of up to 15 years after emplacement, then it begins to deminish. After a period of 50 to 100 years the near-field temperatures are relatively independent of the thermal conductance of the buffer. It is during this initial stage that conditions in the vault would cause the soil and rock moisture to migrate and the buffer to undergo hydro-thermal changes. Sodium bentonites are known to become unstable and undergo structural changes at 100°C and thereby losing some of their swelling potential and imperviousness to water-flow [8].

After about 20 to 30 years a cooling trend in the near-field temperatures is estimated [9] at which time it is reasonable to assume that the repository will start to resaturate. With resaturation or return of moisture the thermal conductivity of buffer and rock will improve and the thermal gradients will further reduce. Thus from the considerations of buffer, the critical period for thermal drying due to moisture migration is soon after the waste emplacement. Since the buffer mass in the borehole placement design is relatively small drying will likely take place in the first year after placement (Figure 7). Scale experiments in construction stages at Stripa Mine in Sweden will verify this estimation. Some laboratory scale heater experiments are in progress at Ontario Hydro to estimate the drying rates of different buffer materials.

5. CONCLUSIONS

The thermal properties of clay based buffer-backfills are dependent on their composition, moisture content and placement density and ambient temperature. A value of 0.5 W/m°C represents the lower limit for the thermal conductivity of bentonite based buffer. Analysis carried out by using a range of thermal property values measured for 3 different clay type buffers suggest that in the near-field, the buffer materials will experience substantial temperature rise of well over 100°C and as well as high thermal gradients. If the buffer and backfills are required to perform satisfactorily as engineered barriers for radionuclide migration, their hydro-thermal stability and the attendent changes in their properties need to be examined closely. Alternatively the thermal loading and geometric arrangement of waste emplacements may have to be adjusted to lower the near-field temperatures below the limits imposed by the buffer materials.

ACKNOWLEDGEMENTS

The authors wish to acknowledge Dr. T. Chan of Lawrence Berkley Laboratory for making the computer code DOT and SAPIV available to Ontario Hydro for thermal analysis of the buffer.

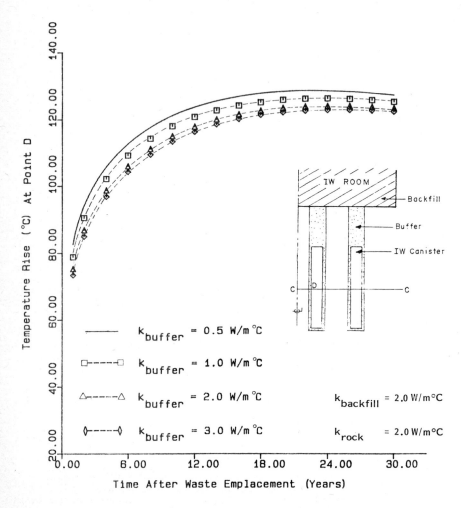

FIGURE 7

TEMPERATURE-TIME HISTORY OF THE BUFFER-CANISTER
INTERFACE FOR AN IW ROOM

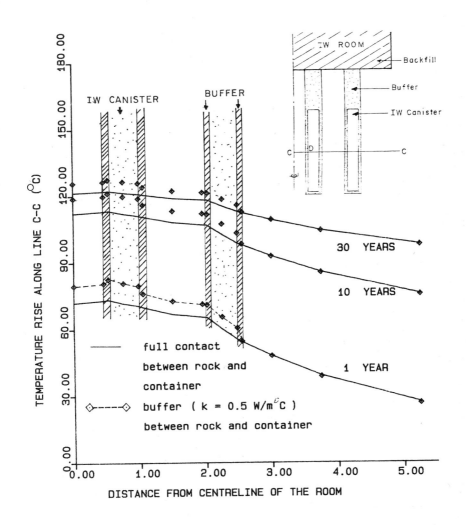

FIGURE 8

TEMPERATURE PROFILE ACROSS THE CENTRE LINE OF
IW CANISTERS FOR 1 TO 30 YEARS AFTER EMPLACEMENT

REFERENCES

1. Burgess, A.S., and P.O. Sandstrom. "Irradiated Fuel and Immobilized Waste Vaults - Preliminary Design Concepts". Atomic Energy of Canada Ltd. Technical Record, TR-48 (1980).

2. Bird, G.W. "Possible Buffer Materials for Use in a Nuclear Waste Vault". Atomic Energy of Canada Ltd. Technical Record, TR-72 (1979)

3. Pusch, R. "Highly Compacted Sodium Bentonite for Isolating Rock-Deposited Radioactive WAste Products". Nuclear Technology 45, 153 (1979).

4. Bell, G.B., and B. Allard. "Chemical Aspects Governing the Choice of Backfill Materials for Nuclear Waste Repositories". Proceedings of Symposium on Backfill as an Engineered Barrier for Radioactive Waste Management, American Chemical Society, 1980 August 24-29.

5. Wood, M.I., and W.E. Coons. "UMTANUM Basalt as a Potential Backfill Component in a Repository Located Within the Columbia River Basalt". Proceedings of Symposium on Backfill as an Engineered Barrier for Radioactive Waste Management, American Chemical Society, 1980 August 24-29.

6. Radhakrishna, H.S. "Heat Flow and Moisture Migration in Cable Back-fills". Ontario Hydro Research Quarterly 20, 2 (1968).

7. Knutsson, S. "Thermal Conductivity of Highly Compacted Bentonite". Karn-Bransle-Sakerhet Teknisk Rapport 11 (1977).

8. Valde, B. "Clay and Clay Minerals in Natural Synthetic Systems". New York, Elseview Science Publishing, 1977.

9. Lee, C.F., and G.R. Simmons. "Thermal Effects on Rock and Fluid Flow". Proceedings, 9th Waste Management Information Meeting, AECL, Toronto, January 1981. (to be published).

TABLE I

PHYSICAL PROPERTIES OF COMPONENTS IN CANDIDATE BUFFERS

Component	Specific Gravity	Size Sand	Fractions (%) Silt	Clay	Air Dry Moisture Content (%)
Wedron Sand (WS)	2.65	100	0	0	0.3
Graded Silica (GS)	2.70	90	10	0	0.5
Blackhills Bentonite (BB)	2.18		18	82	8.3
Avongel Bentonite (AB)	2.21	0	12	88	9.3
Sealbond (SB)	2.76	0	65	35	1.4
Kaolin (K)	2.60	–	30	70	1.1

TABLE II

CHEMICAL CHARACTERISTICS OF BUFFER MATERIALS TESTED

Material	SiO_2	Al_2O_3	Fe_2O_3	MnO	TiO_2	CaO	MgO	K_2O	Na_2O	Loss on Ignition (%)	pH
Wedron Sand (WS)*	99.8	0.10	0.02	-	0.015	0.015	0.005	-	-	0.08	7.0
Graded Silica (GS)*	99.8	0.10	0.02	-	0.015	0.015	0.005	-	-	0.08	7.0
Blackhills Tower Bond Bentonite	65	20.2	2.30	1.0	0.10	1.40	1.00	0.80	3.10	7.0	8.4
Avongel Bentonite (AB)*	60	18.0	2.0	-	-	0.8	2.0	0.4	2.3	5.5	9.3
Sealbond (SB)**	59.1	15.8	5.30	0.12	0.53	4.96	3.99	4.0	0.81	4.1	-
Kaolin (K)**	44.0	36.67	0.31	-	1.0	0.24	0.19	0.17	0.13	13.14	-

* Data supplied by manufacturer
**AECL unpublished data

TABLE III

THERMAL CONDUCTIVITY OF CANDIDATE BUFFER/BACKFILLS

Mix Composition*	Moisture Content Range (5)	Dry Density Range (kg/m³)	Thermal Conductivity in Moist Conditions W/m°C	Thermal Conductivity in Dry Condition W/m°C
WSAB-15	6-17	1100 - 1770	1.3 - 2.8	0.7 - 1.0
WSAB-25	5-16	1200 - 1760	1.2 - 2.7	0.6 - 0.9
WSAB-50	15-29	1330 - 1400	1.2 - 2.0	0.5 - 0.6
GSBB-10	6-10	1850 - 2000	2.0 - 3.0	0.4 - 0.7
GSBB-25	8-12	1500 - 1700	1.2 - 1.7	0.7 - 0.8
GSBB-50	12-20	1490 - 1520	0.7 - 1.2	0.5 - 0.7
GSSB-15	4- 9	2050 - 2175	3.0 - 4.0	1.5 - 1.7
GSK-15	4- 9	2000 - 2200	2.5 - 3.5	1.8 - 2.0
CGBB-15	5-19	1750 - 2000	1.3 - 2.6	0.8 - 1.0
CGBB-25	6-12	1620 - 1710	1.0 - 1.5	0.6 - 0.8
CGBB-50	12-17	1500 - 1550	0.8 - 1.0	0.5 - 0.6
CGBB-100	17-20	1320 - 1400	0.7 - 1.1	0.5 - 0.6
CGSB-15	6-10	1890 - 2200	2.4 - 3.4	1.4 - 1.8
CGSB-25	7-12	1600 - 2050	1.4 - 2.4	0.8 - 1.0
CGSB-50	9-15	1850 - 1980	1.8 - 2.4	0.9 - 1.0
CGSB-100				
CGK-15	6-10	1900 - 2050	2.2 - 3.0	1.7 - 2.0
CGK-25	9-12	1850 - 1980	1.7 - 2.5	0.9 - 1.2
CGK-50	9-15	1770 - 1860	2.1 - 2.2	1.0 - 1.2

*WS = Wedron sand, GS = Graded Silica, CG - Crushed Granite
AB = Avongel Bentonite, BB = Blackhill (Wyoming) Bentonite, SB - Sealbond, K = Kaolin, Z = Zeolite

DISCUSSION

W.R. FISCHLE, Federal Republic of Germany

What sort of samples did you use in your laboratory tests ?

What measurements did they have ? To which diameter did you crush the granite ?

H.S. RADHAKRISHNA, Canada

Thermal conductivity measurements were made on compacted samples of 10 cm diameter by 10 cm height. The soil components were blended dry, mixed with water and then compacted in standard proctor compaction moulds following ASTM Standard procedure of compaction.

The granite rock was crushed to particle sizes of sand and silt to produce a well graded sand. The particle sizes varied from 0.01 mm to 2 mm.

N.A. CHAPMAN, United Kingdom

In your IW disposal concept both the physical effects of complete backfilling and the rewatering and swelling of the clay will have marked effects on the thermal behaviour of the backfill. Did you measure thermal conductivities taking these factors into account, for instance under confined conditions ?

H.S. RADHAKRISHNA, Canada

The thermal conductivity values reported in the paper were measured on compacted samples prior to drying and after drying. Effects of rewetting and swelling are being studied. It is anticipated that the thermal conductivity values should improve on rewetting in a confined condition. However, if the backfill is allowed to swell by large proportions, and then dried, some changes in its thermal con-ductivity could be expected.

M. MAKINO, Japan

You have four holes in line, and the distance between holes is 1.5 m. I feel the distance is rather small as far as thermal impact is concerned. Can you comment ?

H.S. RADHAKRISHNA, Canada

The spacing of canisters or boreholes in the base line design has received several considerations,

    i) thermal effects on rock,

   ii) max canister temperature,

  iii) quality of rock, etc.

In the analysis presented here the proposed spacing of 1.5 m between the canisters does not appear to be critical for either canister or host rock.

The maximum teperature and thermal gradients reached in the buffer certainly require further consideration.

W.R. FISCHLE, Federal Republic of Germany

What temperature do you expect from your canister room ?

How many rooms will you need for the waste ?

H.S. RADHAKRISHNA, Canada

The canister temperature as shown in the paper will depend on the rate of heat production per canister, its time rate of decay and the canister spacing in the rooms. The maximum rise in temperature of the canister for the base-line design discussed in the paper is between 120 to 130°C. This could change if the canister spacing or the thermal conductivity of the buffer/backfills are changed.

K. SHULTZ, Canada

Are you using the temperature rise or the absolute temperature of the canister ?

H.S. RADHAKRISHNA, Canada

The thermal code "DOT" gives the absolute temperatures. However all the plots (Figures 7 & 8) show only the temperature rise as a function of time.

ASSESSMENT OF BARRIER EFFECTS
ON ISOLATION SYSTEM PERFORMANCE
OF GEOLOGIC DISPOSAL OF HIGH LEVEL WASTE

Sumio Masuda and Tadashi Mano
Waste Management Office
Power Reactor and Nuclear Fuel Development
Corporation
Akasaka, Minato-ku, Tokyo, Japan

Masahiko Makino and Takao Ikeda
Nuclear Project Division
JGC Corporation
Bessho 1-chome, Minami-ku, Yokohama, Japan

ABSTRACT

A sensitivity analysis was performed for assessing the
effects of barriers, such as waste form, engineered barrier and
geosphere, on the isolation system performance of the geologic
disposal of high level wastes. For near-field phenomena, (1) time of
initial release and (2) leach time were selected for assessing the
sensitivity of the effects of waste form and engineered barrier,
including waste package, buffer material and backfill.
For far-field phenomena, following parameters were selected:
(3) path length, (4) groundwater flow velocity, (5) retardation
factor of radionuclides and (6) dispersion coeffecient of radionuclides. In this analysis, the values of parameters were selected
in the wide range and for each target value of isolation system
performance, acceptable combinations of each parameter values were
derived.

1.   INTRODUCTION

A geologic repository consists of multiple barriers, such as waste form, engineered-barrier, and natural geologic environment. The phenomena of radionuclide migration from repository can be divided into two parts, the near-field and the far-field. For near-field phenomena, following parameters were selected (1) Time of initial release: time when the engineered system loses its confinement function by attack of surrounding water and (2) Leach time: time duration of constant release of radionuclide from the waste form into the geologic environment. For far-field phenomena, following parameters were selected: (3) Path length: transport distance of radionuclides with groundwater, (4) Groundwater flow velocity, (5) Retardation factor of radionuclides, and (6) Despersion coefficient of radionuclides.

This study aimed at the sensitivity analysis of above six parameters on isolation system performance, and it was recognized that the near-field phenomena plays important part for isolating radionuclide in the geologic disposal of high-level wastes from spent fuel reprocessing. Subsequently, this study will move to the next step to develop the realistic model of radionuclide migration in the near-field and to propose the research and development target for geologic disposal system.

2.   DESCRIPTION OF THE STUDY

2.1 Potential hazards involved in high-level waste

Since the high-level waste from spent fuel reprocessing has large amount of radioactivity and contains long-lived radionuclides, it is expected to be hazardous for a long period of time. Fig. 1 shows variation in the ingestion toxicity index of high-level waste with time. The term, "ingestion toxicity index" is defined for an individual radionuclide as radioactivity devided by the maximum permissible concentration in drinking water. The ingestion toxicity index is normalized by the amount of radioactivity produced from 1,000 MWe nuclear power plant for a year, and its unit becomes "$m^3H_2O/GWeY$". As seen from Fig. 1, the hazard of fission products (FP) prevails for 1,000 years after reprocessing, and then it levels with that of actinides. After 10^4 years, the hazard of actinides increases due to the decay of their parent nuclides. The hazard of high-level waste expressed by the ingestion toxicity index is merely relative, however, the potential hazard will lurk for over 1,000 years.

2.2 Radionuclide migration scenario

2.2.1 Assessment method

To assess the isolation system performance of geologic disposal, it is necessary to provide the nuclide migration scenario in the disposal system. Fig. 1 does not include the effects of barriers, such as waste form, engineered barrier, and natural environment.

The isolation system performance might be represented by the potential hazard of radioactivity released to the biosphere through radionuclide migration along with underground water. In this study, the isolation system performance is expressed by means of ingestion hazard index, which is defined for an individual radionuclide release rate to the biosphere divided by the maximum permissible concentration in drinking water [1], being normalized by radioactivity produced in the 1,000 MWe nuclear power plant for a year. The unit of ingestion hazard index for geologic disposal is "$m^3H_2O/GWeY^2$", which corresponds to the quantity of water required to dilute the underground water to the permissible level [2][3].

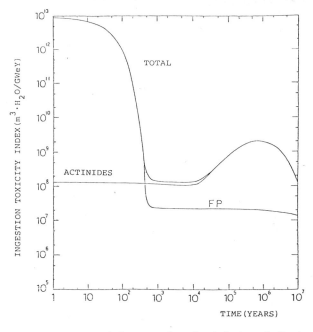

Fig. 1 Potential Hazards of High-Level Waste

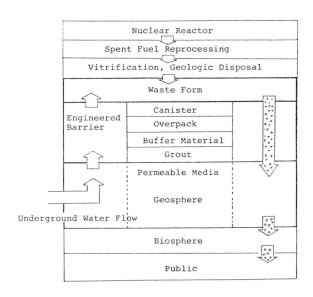

Fig. 2 Radionuclide Migration Scenario

2.2.2 Repository model and radionuclide migration scenario

The geologic disposal system consists of three barriers. The first barrier is "engineered barrier" which consists of the canister, overpack, backfill material and grout and it serves to delay the contact of the waste form with underground water by its confinement function [4].

The second barrier is the waste form itself. The waste form is chemically stable and radionuclides in the waste form are released very slowly into the surrounding water after the engineered barrier loses its confinement function.

The third barrier is the geosphere. During and after leaching out from the waste form, radionuclides will migrate along with the underground water flow in the geosphere. In the course of the migration, the nuclides undergo adsorption and absorption by geospherical material, therefore they will move slower than the underground water flow.

Fig. 2 shows the migration scenario. The following equation was used as a nuclide migration equation for the scenario:

$$K_i \frac{\partial N_i}{\partial z} = D \frac{\partial^2 N}{\partial z^2} - V \frac{\partial N_i}{\partial z} + K_{i-1} \lambda_{i-1} N_{i-1} - K_i \lambda_i N_i + F(t) \delta(z)$$

Where,

N_i = radionuclide concentration in the underground water
z = path length
K_i = retardation factor (ratio of underground water flow velocity to nuclide migration velocity)
D = dispersion coefficient
V = underground water flow velocity
λ_i = decay constant
$F(t)$ = radionuclide release rate from the waste form
$\delta(z)$ = Delta function

Fig. 3 shows a relation between a repository model and migration equation. Various release modes are conceivable, however, the "band release mode" was adopted in this study. In Fig. 3, Ts means the time of initial release and LT means leach time. Radionuclide migration in the geosphere and its ingestion hazard index were calculated by MGRAT-3 code [5].

Table I summarizes parameters in each barrier and variation in each parameter value for performing sensitivity analysis in this study. Table II shows the calculation conditions of the generation of high-level waste, and Table III shows the calculation conditions on radionuclide.

2.3 Sensitivity analysis on barrier effects

2.3.1 Standard migration scenario

The standard migration scenario was selected as a reference among cases to assess the sensitivity of barrier effects on the isolation system performance. Fig. 4 shows the ingestion hazard index change of the typical radionuclides along with the time passed, in the standard migration scenario shown in Table I. As seen from Fig. 4, the hazard of ^{99}Tc and ^{129}I, whose migration velocity is equal to the underground water flow velocity, appear first, then the hazard of ^{90}Sr appears. The hazard of ^{226}Ra, which are formed by the decay of their precursor, U and Np, appear last. The hazard of Ra becomes dominant 10^5 years after the start of disposal and lasts

Table I Parameters in Each Barrier and Variation in Each Parameter Value

Barrier	Parameter		Parameter value	
	Sym.	Description	Standard scenario	Variation
Engineered Barrier	Ts	Time of initial release Time when the engineered barrier loses its confinement function and the waste form is assumed to be in contact with groundwater	1 Year	$1 \sim 10^7$ Years
Waste form	LT	Leach time Time duration of constant release of nuclides from the waste form into the geosphere	10^5 years: Constant for all nuclides	$10 \sim 10^7$ Years
	$F(t)$	Release rate	Band release	Band release
Geosphere	Z	Path length Distance from repository to biosphere along with underground water flow	100 m	$1 \sim 10^4$ m
	V	Underground water flow velocity	10 m/Year	$1 \sim 10^3$ m/Year
	D	Dispersion coefficient	100 m²/Year	$1 \sim 10^3$ m²/Year
	K_L	Retardation factor Ratio of underground water flow velocity to nuclide migration velocity which has different value among nuclides	Experimental value in Western U.S. desert soils [6]	1/10 ~ 10 times of the standard scenario case and the experimental value in granite [4]

Table II Calculation Conditions of the Generation of High Level Waste

Item	Description
Reactor type	PWR
Burn-up Specific power U enrichment Electric power	35,000 MWD/MTU, 1,000 days continuous operation 35 MW/MTU 3.4 % 1,000 MWe, 33 % thermal effecient
Time of reprocessing, geologic disposal	Reprocessing after 180 days cooling; Immediate vitrification and geologic disposal after reprocessing
Incorporated fraction of nuclide into waste form	Incorporated fraction of Pu and U in the high-level waste at reprocessing is assumed to be 0.5 %. Other nuclides are all incorporated in the waste.
Inventory of radionuclide	Radionuclide inventory through 1,000 MWe nuclear power plant operation, expressed in Ci/GWeY. See Table III.
Burn-up calculation	ORIGEN code

Table III Calculation Conditions on Radionuclides

Nuclide	Half life [Y]	Retardation factor [6] [-]	M.P.C. in water $[Ci/m^3 H_2O]$ [1]	Inventory [Ci/GWeY]
^{90}Sr	2.88×10^1	1.0×10^2	3.0×10^{-7}	2.68×10^6
^{99}Tc	2.12×10^5	1.0	2.0×10^{-4}	4.76×10^2
^{129}I	1.7×10^7	1.0	6.0×10^{-8}	1.21
^{137}Cs	3.0×10^1	1.0×10^3	2.0×10^{-5}	3.54×10^6
^{234}U	2.47×10^5	1.43×10^4	3.0×10^{-5}	9.02×10^{-1}
^{230}Th	9.0×10^4	5.0×10^4	2.0×10^{-6}	4.97×10^{-6}
^{226}Ra	1.62×10^3	5.0×10^2	3.0×10^{-8}	2.27×10^{-9}
^{237}Np	2.14×10^6	1.0×10^2	3.0×10^{-6}	1.28×10^1
^{233}U	1.62×10^5	1.43×10^4	3.0×10^{-5}	8.87×10^{-6}
^{225}Ra	4.1×10^{-1}	5.0×10^2	5.0×10^{-7}	3.57×10^{-7}
^{243}Am	7.95×10^3	1.1×10^4	4.0×10^{-6}	4.60×10^2
^{239}Pu	2.44×10^4	1.0×10^4	5.0×10^{-6}	5.01×10^1

until 10^6 years.

Fig. 4 is shown in two-dimensional, i.e., hazard vs time, however, Fig. 5 is expressed three-dimensionally, i.e., hazard vs time and path length. The hazard of ^{90}Sr reaches the peak in a small range of time and path length, and it disappers in over 1,000 years. The hazard which follows ^{90}Sr is caused by ^{129}I and 99Tc and it remains for a long period of time. The peak of the hazard in $10^5 \sim 10^6$ years is that of Ra which continues within path length of 1000m.

It can be seen from Figs. 4 and 5 that the hazard index of each nuclide shows different behaviors, depending on the types of nuclides. Each nuclide has the maximum value of the ingestion hazard index at different time after the start of the disposal. In this study, the concept of maximum ingestion toxicity index (hereafter MIHI) was introduced as the maximum value of the sum of the ingestion toxicity index of each nuclide along with time.

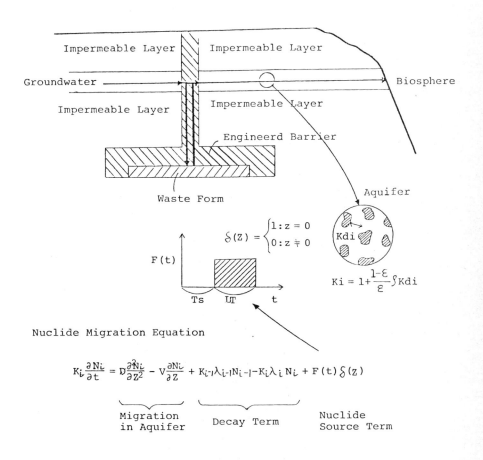

Fig. 3 Repository Model and Nuclide Migration Equation

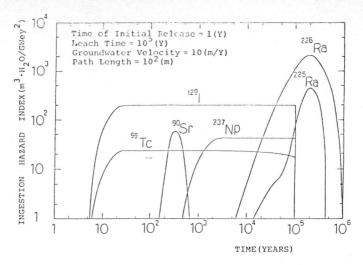

Fig. 4   Ingestion Hazard Index of Nuclide
(Standard Migration Scenario)

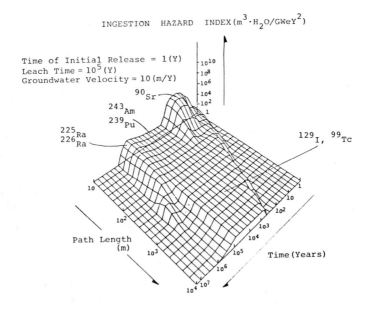

Fig. 5   Ingestion Hazard Index vs Time and Path Length
(Standard Migration Scenario)

2.3.2 Effects of engineered barriers

The integrity of the engineered barrier is represented by the time of initial release (Ts), therefore, the effects of the engineered barriers can be assessed from the sensitivity of the maximum ingestion hazard index with respect to the value of Ts.

Fig.6 shows the effects of the engineered barrier, normalized by the standard migration scenario at a specific path length (Z/V): travel time. As shown in Fig. 6, in the Z/V range of $1 \sim 2$ years, the MIHI decreases to $1/10^2 \sim 1/10^3$ if the time of initial release is increased up to 300 years. In this case, nuclide with strong toxicity, such as ^{90}Sr and ^{137}Cs will decay during the period, and accordingly the MIHI will lower. However, if the travel time, Z/V, is above 10 years, the MIHI will remain almost unchanged even if the time of initial release increases. In this case, ^{90}Sr and ^{137}Cs will decay before they reach to the travel time (Z/V) of 10 years, and the effects of the engineered barrier is not so important. The MIHI slightly increases in Ts of $10^5 \sim 10^6$ years, because the Ra inventory increases due to the decay chain of ^{234}U $\rightarrow$ ^{230}Th $\rightarrow$ ^{226}Ra, ^{237}Np $\rightarrow$ ^{233}U $\rightarrow$ ^{229}Th $\rightarrow$ ^{225}Ra.

2.3.3 Effects of waste form

The integrity of the waste form is represented by the leach time. Fig. 7 shows the effects of the waste form. As shown in the figure, there is a nearly reverse proportional relationship between the MIHI and the leach time. At the travel time (Z/V) of 1 year the dominant nuclide is ^{90}Sr, and at the Z/V of 10^2 years the dominant nuclide is ^{129}I, and behaviors of such nuclides are simple. However, at the Z/V of 10 years dominant nuclides are ^{129}I up to the leach time of 10^4 years and Ra above that of 10^5 years. The reason why the MIHI value at Z/V = 1 year does not vary during the period of $10^4 \sim 10^5$ years is that the hazard of Ra increases.

2.3.4 Effects of path length

Fig. 8 shows the effects of nuclide migration path length. In the assessment of the effects of it, not only the path length but also the underground water flow velocity must be taken into consideration. Even if the path length is increased, the migration travel time will not increase if the water flow velocity increases. Hence, in this study, the path length is expressed by the travel time (Z/V), which the underground water requires to reach the biosphere from the repository.

As seen in Fig. 8, the MIHI will decrease until Z/V = 50 years, then it will remain constant. The dominant nuclide are ^{90}Sr and ^{137}Cs until Z/V of 10 years, then Ra becomes dominant up to Z/V of 50 years. Finally long-lived ^{129}I and ^{99}Tc will prevail, remaining at a fixed level.

2.4 Required conditions for each barrier

The maximum ingestion hazard index method for the assessment of the isolation system performance of the geologic disposal system is superior in parametric survey, however, it it somewhat inconvenient for the assessment of absolute hazards, such as environmental dose. In this study, however, the target value of the MIHI was selected, and the required conditions for each barriers were derived for it. Herein, the target value of the MIHI is selected at 10^4 m^{3}H$_2$O/GWeY^2. Water flow of 10^4 m^{3}H$_2$O/year corresponds to such a small creek whose flow rate is about 1 m^3/hr.

2.4.1 Case: leach time is fixed

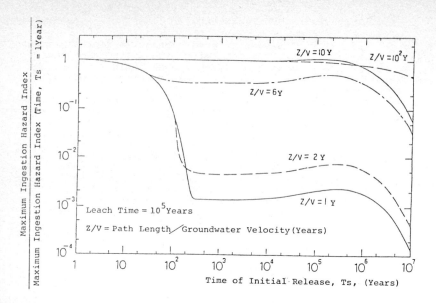

Fig. 6   Relative Effects of Engineered Barrier

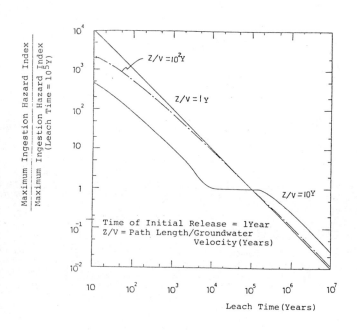

Fig. 7   Relative Effects of Waste Form

Fig. 9 shows the required combination of the time of initial release and travel time (Z/V) at the fixed leach time, i.e., 10^3, 10^4, 10^5, 10^6 and 10^7 years, for satisfying the isolation system performance. In the figure, the conditions inside the curve (origin side) involves hazards higher than 10^4 m3H2O/GWeY^2, and the conditions outside the curve are less hazardous than tha value. The figure shows a trend that the range of acceptable conditions increases with an increase in the leach time. The result is summarized in Table IV.

2.4.2 Case: time of initial release is fixed

Fig. 10 shows the required combination of the leach time and the travel time (Z/V) at the fixed time of initial release, i.e., 1, 10, 10^2, 10^3, 10^4, 10^5, 10^6 and 10^7 years. As shown in the figure, the effects of the time of initial release is strong when the Z/V is under 10 years, but is not strong when the Z/V is over 10 years. It is note worth that the required conditions become severer when the time of initial release is 10^5 years than when the time of initial release is 10^3 and 10^4 years. The reason for this is that the Ra inventory increases. The result is summarized in Table IV.

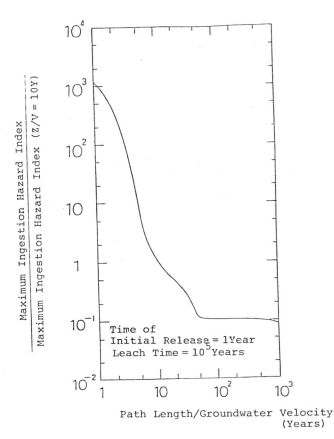

Fig. 8 Relative Effect of Travel Time

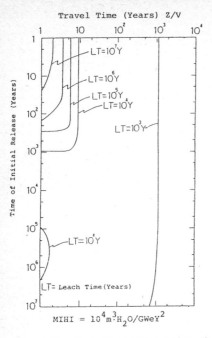

Fig. 9 Required Combination of Barrier Conditions (LT: fixed)

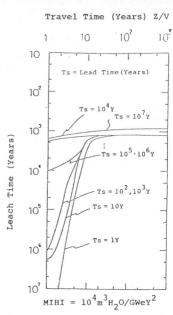

Fig. 10 Required Combination of Barrier Conditions (Ts: fixed)

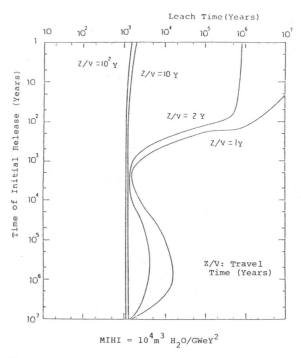

Fig. 11 Required Combination of Barrier Conditions (Z/V: fixed)

Table IV Acceptable Combination of Each Parameter Values Satisfying MIHI of 10^4 [$m^3 H_2O/GWeY^2$]

Fixed parameter	LT (Years)	10^3	10^4	10^5	10^6	10^7
Acceptable conditions	Ts (Years)	10^7	1000	300**	200**	30**
	Z/V (Years)	1000	10	6**	4**	2**

Fixed parameter	Ts (Years)	1	10, 10^2, 10^3	10^4	10^5, 10^6	10^7
Acceptable conditions	LT (Years)	3000	3000	2000	2000	2000
	Z/V (Years)	10	7	1	10	1

Fixed parameter	Z/V (Years)	1	2	10	10^2	10^3
Acceptable conditions	Ts (Years)	500	500	1	1	1
	LT (Years)	2×10^4	5000	2000	2000	1000

 * Maximum Ingestion Hazard Index
** MIHI of 10^4 will be obtained where either of the two conditions is satisfied

2.4.3 Case: travel time is fixed

Fig. 11 shows the required combination of the time of initial release and the leach time, at the fixed Z/V value, i.e., 1, 2, 10 and 10^2 years. As shown in the figure, the effects of the time of initial release and leach time are large when the Z/V is 1, 2 years, but they are not effective when the Z/V is over 10 years. It can be said that the requirement on the engineered barrier can be reduced by increasing the Z/V up to 10 years. The result is summarized in Table IV.

3.   CONCLUSION

This study shows the effectiveness of the geologic disposal of high-level waste to reduce the potential hazard of the waste. Different nuclides show a different behavior in the geologic disposal system. The environmental impact, i.e., ingestion hazard, is dominated by short-lived FPs such as ^{90}Sr at first. Thereafter it is controlled by non-sorptive and long-lived FPs such as [129]I, followed by actinides such as [239]Pu and finally Radium. In this study, the characteristics of each barrier and the barrier effects on the maximum ingestion hazard index were clarified. The proper selection of the conditions of the barriers can reduce the hazards in geologic disposal system effectively. In order to limit the radionuclide concentration in the biosphere to such levels below the maximum permissible concentration, there are a variety of combination of barrier conditions, time of initial release, Ts, leach time, LT, and travel time, Z/V.

U.S. NRC proposed the required conditions of geologic disposal system in the 10 CFR Part 60, in which they recommend that the life time of engineered barriers is to be longer than 1,000 years, the leach time is to be longer than 10^5 years, and radio-nuclide travel time is to be longer than 1,000 years[7]. The re-commended elements seem to be equivalent to the time of initial release Ts, leach time LT, and travel time Z/V in this study. As a result of this study, the following combinations of barrier con-ditions were derived for obtaining the certain isolation system performance:

leach time	<u>100,000 years</u>	3,000 years	1,000 years
time of initial release	300 years	<u>1,000 years</u>	1 year
travel time	6 years	7 years	<u>1,000 years</u>

Compared to the result of this study, the NRC recommendation seems to be enough for reducing the environmental impact of high-level waste to the permissible level.

This study aimed at the sensitivity analysis of barrier conditions on isolation system performance, and it was recognized that the near-field phenomena plays important part. Actual near-field phenomena must be far more complicated than the model here applied. Consequently this study will move to the next step to assess the realistic mode of radionuclide migration in the near- and far-field, including thermal, thermomechanical, thermochemical, thermohydrogic and radiochemical effects, and to propose the specifi-cation of each element in barriers in accordance with the experi-mental data to be obtained.

4.   REFERENCES

[1] US Nuclear Regulatory Commission,: "10 CFR Part 20, Standards for Protection against Radiation"

[2] T. H. Pigford, S. Masuda,: "Feasibility of Geologic Disposal of High Level Waste", Nuclear Engineering, Vol. 27, No. 2, pp. 31 - 36, in Japanese

[3] S. Masuda,: "Hazard from Deep Geologic Disposal of Radioacitive Waste", N141-80-07 Power Reactor & Nuclear Fuel Development Corporation, August 1980

[4] XXX,: "Handling of Spent Nuclear Fuel and Final Storage of Vitrified High Level Reprocessing Waste", General, pp. 29 - 36, KBS Report (1978)

[5] M. Harada, P. L. Chambré, M. Foglia, K. Higashi, F. Iwamoto D. Leung, T. H. Pigford, D. Ting: "Migration of Radio-nuclides Through Sorbing Media Analytical Solution - I" LBL-10500 UC 70, February 1980

[6] H. C. Burkholder,: "Method and Data for Predicting Nuclide Migration in Geologic Media", BNWL-SA-5822 July 1976

[7] US Nuclear Regulatory Commission,: "10 CFR Part 60, The Technical Criteria for Regulating Geologic Disposal of High-Level Radioacitive Waste"

DISCUSSION

H.C. BURKHOLDER, United States

The results in Figure 6 of your paper seem to suggest that overall isolation system performance as measured by the maximum hazard index is independent of the value chosen for the time of initial release (i.e., the container lifetime) as long as the groundwater travel time from the repository to the biosphere is equal to or greater than 10 years. Given these results, how much benefit or value would you attach to the inclusion of long-lived containers in the isolation systems ? What fraction of the potential isolation sites in Japan would you expect to have groundwater travel times of less than 10 years ? If you expected the ground water travel time to be less than 10 years at sites in Japan, would you choose geologic disposal as Japan's disposal option ?

T. MANO, Japan

Your understanding is correct. Yes, when the groundwater travel time from the repository to the biosphere is equal to or greater than 10 years, long-lived containers are not required for the total isolation system. We need to get site data for evaluating the fraction of the potential isolation sites in Japan which have groundwater travel times of less than 10 years.

Yes, we will make a required combination of conditions for each barrier, a condition for Japan's disposal option.

M. MAKINO, Japan

Dr. Burkholder pointed out that the effects of engineered barriers are not so important at $Z/V = 10$ years. This question is very important and I would like to explain this to you. If land availability is restricted, in such a case the integrity of the engineered barrier may become important, however, if we can get a high enough Z/V value, we do not need the integrity of the engineered barrier so much. We do not intend to handle these evaluation results in any specific way.

F. GERA, Italy

I believe that the Hazard Indexes you have shown are based on the old ICRP limits. If you recalculate the indexes on the basis of the limits given in ICRP Publication 30 you will see that the relative importance of several radionuclides is drastically changed. For example, ^{237}Np will become one of the critical radionuclides.

T. MANO, Japan

You are right. The maximum ingestion hazard indices will be changed if the MPC values are changed. We heard that the MPC of ^{237}Np was lowered, but this has not yet been employed in Japan and USA.

E.K. PELTONEN, Finland

In your study you have used an assumption that the MPC values are acceptable concentrations for radionuclides in surface waters. Is this a subjective choice or is it recommended by your authorities ?

T. MANO, Japan

      We have temporarily adopted MPC as acceptable concentrations for radionuclides in surface water for the purpose of this sensitivity study.

E.K. PELTONEN, Finland

      What kind of soil or rock have you assumed in your analysis ?

T. MANO, Japan

      This analysis did not investigate any specific rock or soil. The variations of parameters will cover the range of all kinds of rock.

E.K. PELTONEN, Finland

      You have used only one set of constant values for retardation factors, which means that you have a constant porosity (in the case where k_d values have been employed). Is this consistent with the use of different permeability values ?

T. MANO, Japan

      We used ground water flow instead of permeability. Also retardation factors are varied over a range of a factor of two. The discussion on sensitivity to retardation factors is not included in this paper.

R.E. WILEMS, United States

      If one were to accept a maximum ingestion hazard index of 10^4 m^3 $H_2O/GWeY^2$ as indicated in Figure 9 of your paper, then since nearly any site would have a travel time of 10 years, does not your analysis support a leach time of 10^4 years and indicate only slight improvement by developing a waste form of 10^5 yrs ?

T. MANO, Japan

      You are right.

R.E. WILEMS, United States

      How sensitive are your results presented in Figure 9 to the value of Maximum Ingestion Hazard Index, for example, if a MIHI of 10^3 were required what Travel Time would be required for LT = 10^4 y and for LT = 10^5 y, if Ts is between 1 and 100 yrs ?

T. MANO, Japan

      In case of MIHI = 10^3, the necessary travel time is 30 years for LT = 10^5 and more than 10^3 years for LT = 10^4.

WASTE PACKAGE PERFORMANCE ASSESSMENT

D.H. Lester
Science Applications, Inc.

ABSTRACT

This paper describes work undertaken to assess the life-expectancy and post-failure nuclide release behavior of high-level and TRU waste packages in a geologic repository. The work involved integrating models of individual phenomena (such as heat transfer, corrosion, package deformation, and nuclide transport) and using existing data to make estimates of post-emplacement behavior of waste packages.

A package performance assessment code was developed to predict time to package failure in a flooded repository and subsequent transport of nuclides out of the leaking package. The model has been used to evaluate preliminary package designs. The results indicate, that within the limitation of model assumptions and data base, packages lasting a few hundreds of years could be developed. Very long lived packages may be possible but more comprehensive data are needed to confirm this.

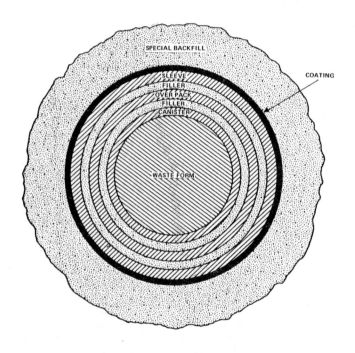

Figure 1. Cross-Section of Maximum Protection Waste Package.

1. INTRODUCTION

Early in the efforts to develop waste packages by the U.S. Department of Energy and its contractors it was recognized that a means of providing some measure of design progress was needed. Some approximate estimate of post-emplacement performance was required to determine what design feature were desirable and what materials problems required attention. The work described in this paper was undertaken to develop such an estimating technique.

Several limitations were necessary. Since results were needed in the near-term it was necessary to base the assessment model on the type of data currently available. At the same time it was desirable to assemble the model in a way that would allow for later refinements in data and phenomenological understanding. Part of the effort in developing the model was directed toward identification of data and modeling needs in order to produce more rigorous package and near-field models in the future.

1.1 Assumption

The performance assessment model (called "BARIER") considers a cylindrical waste package consisting of a series of containers of a variety of materials including a backfill between the emplaced package and the host rock of the repository. All phenomena are considered uniform over the package length. Figure 1 is a diagram of a package cross-section.

The model is a deterministic description of sure, slow process occurring after the package is emplaced in the repository. Key assumptions of the model are:

- Flooded repository at time = 0 with very low water flow velocities near the package.

- Normal package and repository history - no events such as earthquake, intrusion of persons, or package damage.

- Constant corrosion rates (function of temperature and radiation),

- Full lithostatic, hydrostatic pressure developed at time = 0.

- Constant groundwater or brine chemistry.

1.2 Repository Conditions

Conditions used for the repository in which the packages are located have been modeled after existing environmental impact statements (DOE, 1980) and draft repository reference conditions documents. Currently a hydrostatic pressure of 850 psi and lithostatic rock pressure of 2500 psi is being used for the salt repository. A hydrostatic pressure of 450 psi is being assumed for hardrock repositories (basalt, granite or tuff). The model is equipped to fit repository temperature versus time to a log function. Temperature versus time from the U.S. commercial wastes (DOE, 1980) have been used in the past. Current investigations have been based on constant temperature versus time for two parametric extremes: $250^{\circ}C$ and $165^{\circ}C$. These temperatures and pressures are assumed to exist at the boundary of the package backfill and the host rock.

2. MODEL DESCRIPTION

This section describes the various submodels in the performance assessment model and the manner in which they are interfaced. The submodels have been programed in the BARIER code as subroutines to facilitate future upgrading of individual submodels.

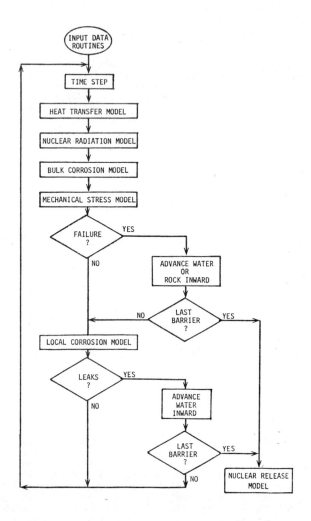

Figure 2.   Schematic of Performance Model.

2.1 Overview

Figure 2 is a simplified diagram of the performance model. The package degradation portion of the model is driven by a basic time step routine. At each increment of time a heat transfer model calculates the package temperature profile. A nuclear radiation model then calculates the radiation level at the surface of all wetted barriers. Bulk corrosion of all wetted barriers is then assessed for that time step in the corrosion routine. Mechanical stresses are then checked for all barriers subjected to lithostatic or hydrostatic forces. Barrier failure is then tested. In the event of failure from pressure the water and/or rock forces are then advanced inward. The local corrosion routine then is used to assess the possibility of water advancing inward due to localized cracks, or pits. Time is advanced until the last barrier leaks and leaching is initiated. A nuclide release model is then used to assess the release versus time profile.

Figure 3 is a schematic of a barrier element in the model. A one-half longitudinal cross section through the cylindrical element is shown. The element may consist of some combination of all the possible components shown. The solid wall may be composed of up to two materials: a no-strength cladding (material #2) and a structural base material (material #1). The presence or absence of the individual components of one element is conveyed by setting the diameter boundaries of each component. Each of these barrier elements is treated as a unit in the time sequenced degradation model described above.

2.2 Temperature Model

The temperature model is a standard conduction, radiation calculation for concentric cylinders in steady-state heat transfer. It is assumed that temperature profiles reach steady-state for each time step. This is reasonable since the package is small and the time steps are generally several years.

2.3 Nuclear Radiation Model

The radiation model calculates gamma fields as R/hr at the outer surface of wetted barriers. The model contains four types of photon sources: High Level Waste Glass from Commercial Reprocessing, one spent PWR fuel bundle, tightly packed rods for three spent PWR fuel bundles, and High Level Waste Glass from government activities. The sources are characterized by energy groups as a function of time stored as tables in the data base.

The calculation of the gamma flux at a particular location is given by:

$$\text{Flux} = B \times S_v \times \frac{R_o^2}{a + x} \times F(b_2)$$

where

B = buildup factor (dimensionless)

S_v = source intensity (Photons/cm^3/sec)

R^o = radius of the cylinder, (cm)

a = distance to the point of interest from the edge of the cylinder, (cm)

z = self-shielding distance factor, (cm)

b_2 = number of mean free paths to the point of the interest, (dimensionless)

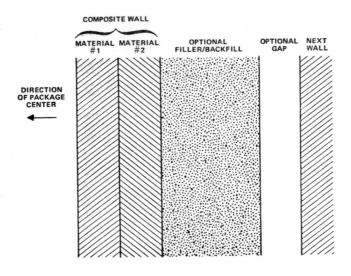

Figure 3. Barrier Element in the Performance Model.

F = function defined by

$$F(b_2) = \int_0^{\pi/2} e^{-b_2 \sec\theta} \, d\theta$$

2.4   Corrosion Models

Separate corrosion rates for bulk and local corrosion are calculated in the performance model. Both corrosion use a similar form but with different numerical parameters. Corrosion rate, R_c, is given by

$$R_c = K(T) \, M_r(\phi)$$

where

R_c = rate of penetration as inches/year

$K(T)$ = rate constant as function of temperature, T

M_r = multiplier as a function of radiation dose, ϕ

Currently the function K is

$$K = K_1 \qquad T \leq 100°C$$

$$K = K_2 \qquad T > 100°C$$

where K_1, K_2 are constants. M_r is correlated as

$$M_r = A \log_{10}\phi + B$$

where A,B are constants.

Values of K_1, K_2 have been developed from empirical data in the literature for both oxic and anoxic conditions for brine and hardrock groundwater. In the local corrosion model the mechanism with the largest values for K_1, K_2 are picked as the dominant mechanisms.

In the progression of the package toward failure the wall is considered to be weakened only by bulk corrosion but leaks can develop from local corrosion. Thus it is possible for water to be present inside a barrier which has not yet been crushed due to rock creep.

2.5   Mechanical Stress Model

The package is seen as a series of cylinders, each composed of a multi-element stress member composed of parts depicted in Figure 3 (Section 2.1). Each element is modeled as a backfill or filler (if preset) composed of a packed particulate material resting against an elastic wall. At each time step it is assumed that stress-strain is in equilibrium and pressure profiles are calculated. The backfill is treated as a stress member with compressive strength. The wall is said to fail when plastic deformation is triggered. At this time all external forces (and water) are transferred to the next inner barrier.

The stress model is simplified in its geometric treatment and assumes that forces are non-directional and uniform along the cylinder length. The ends of the cylinder are assumed to be under a constant stress and free to move. Effects of non-uniform shearing or localized forces are not accounted for. Such sophistication would be a worthwhile improvement but was outside the scope of the work.

2.6   Nuclide Release Model

The nuclide release model has the following key assumptions:

● The backfill is intact and homogeneous

● The nuclide concentration in the repository = 0

● Nuclide sorption is represented by solid-liquid equilibrium

● Movement of water through the backfill is by diffusion only

Failed barrier walls are considered to be porous media with flow resistance but no capacitance (sorption).

The equation describing is the release model is

$$\frac{\partial c}{\partial t} = \left[\frac{D}{1 + \frac{K_d \rho}{\varepsilon}} \right] \frac{\partial^2 c}{\partial x^2} - \frac{\lambda}{\left[1 + \frac{K_d \rho}{\varepsilon} \right]} c$$

where the backfill is considered to be a slab of thickness X (unrolled cylinder) and

C = concentration in the backfill solution

D = effective diffusivity

K_d = sorption retardation factor

ρ = bulk density of backfill

ε = porosity of backfill

λ = radioactive decay constant

x = 0 at the canister, $x = \ell$ at the geology

The boundary conditions are:

Before canister inventory depletion

$$\frac{\partial c}{\partial x}\Big|_{x=0} = h_o(C - C_o) \qquad\qquad \text{for all time}$$

where C_0 = solubility of the nuclide in water

$C = 0$, $t = 0$ for all x

$$\left.\frac{\partial c}{\partial x}\right|_{x=\ell}, \ t = 0 \hspace{4cm} \text{for all time}$$

After canister inventory depletion

$C = 0$, $x = 0$,   all time

$$\left.\frac{\partial c}{\partial x}\right|_{x=\ell} = - h_\ell(c) \hspace{3cm} \text{all time}$$

Fourier transform solutions to these equations have been developed and programmed in the backfill release model. Release rates are calculated based on the concentration profiles.

3.   RESULTS

Table I shows results for some packages in a salt repository. The same cases are shown in Table II for a hardrock repository. The model is currently not sufficiently refined to make significant distinctions between different types of hardrock.

The stabilizer is the term used for material in the spent fuel canister in the space between fuel rods and the canister inside diameter. Packages are crushable or rigid depending on whether or not there are any voids within the canister contents.

Two types of package concepts are shown in the results. The "A" concept has a crushable canister and relies on a heavy iron (13 inches thick) overpack to protect the package against cracking by rock or hydrostatic pressure. The "B" concept relies on the rigidity of the waste form and the corrosion resistance of a titanium alloy (TICODE-12). In the case of spent fuel the canister is made rigid by using lead cast in all void spaces around the fuel. In the case of commercial waste the canister is rigid if filled completely. A and B concepts also differ in backfill thickness; A uses 52 inches of backfill and B uses 13 inches.

In the salt repository the B package performs better than A since the combination of high corrosion resistance and rigid waste form work together to prevent early crushing and delay the contact of waste with water. The A package does not do as well because the iron corrodes too rapidly and eventually the package crushes from lithostatic pressure. The effect on breakthrough time from package A (52 inch backfill) to package B (13 inch backfill) is insignificant compared to the difference between the two nuclides with different k_d factors (U-238, k_d = 1400 and Tc-99, k_d = 1). Sorption retardation factor has orders of magnitude effect compared to backfill thickness. The decreased life from spent fuel to HLW is due mainly to the much higher heat loading in the HLW causing very high temperatures which accelerate corrosion and weaken the package materials.

The trends in hardrock are essentially the same but package A life is generally longer since the influence of lithostatic pressure is not present. Both package types last longer because corrosion rates are less in hardrock than in salt.

Table I. Performance of Waste Packages in Salt Repository.

Repository Temperature = 165°C Hydrostatic Pressure = 850 psi
Lithostatic Pressure = 2500 psi Anoxic Conditions

PACKAGE TYPE	WASTE FORM	HEAT LOADING (WATTS)	CRUSHABLE OR RIGID	PACKAGE DESCRIPTION BARRIER ELEMENT	MATERIAL	THICKNESS (IN)	TIME TO WASTE-WATER CONTACT (YRS)	U-238 RELEASE BREAKTHROUGH TIME (YRS)	STEADY-STATE RELEASE BEGIN (YRS)	END (YRS)	RATE (Ci/YR)	Tc-99 RELEASE BREAKTHROUGH TIME (YRS)	STEADY-STATE RELEASE BEGIN (YRS)	END (YRS)	RATE (Ci/YR)
A	Spent Fuel (PWR) 1 bundle	550	Crushable	Backfill	Sand/Bentonite	52.00	310	8.6×10^5	8.6×10^6	2.2×10^{10}	1.6×10^{-9}	1080	8000	1.9×10^6	3.9×10^{-8}
				Overpack	Iron	13.00									
				Canister	Mild Steel	0.25									
				Stabilizer	Air	--									
B	Spent Fuel (PWR) 1 bundle	550	Rigid	Backfill	Sand/Bentonite	13.00	1400	3.4×10^4	3.3×10^5	1.8×10^{10}	1.3×10^{-12}	1420	1690	1.8×10^6	6.7×10^{-8}
				Overpack	Ticode-12	0.25									
				Canister	Mild Steel	0.25									
				Stabilizer	Cast Lead	--									
A	Commercial HLW Glass	2160	Crushable	Backfill	Sand/Bentonite	52.00	200	8.7×10^5	8.7×10^6	4.7×10^9	4.8×10^{-13}	980	8000	2.5×10^6	2.4×10^{-8}
				Overpack	Iron	13.00									
				Canister	Mild Steel	0.25									
B	Commercial HLW Glass	2160	Rigid	Backfill	Sand/Bentonite	13.00	530	3.5×10^4	3.5×10^5	3.1×10^9	8.1×10^{-13}	560	840	2.3×10^6	4.0×10^{-8}
				Overpack	Ticode-12	0.25									
				Canister	Mild Steel	0.25									

Table II. Performance of Waste Packages in Hardrock Repository.

Repository Temperature = 165°C   Anoxic Conditions
Hydrostatic Pressure = 450 psi

PACKAGE TYPE	WASTE FORM	HEAT LOADING (WATTS)	CRUSHABLE OR RIGID	PACKAGE DESCRIPTION BARRIER ELEMENT	MATERIAL	THICKNESS (IN)	TIME TO WASTE-WATER CONTACT (YRS)	U-238 RELEASE BREAKTHROUGH TIME (YRS)	STEADY-STATE RELEASE BEGIN (YRS)	END (YRS)	RATE (CI/YR)	Tc-99 RELEASE BREAKTHROUGH TIME (YRS)	STEADY-STATE RELEASE BEGIN (YRS)	END (YRS)	RATE (CI/YR)
A	Spent Fuel (PWR) 1 bundle	550	Crushable	Backfill	Sand/Bentonite	52.00	1670	1.2×10^6	1.2×10^7	2.2×10^{10}	7.8×10^{-13}	2450	9.4×10^3	1.9×10^6	3.9×10^{-8}
				Overpack	Iron	13.00									
				Canister	Mild Steel	0.25									
				Stabilizer	Air	--									
B	Spent Fuel (PWR) 1 bundle	550	Rigid	Backfill	Sand/Bentonite	13.00	1890	4850	4.7×10^5	1.8×10^{10}	1.3×10^{-12}	1900	2.2×10^3	1.8×10^6	6.7×10^{-8}
				Overpack	Ticode-12	0.25									
				Canister	Mild Steel	0.25									
				Stabilizer	Cast Lead	--									
A	Commercial HLW Glass	2160	Crushable	Backfill	Sand/Bentonite	52.00	1630	1.2×10^6	1.2×10^7	4.6×10^9	4.8×10^{-13}	2410	9430	2.5×10^6	2.4×10^{-8}
				Overpack	Iron	13.00									
				Canister	Mild Steel	0.25									
B	Commercial HLW Glass	2160	Rigid	Backfill	Sand/Bentonite	13.00	530	5×10^4	4.9×10^5	3.1×10^9	8.1×10^{-13}	560	840	2.3×10^6	4.0×10^{-8}
				Overpack	Ticode-12	0.25									
				Canister	Mild Steel	0.25									

4.   CONCLUSIONS

Only a sampling of results has been presented in this paper but most other results are similar. In general, the results indicate the construction of packages which will last several hundreds of years can be achieved. However, such conclusions must be taken with caution since the data are limited and the models used are relatively simple. In all cases an effort was made to bias uncertainties in a conservative direction but further work is needed to confirm this. Better corrosion and materials properties data would greatly facilitate more accurate predicting of package performance.

5.   REFERENCES

U. S. Department of Energy, Final Environmental Impact Statement, Management of Commercially Generated Radioactive Waste, DOE/EIS-0046F, October 1980.

DISCUSSION

J.D. BLACIC, United States

How do you rationalize as conservative your assumption that a hard rock repository waste package will only ever see a hydrostatic stress ?

D.H. LESTER, United States

The model does not account for non-homogeneous or discontinuous stress such as motion along a fault line. It may indeed not be conservative if such events occur. We have dealt only with sure, certain processes. We could account for this better by assuming a hard rock lithostatic pressure equal to say the overburden pressure.

P. UERPMANN, Federal Republic of Germany

I would like to comment on the use of labeling brines with letters A, B, etc. which already have a specific definition in some countries. A brine labeled "A" is often interpreted as a solution which is saturated with both NaCl and $MgCl_2 \cdot 6H_2O$; a brine labeled "B" interpreted as a solution in equilibrium with NaCl and KCl. In order to avoid confusion it seems reasonably to label the reference brine "WIPP brine B" with another name, especially since your reference brine is not even saturated with NaCl.

D.H. LESTER, United States

This sands like a good suggestion for the WIPP project. Our intention is to use a dissolution intrusion brine as an reference solution. In most cases the corrosion data we have does not support any attempt to make meaningful distribution between different brine chemistries.

C. McCOMBIE, Switzerland

In the nuclide release model what kind of sorption behaviour was assumed for the backfill ? For example, is potential reduction in sorption due to the presence of corrosion products (cf. Belgian paper) taken into account ?

D.H. LESTER, United States

We used very low k_d's for the nuclides in question. Our results indicate that the diffusion barrier behavior of the backfill is much more useful than the sorption delay anyway.

A related point is the role the backfill could have in limiting corrosion rates. We have have used immersion type corrosion data but a good backfill diffusion barrier would likely limit the corrosion rates to a diffusion controlled regime.

DIFFUSION MODELLING OF THE BOREHOLE EMPLACEMENT
CONCEPT FOR A NUCLEAR FUEL WASTE DISPOSAL VAULT

S. Cheung, G.W. Bird and C.B. So
Atomic Energy of Canada Research Company
Whiteshell Nuclear Research Establishment,
Pinawa, Manitoba ROE 1L0

ABSTRACT

An analytical solution has been developed for ionic diffusion in two dimensions from a cylinder-within-a-cylinder. The resulting equations have been used to model the borehole emplacement concept for a nuclear fuel waste disposal vault. The calculations show that the diffusive mass flux at the radial buffer-rock interface is approximately 10% higher than that calculated for diffusion in the radial direction using a one-dimensional infinite cylinder model. Similarly, it is shown that increasing the radial thickness of the buffer material beyond 0.45 m and the vertical thickness above the waste package beyond 1 m will have little additional effect in reducing the radionuclide flux from a failed waste container of 0.35 m diameter and 3 m long.

* Issued as AECL-7351

1.   INTRODUCTION

The Canadian research program for nuclear fuel waste management is based on the concept that the waste can be effectively isolated by deep underground disposal in stable geological formations.

At present, disposal in vaults mined into crystalline hard rock formations, such as granite batholiths of the Precambrian shield, is considered to be the most favorable option [1]. In such a vault the only potentially significant mechanism of radionuclide release is for circulating groundwater to penetrate to the waste, leach out the radionuclides and carry them back to the surface. A number of protective barriers will be used to isolate the waste and minimize the probability of any significant escape. These are the waste form itself, the waste container, the buffer material surrounding the container, backfill and sealing materials which fill the remainder of the vault, the massive geological formation, and finally dispersion, dilution and retention in the biosphere.

In this paper we discuss mass transport through the buffer material emplaced between the waste form and the host rock. Desirable buffer properties have been discussed elsewhere [2,3] and two alternative emplacement methods have been suggested [3], as illustrated schematically in Figure 1.

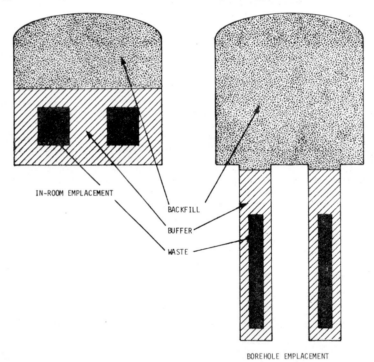

IN-ROOM EMPLACEMENT

BACKFILL

BUFFER

WASTE

BOREHOLE EMPLACEMENT

Figure 1   Schematic illustrations of two possible methods of waste package emplacement. Not to scale.

A number of possible buffer materials have been suggested, and one composed of bentonite (montmorillonite), montmorillonite/illite mixed layer clays, illite clays, or some mixture of these minerals, may be used. The montmorillonite or mixed layer clays are particularly attractive because their inherent swelling properties give such low hydraulic conductivities that diffusion is the primary mechanism of mass transport. Here we have considered only the borehole emplacement option (Figure 1b) and have developed a model for mass transport from the waste to the surrounding rock or backfill. Calculations based on this model are presented to show how effectively the buffer will retard the migration of radionuclides.

2.   MASS TRANSPORT THROUGH THE BUFFER

An analytical two-dimensional mass-transport model has been developed that describes the radial (horizontal) and axial (vertical) distributions of radionuclide concentration in the space between the waste form and the confining rock mass and backfill material. The assumptions used and the mathematical development of the equations are detailed in Appendix A.

The resulting equations [(3) and (6) of Appendix A] are:

$$C_1(r,Z) = A(r,Z) + \sum_{i=1}^{m-1} B_i(r,Z) \, C_a(r_i,Z_1) \qquad (1)$$

and

$$C_2(r,Z) = C(r,Z) + \sum_{j=1}^{\ell-1} D_j(r,Z) \, C_b(R_1,Z_j) \qquad (2)$$

for the radionuclide concentration distributions between the waste form and the backfill, and between the waste form and the rock, respectively.

The mass flux (concentration gradient) can be obtained by partial differentiation of equations (1) and (2) with respect to r and Z.

Equations (1) and (2) and their partial derivatives were solved for several buffer dimensions and two different coefficients of buffer diffusivity. The computations were done for a vertical quarter-plane of a waste form emplaced in a borehole, as illustrated by the shaded segment of Figure 2.

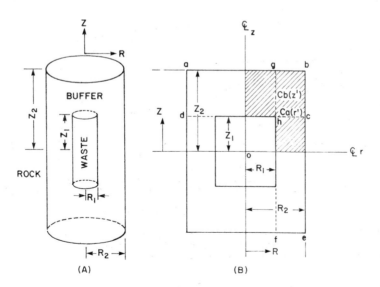

(A)   (B)

Figure 2   Geometric configuration of the waste form for the borehole-emplacement method. In this paper a waste container is assumed to be 0.35 m diameter by 3.0 m long, i.e. $R_1 = 0.175$ m, $Z_1 = 1.5$ m. An equivalent thickness of buffer is assumed above and below the waste container. The calculations were carried out for the shaded area of Figure 2(B).

In this study, we are computing the flux of radionuclides at any point along the buffer-rock or buffer-backfill interfaces as well as the integrated total flux across these interfaces.

To solve equations (1) and (2), the following boundary conditions are assumed:

1. The concentration of diffusing species at the buffer-container interface is considered to be constant. This implies an infinite source.

2. The concentrations of diffusing species at the buffer-rock and buffer-backfill interfaces are taken to be zero. This implies that when the radionuclides arrive at these interfaces, they are flushed away immediately.

For this analysis, the length ($2Z_1$) and the diameter ($2R_1$) of the waste container are taken to be 3 m and 0.35 m respectively.

3.   RESULTS AND DISCUSSION

Figure 3 shows the concentration contours of a diffusing species for the one-quarter vertical plane of the waste container and buffer illustrated in Figure 2. The contours are expressed in per cent of the concentration at the container-buffer interface (100%). The radial scale has been expanded by a factor of 10 to better illustrate the contour lines.

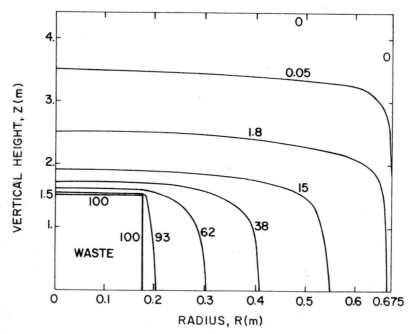

Figure 3   Concentration contours for a quarter-plane of a cylinder-within-a-cylinder (shaded area of Figure 2(B)). The contours are expressed as % of that at the container-buffer interface.

The concentration of diffusing species decreases from the container-buffer interface to the buffer-backfill, or buffer-rock interface, as expected. In the radial direction, below the height where end effects are noted (approximately 0.5 m below the top of the container, see Figure 3), the concentration is almost a linear function of radial distance (up to 0.55 m). The concentration

in the axial direction decreases abruptly from the end surface (100%) to
approximately 2% at a distance of 1 m above the container.

Figure 4 shows the effect of the radial thickness of the buffer (R_2-R_1)
on the concentration gradient (dC/dR) at the buffer-rock interface, as a
function of Z. The concentration gradient is expressed as a fraction of the
concentration at the container-buffer interface per metre. This concentration
gradient is directly related to mass flux. For comparison, the concentration
gradients at the buffer-rock interface calculated using a one-dimensional
infinite cylinder model are shown in Figure 4 as dotted lines. In all cases the
concentration gradient remains almost the same from the container centreline
(Z=0) to a vertical height (Z) of 1 m (0.5 m below the top of the container) for
both the one- and two-dimensional models. The concentration gradient for the
two-dimensional model then decreases abruptly, and at Z=3m is insignificant.
The concentration gradient decreases as expected, with increasing radial
thickness of the buffer.

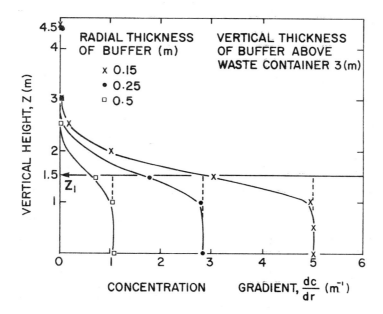

Figure 4   The effect of increasing the radial thickness of the buffer (R_2-R_1) on
the radial concentration gradient at the buffer-rock interface. Point
Z_1 shows the half length of the waste container. The vertical thickness of the buffer was 3.0 m. The dotted lines are the concentration
gradients calculated for a one-dimensional diffusion model.

Figure 5 shows the effect of the vertical thickness of the buffer
(Z_2-Z_1) above the container on the concentration gradient (dC/dR) at the
buffer-rock interface as a function of Z, for a radial buffer thickness of 0.25
m. It is apparent that for a vertical thickness of buffer above the waste
container of 0.5 m or greater, the concentration gradients at the buffer-rock
interface are practically unaffected.

Figure 6 shows the effect of the vertical thickness (Z_2-Z_1) of the
buffer above the container on the vertical concentration gradient as a function
of radial distance (R), for a radial buffer thickness of 0.25 m. It can be seen
that the concentration gradient decreases from the centreline (R=0) to the
buffer-rock interface in all cases. If the vertical thickness of the buffer is
increased from 0.5 m to 1.0 m (from Z_2=2 m to Z_2=2.5 m) the vertical gradient at
1 m above the container is approximately 40 times less than that at 0.5 m.

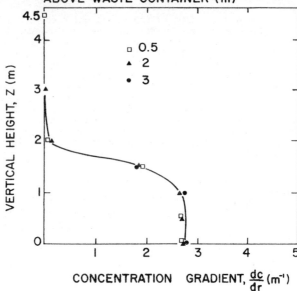

RADIAL THICKNESS OF BUFFER 0.25 m
VERTICAL THICKNESS OF BUFFER
ABOVE WASTE CONTAINER (m)

Figure 5   Effect of the vertical thickness of the buffer (Z_2-Z_1) on the radial
concentration gradient profile at the buffer-rock interface, for a
radial buffer thickness of 0.25 m.

In Figure 7, the total radial mass transport through the buffer given
by the one-dimensional infinite-cylinder diffusion model [4] is compared with
that given by our two-dimensional model, for similar conditions. We have used
uranium in our example and assumed a solubility of 10^{-10} molar as the concentration at the container-buffer interface (a reasonable concentration for reducing
groundwater conditions [5]). The diffusion coefficient of dissolved uranium in
soil was assumed to be 2×10^{-10} m^2/s. This value is obtained by multiplying the
diffusivity of ions in water ($\simeq 2 \times 10^{-9}$ m^2/s) by an assumed effective buffer
porosity of 0.1 [6]. It is recognised that the diffusivity of uranium should be
less than that of simple ions in water.

The total radial flux of uranium is obtained by integrating the flux
over the area for ionic diffusion. For the one-dimensional model the area is:
$(2\pi R_2)$ $(2Z_1)$. For the two-dimensional model the length of the buffer is greater
than the container length and diffusion occurs in both axial and radial
directions.   Therefore, the effective area is: $(2\pi R_2)$ $(2Z_2)$.   For the
two-dimensional model, the total radial flux of diffusing ions is approximately
10% more than for the one-dimensional model, at a buffer thickness of 0.15 m.
Using the two-dimensional model, the total flux is slightly less,, assuming the
same area $(4\pi RZ_1)$ as for the one-dimensional model.

These results can be explained using Figure 4.   The concentration
gradients plotted on Figure 4 show that there is a small radial flux contribution through the buffer beyond the length (Z_1) of the container. The integration for the two-dimensional model includes this contribution beyond the end of
the container.

Similarly, from Figure 4 it can be seen that integrating the concentration gradients for the one- and two-dimensional models only over the length
of the container (Z_1) will result in a lower total flux for the two-dimensional
model.   This occurs because some of the diffusing ions from the radial surface
of the container move beyond the length of the container.

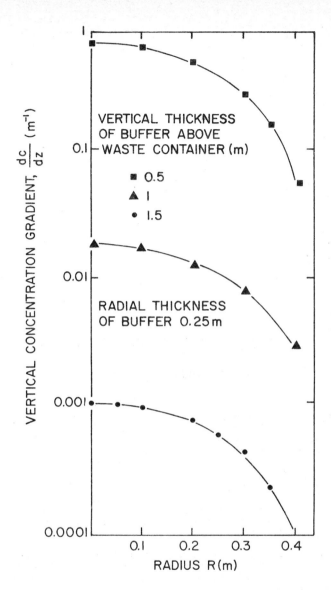

Figure 6   Effect of vertical thickness of the buffer $(Z_2 - Z_1)$ on the vertical concentration gradient profile, for a radial thickness of 0.25 m.

It can be seen from Figure 7 that the total radial flux decreases as the radial buffer thickness increases, and the total flux tends to level off at approximately 0.45 m.   This is due to the dominant effect of the decreasing concentration gradient rather than the increasing surface area with increasing radial thickness of the buffer.

The total axial fluxes at the buffer-backfill and buffer-rock interfaces are estimated to be 4 per cent of the total radial flux when the vertical buffer thickness above the waste container is 0.5 m (Z_2=2 m).   Increasing this thickness to 1 m (Z_2=2.5 m) decreases the total axial fluxes to 0.1% of the total radial flux.

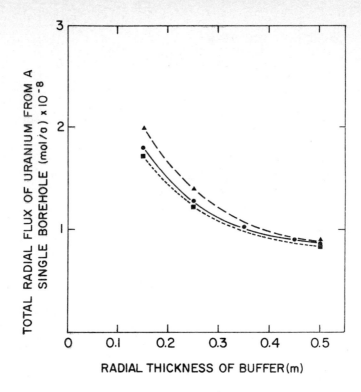

Figure 7 Comparison of the total radial flux derived from one-dimensional and
         two-dimensional diffusion models. ▲————▲ Two-dimensional diffusion
         model, using the total effective area for diffusion. This calculation
         included the contribution from the end caps. ●————● One-dimensional
         diffusion in the radial direction only. ■————■ Two-dimensional
         diffusion for the same area of buffer as used in the one-dimensional
         model.

     The retention time for radionuclides in the buffer can be calculated if
the diffusion coefficients and retardation factors in the buffer material are
known. We have calculated retention times for three buffer thicknesses using
the approximate equation of Neretnieks [7].

     For these calculations our assumption was a radionuclide diffusion co-
efficient of 2×10^{-10} m^2/s. The results are given in Table 1. They show that
the retardation factor for the buffer is an important variable in determining
the time to breakthrough for the radionuclides. Assuming retardation factors
similar to those in Table 1 [8], most of the short-lived fission products, such
as Sr^{90} and Cs^{137}, will completely decay before breakthrough. A tenfold decrease in the effective porosity of the buffer to 0.01 results in a tenfold
increase in retention time. An effective porosity of 0.01 seems reasonable for
a high-density bentonite-based buffer [6].

     The work described here results from the initial part of a continuing
program dealing with the mathematical modelling and experimental investigation
of mass-transport phenomena in the buffer and backfill regions of a nuclear
waste-disposal vault.   The modelling studies will be extended to include
analysis of the in-room disposal option and the effect of different boundary
conditions, such as an impermeable rock along the borehole wall, on the current
model.   An experimental program is planned to determine whether the ionic
diffusion coefficients in possible buffer materials are similar to those used in
these calculations.

TABLE 1   Retention Time in Buffer for Various Radionuclides Calculated with the Equation of Neretnieks [7] for a Buffer Density of 2.1 g cm^{-3}.

Nuclide	Half-life (years)	Retardation factor (from reference [8])	Retention time in Buffer (years) radial thickness in (metres)			
			0.125	0.25	0.35	0.45
Sr^{90}	27.7	600	156	625	1227	2029
Tc^{99}	2.1x10^5	1	0.3	1.0	2.0	3.4
I^{129}	1.7x10^7	1	0.3	1.0	2.0	3.4
Cs^{135}	3.0					
		400	104	417	818	1352
Cs^{137}	30.0					
Ra^{226}	1602	800	208	834	1636	2704
Th^{229}	7340	> 1000	260	1000	2000	3400
Np^{237}	2.1x10^6	200	52	208	409	676
Pu^{239}	2.4x10^4	1200	313	1253	2455	4058
Am^{241}	458					
		> 4000	1040	4170	8180	13520
Am^{243}	7370					

4.   SUMMARY AND CONCLUSIONS

An analytical solution has been developed for radionuclide diffusion in two dimensions from a cylinder-within-a-cylinder. The equations can be used to calculate the concentration of diffusing species and the mass fluxes (concentration gradients) as a function of the vertical and horizontal thicknesses of buffer material for the borehole-emplacement concept of nuclear fuel waste disposal. The calculations show that the diffusive flux at the outer edge of the buffer is approximately 10 per cent higher than that obtained assuming diffusion in the radial direction only. Increasing the radial thickness of the buffer beyond 0.45 m and increasing the vertical thickness above the waste container beyond 1 m will have little additional effect in reducing radionuclide flux from a failed waste container of 0.35 m diameter and 3 m long.

Our calculations show that if the buffer thickness above the waste container is greater than 1 m, a one-dimensional diffusion model will be sufficiently accurate for practical purposes.

Calculation of retention times for various radionuclides in the buffer shows that the buffer material can play an important role in retarding migration from the waste container, particularly for those radionuclides with short half-lives, or high retardation factors.

An experimental program is required to confirm the diffusion coefficients and retardation factors assumed for the buffer material in these calculations.

ACKNOWLEDGEMENTS

Reviews by J.R. Dryden, T.W. Melnyk, G.L. Rigby and B.J.S. Wilkins substantially improved this manuscript.

REFERENCES

1. Boulton, J, (Editor) "Management of Radioactive Fuel Wastes: The Canadian Disposal Program", Atomic Energy of Canada Limited Report, AECL-6314 (1978).

2. Pusch, R, "Highly Compacted Sodium Bentonite for Isolating Rock-Deposited Radioactive Waste Products", Nuclear Technology 45, 153 (1979).

3. Bird, G.W, "Selection and Evaluation of Buffer and Backfill Materials for use in a Nuclear Fuel Waste Disposal Vault: The Canadian Program", Proceedings of Workshop on Research and Development Relating to Backfill for Underground Nuclear Waste Management. U.S. National Bureau of Standards, Gaitherburg Maryland (1981, April) in preparation.

4. Carslaw, H.S., and Jaeger, J.C., "Conduction of Heat in Solids", Oxford University Press 1959.

5. Goodwin, B.W., "Maximum Total Uranium Solubility Under Conditions Expected in a Nuclear Waste Vault", Atomic Energy of Canada Limited Technical Record TR-29, September 1980, unpublished work available from Atomic Energy of Canada Research Company, Chalk River, Ontario, KOJ 1JO.

6. Nowak, E.J., "The Backfill Barrier as a Component in a Multiple Barrier Nuclear Waste Isolation System", Sandia National Laboratories Publication, SAND 79-1109 (1980).

7. Neretnieks, I., "Transport of Oxidants and Radionuclides Through a Clay Barrier", KBS-Teknisk Rapport 79 (1978).

8. Wuschke, D.M., Mehta, K.K., Dormuth, K.W., Andres, T., Sherman, G.R., Rosinger, E.L.J., Goodwin, B.W., Reid, J.A.K., and Lyon, R.B., "Environmental and Safety Assessment Studies for Nuclear Fuel Waste Management, Volume 3, Post Closure Assessment", Atomic Energy of Canada Limited Technical Record TR-127, Volume 3 (in preparation), to be available from Atomic Energy of Canada Research Company, Chalk River, Ontario, KOJ 1JO.

APPENDIX A

An analytical solution is developed for a two-dimensional mass-transport equation that describes the axial and radial mass flux in the space between two co-axial cylinders (see Figure 2). The following assumptions are made:

1. The waste container is placed symmetrically in the buffer material. The waste container is assumed to have failed and to be uniformly permeable to radionuclides over its entire surface.

2. Mass transport is by diffusion only.

3. The medium is isotropic and of constant diffusivity.

4. There are no chemical reactions that retard migration.

5. Steady-state transport is assumed.

6. Radionuclide concentrations at the buffer-waste interface (the surface of the inner cylinder) and at the buffer-rock and buffer-backfill interfaces (the surface of the outer cylinder) are constant.

The solution is obtained by combining the analytical solutions derived for sections abcd and befg of Figure 2. The analytical solution for section abcd under steady-state conditions is obtained from Carslaw and Jaeger [4] as:

$$C_1(r_1,z) \sim \frac{2}{R_2^2} \sum_{n=1}^{\infty} \frac{J_o(r\alpha_n) \sinh[(Z_2-z)\alpha_n]}{J_1^2(R_2\alpha_n) \sinh[(Z_2-Z_1)\alpha_n]} \times \left[\int_o^{R_1} r'C_oJ_o(r'\alpha_n)dr' \right]$$

$$+ \sum_{i=0}^{m} r_i W(r_i) J_o(r_i \alpha_n) C_a(r_i, Z_1) \Bigg] \tag{1}$$

where C_1 = the radionuclide concentration at (r,z) of section abcd minus the concentration at the buffer-rock interface,

$\quad C_o$ = the constant radionuclide concentration at the buffer-waste interface minus the concentration at the buffer-rock interface,

$\quad J_o, J_1$ are Bessel functions of zero and first order, respectively,

$\quad \alpha_n$ is the nth root of $J_o(R_2 \alpha_n) = 0$,

$C_a(r_i, Z_1) = C_1(r_i, Z_1)$ $\qquad\qquad R_1 \le r_i \le R_2$.

The following approximation has been used in the second term (in square brackets) of equation (1):

$$\sum_{i=0}^{m} r_i W(r_i) J_o(r_i \alpha_n) C_a(r_i, Z_1) \approxeq \int_{R_1}^{R_2} r' C_a(r', Z_1) J_o(r' \alpha_n) dr' \tag{2}$$

where the segment (R_1, R_2) is represented by m segments and $W(r_i)$ is the weight function of the numerical integration scheme. In this paper Simpson's Rule is used.

Using the facts that

$$C_a(R_1, Z_1) = C_o$$

and

$$C_a(R_2, Z_1) = 0$$

we can express C_1 as

$$C_1(r,Z) = A(r,Z) + \sum_{i=1}^{m-1} B_i(r,Z) C_a(r_i, Z_1) \tag{3}$$

where $\quad A(r,z) = \dfrac{2}{R_2^2} C_o \sum_{n=1}^{\infty} \dfrac{J_o(r\alpha_n)\, \sinh[(Z_2-z)\alpha_n]}{J_1^2(R_2\alpha_n)\, \sinh[(Z_2-Z_1)\alpha_n]}$

$$\times \left[\int_0^{R_1} r' J_o(r'\alpha_n) dr' + W(r_o) R_1 J_o(R_1\alpha_n) \right]$$

$W(r_o) = \dfrac{R_2 - R_1}{6m}$

and

$B_i(r,z) = \dfrac{2r_i W(r_i)}{R_2^2} \sum_{n=1}^{\infty} J_o(r_i\alpha_n) \dfrac{J_o(r\alpha_n)\, \sinh[(Z_2-z)\alpha_n]}{J_1^2(R_2\alpha_n)\, \sinh[(Z_2-Z_1)\alpha_n]}$

For section befg, the analytical solution is also obtained from Carslaw and Jaeger [4] as:

$$C_2(r,Z) \simeq \frac{2}{Z_2} \sum_{n=1}^{\infty} \frac{F_o\left(\frac{(2n-1)\pi r}{2Z_2} \; ; \; \frac{(2n-1)\pi R_2}{2Z_2}\right)}{F_o\left(\frac{(2n-1)\pi R_1}{2Z_2} \; ; \; \frac{(2n-1)\pi R_2}{2Z_2}\right)} \sin\left(\frac{(2n-1)\pi(z+Z_2)}{2Z_2}\right)$$

$$\times (-1)^{n+1} \times \left[\frac{2Z_2}{(2n-1)\pi} C_o \sin\left(\frac{(2n-1)\pi Z_1}{2Z_2}\right) \right.$$

$$\left. + \sum_{j=0}^{\ell} W(z_j) \cos\left(\frac{(2n-1)\pi z_j}{2Z_2}\right) C_b(R_1, Z_j) \right] \tag{4}$$

where $C_2(r,Z)$ = the radionuclide concentration at (r,Z) of section befg minus the concentration at the buffer-rock interface,

$F_o(x,y)$ = $I_o(x)K_o(y) - K_o(x)I_o(y)$,

I_o, K_o = The zero order of modified Bessel functions of the first and second kind, respectively,

$C_b(R_1, Z_j)$ = $C_2(R_1, Z_j)$ $\qquad\qquad Z_1 \leq Z_j \leq Z_2$.

The following approximation has been used in the second term (in square brackets) of equation (4):.

$$\sum_{j=o}^{\ell} W(z_j) \cos\left(\frac{(2n-1)\pi z_j}{2Z_2}\right) C_b(R_1, z_j) \simeq \int_{Z_1}^{Z_2} dz' \cos\left(\frac{(2n-1)\pi z'}{2Z_2}\right) C_b(R_1, Z') \tag{5}$$

where the segment (Z_1, Z_2) is represented by ℓ segments and $W(Z_j)$ is the weight function of the numerical integration scheme using Simpson's Rule.

Using the facts that

$C_b(R_1, Z_1) = C_o$

$C_b(R_1, Z_2) = 0$

we can express C_2 as

$$C_2(r,Z) = C(r,Z) + \sum_{j=1}^{\ell-1} D_j(r,Z) C_b(R_1, Z_j) \tag{6}$$

where $C(r,z)$ = $\frac{2C_o}{Z_2} \sum_{n=1}^{\infty} (-1)^{n+1} \frac{F_o\left(\frac{(2n-1)\pi r}{2Z_2} \; ; \; \frac{(2n-1)\pi R_2}{2Z_2}\right)}{F_o\left(\frac{(2n-1)\pi R_1}{2Z_2} \; ; \; \frac{(2n-1)\pi R_2}{2Z_2}\right)}$

$$\times \sin\left(\frac{(2n-1)\pi(z+Z_2)}{2Z_2}\right)$$

$$\times \left[\frac{2Z_2}{(2n-1)\pi} \sin\left(\frac{(2n-1)\pi Z_1}{2Z_2}\right) + \left(\frac{Z_2 - Z_1}{6\ell}\right) \cos\left(\frac{(2n-1)\pi Z_1}{2Z_2}\right) \right]$$

and

$$
D_j(r,z) = \frac{2W(z_j)}{Z_2} \sum_{n=1}^{\infty} (-1)^{n+1} \frac{F_o\left(\frac{(2n-1)\pi r}{2Z_2} \; ; \; \frac{(2n-1)\pi R_2}{2Z_2}\right)}{F_o\left(\frac{(2n-1)\pi R_1}{2Z_2} \; ; \; \frac{(2n-1)\pi R_2}{2Z_2}\right)}
$$

$$
x \sin\left(\frac{(2n-1)\pi(z+Z_2)}{2Z_2}\right) \quad x \cos\left(\frac{(2n-1)\pi z_j}{2Z_2}\right)
$$

To determine C_a and C_b, we note that the regions of validity of the solutions C_1 and C_2 in equations (3) and (6) overlap at cbgh. For the line segment gh, we have, from equation (3),

$$
C_b(R_1,z_j) = A(R_1,z_j) + \sum_{i=1}^{m-1} B_i(R_1,z_j)C_a(r_i,Z_1) \tag{7}
$$

For the line segment hc, we have, from equation (6),

$$
C_a(r_i,Z_1) = C(r_i,Z_1) + \sum_{j=1}^{\ell-1} D_j(r_i Z_1)C_b(R_1,z_j) \tag{8}
$$

Equations (7) and (8) form a set of linear simultaneous equations and can be solved readily for C_a and C_b.

Once C_a and C_b are determined, the concentration at any point within the buffer material can be obtained from equation (3) or (6). This solution is unique within the buffer material. Therefore, the concentration gradients (mass fluxes) are vigorously defined by the partial differentiation of equation (3) or (6) with respect to r and Z.

It has been verified that the results for the limiting case ($2Z_1 \to \infty$) of this two-dimensional model approach those calculated for diffusion in the radial direction using a one-dimensional infinite-cylinder model.

DISCUSSION

D.H. LESTER, United States

I noticed you have use a diffusion coefficient which is molecular diffusivity x porosity. We have used :

$$\frac{\text{diffusivity x porosity}}{\text{tortuosity}}$$

tortuosity = from 1 to 6.

S. CHEUNG, Canada

The diffusion coefficient we used is diffusivity x effective porosity. This diffusion coefficient is in the same form as yours. The effective porosity has taken into account the tortuosity which is considered to be 1 to give a conservative value.

M. MAKINO, Japan

How can we evaluate the transient effect, such as decay of radionuclides which have a relatively long half life, when the retention time as a buffer for some nuclides is rather long ? Do you intend to handle transient effects in your calculations in the future ?

S. CHEUNG, Canada

The analytical solutions we developed do not take into account radioactive decay. We expect that analytical solutions will not be found if decay is included. The effect of decay can be approximated in our equations by systematically adding decay terms (this would ignore decay chain effects). Alternatively a more general description will probably require numerical solution based on a finite difference/finite element approach.

P. UERPMANN, Federal Republic of Germany

What are the pH-Eh values you assumed for your solution in contact with the buffer material ?

S. CHEUNG, Canada

At neutral pH magnetite-hematite Eh condition /see reference (5)7.

Session 7

WORKING GROUPS DISCUSSIONS

WORKING GROUP ON GRANITE AND CRYSTALLINE HOST ROCK

Chairman : G.R. SIMMONS (Canada)
Rapporteur : P. JOHNSTON (NEA)

WORKING GROUP ON SALT FORMATIONS

Chairman : R.H. KOSTER (Federal Republic of Germany)
Rapporteur : H. BURKHOLDER (United States)

WORKING GROUP ON OTHER HOST MEDIA AND ENGINEERED BARRIERS

Chairman : R. HEREMANS (Belgium)
Rapporteurs : N. CHAPMAN (United Kingdom), F. GERA (Italy)

WORKING GROUP ON GRANITE AND CRYSTALLINE HOST ROCK

Chairman : G.R. SIMMONS, AECL, Canada
Rapporteur : P. JOHNSTON, NEA

The environmental and safety assessment of geologic disposal is the main objective of research programmes. These assessments include the vault, the geosphere and the biosphere ; all these components must be considered as parts of the isolation system, not separately. It was recognized in the discussions that investigations of near-field phenomena are only a part of the overall repository safety assessment, and that the importance of near-field phenomena must be seen in the context of a broader study. The emphasis that needs to be placed on investigations of near-field effects can be derived from sensitivity analyses of radionuclide releases. The presentation by Mr. Mano has given an example of such a sensitivity study, but more understanding will be needed before a realistic assessment of the importance of specific near-field effects can be made. At this stage it appeared that the near-field fundamental phenomena are not sufficiently well understood to allow detailed sensitivity analyses and the assigning of priorities to the different aspects of investigations reported at the meeting.

Chemical species in groundwater in granite were discussed both in contributed papers and in the working group. Knowledge of the chemical species of radionuclides in the region of a repository was seen as essential to understanding retention and migration of radionuclides, and to calculate the source term for safety assessments. The modelling of sorption and solubilities of radionuclides is however very difficult in the near-field where the chemical behavior may be strongly affected by thermal alterations of the host rock surfaces and by the canisters, buffer and backfill materials.

It was felt that in the near-field, the concept of a unique k_d value for each radionuclide would not be adequate for a realistic understanding of retention mechanisms. The migration of radionuclides can only be accurately modelled with an understanding of the thermo-dynamic mechanisms involved, and an accurate characterization of the geochemical environment. Radiolysis near the waste package may also have an important impact on the chemical species and oxidation states of radionuclides. In addition, corrosion products from the waste canister and package may influence the chemical species and the sorption capacity of buffer and host rock for radionuclides in the near-field.

Because of the difficulty of modelling radionuclide retention in a complicated environment, it was suggested by a number of participants that materials having an unpredictable behavior in the hydro-thermal environment of the near-field should be avoided in repository and waste package design. This may be achieved by controlling the materials used or by controlling the hydro-thermal environment to avoid material alterations.

The working group agreed that construction of a waste repository in granite is feasible with existing civil and mining engineering technology. The construction and operation of a repository would be similar to existing mining experience. The discussions concerned safety assessment, and specifically addressed slow crack growth and rock creep associated with excavation of cavities and heating from waste. These phenomena may affect rock mass strength and radionuclide migration through fissuring. Several contributed papers presented thermo-mechanical analyses of repositories, however none of these models included constitutive equations for creep. Very little experimental data is available to judge the importance of this phenomenum in hard rock.

It was suggested that there will be an envelope around a repository within which significant repository-induced modification of the host rock structure occurs, and beyond which the host rock is not significantly disturbed. Attempts should be made to quantify the extent of significant structural modification.

The development of dynamic response models and seismic criteria were discussed. Some work is underway in Canada, and studies are being done in Japan and the USA. These studies suggest that seismic events should not seriously affect the stability of underground repositories.

The desirability of international exchange of near-field computer programs and experimental results was discussed. The possibility of using established code exchange routes as in other areas of nuclear safety and modelling was mentioned. Some code comparisons already exist in the United States, Sweden and Canada, and the working group recognized that these activities may become more important as codes become more widely established.

The working group agreed that international co-operation is essential in the research directed at understanding near-field phenomena. This co-operation should be initiated at the engineering and scientific level, and may subsequently be formalized. Facilities currently available for in-situ research in granite include the Stripa mine in Sweden, the Climax and Colorado School of Mines facilities in the United States, and hard-rock facilities in Japan and the United Kingdom. Granite facilities will be constructed in the near future in Switzerland and Canada. The working group emphasized that an effort must be made to encourage co-ordination, and to ensure efficient use of limited resources.

WORKING GROUP ON SALT FORMATIONS

Chairman : R.H. KOSTER, Federal Republic of Germany
Rapporteur : H. BURKHOLDER, ONWI, USA

The working group discussed the objectives and the topics placed before it by the secretary of the meeting. The discussions and conclusions of the group regarding these items are summarized in the sections that follow.

IDENTIFICATION OF RESEARCH NEEDS AND PRIORITIES

The group considered the data base for near-field phenomena associated with certain waste forms and the host rock to be generally adequate, but they considered the data base for the canister, backfill, and seals to need improvement. The group felt that the waste form, canister, backfill,host rock, and surrounding geologic media should be treated as a multibarrier system. However, the group questioned the usefulness of canisters and backfills as barriers in the isolation system after complete closure of the repository. Specific areas where additional data seemed needed included the decomposition, leaching, and solubility of high-level wastes, spent fuel, and medium level wastes ; radiation effects on package components, and interstitial host rock fluids and thermochemical and thermomechanical effects on waste forms, containers, and impurities in the host rock.

The group generally considered that the effort needs to be greater in the near-field model integration area than in the new model development area. Specific areas where improved models were needed included the thermochemical interactions between the waste form, container, and host rock during the post-closure period and the thermo-hydrologic effects in the repository during the operational period.

The group considered the development of a near-field system performance model and the integration of that model with a far-field system performance model to be of paramount importance. The group did not believe that rational decisions about R & D priorities in the various near-field areas could be made without the development and use of such a model. The so-called "systems approach" appeared to be the only way to trade-off the importance of various near-field phenom-ena to overall disposal system performance and to evaluate the relative effectiveness of various disposal system components in providing that performance.

PROMOTION OF INTERNATIONAL COOPERATION

The group concluded that the OECD/NEA could best promote international cooperation by organizing additional future meetings concerning near-field topics. Cooperative work on developing an integrated near-field performance model seemed impractical because of the different disposal philosophies and decision environments in various countries. Also, small countries cannot afford to give up individuals to such activities. The group thought that a meeting which focused on integrated near-field performance models (i.e., one that focused on the interactions of various models or disposal system phenomena) would be a good topic for a future meeting. The group identified the need for more efficient information exchange. Agreements to exchange preliminary information seemed essential to satisfying that need.

THE IMPORTANCE OF NEAR-FIELD PHENOMENA

The group felt that the investigation of near-field phenomena should continue but concluded that the importance of near-field phenomena relative to far-field phenomena could not be quantitatively determined without an overall disposal system performance model. The importance seemed to be strongly influenced by disposal philosophy (e.g., temperature limit for the host rock) and the regulatory environment (e.g., container lifetime standards). Phenomena that would ordinarily be unimportant can be made important by decisions in these areas.

EXPERIMENTAL AND MODELING WORK CONCERNING SALT CREEP AND FRACTURING

The group concluded that on-going R & D activities were adequate if followed to completion.

EXPERIMENTAL AND MODELING WORK OF WATER AND GAS MOVEMENT

The group concluded that on-going R & D activities should be continued but that the emphasis on work in this area needs to be re-examined using an integrated near-field performance model.

COMPUTER CODE VALIDATION

The group concluded that computer codes should be validated with experimental data. Such data can be obtained by laboratory experimentation and field testing. The choice of experimental approach to validation depends strongly on the phenomenon being modeled. Some phenomena occur so rapidly that field testing is possible. Often the phenomena occur so slowly that accelerated testing must be employed. However, such tests are very difficult to interpret and extrapolation to long-times is necessary.

WORKING GROUP ON OTHER HOST MEDIA AND ENGINEERED BARRIERS

Chairman : R. HEREMANS, Belgium
Rapporteurs : N. CHAPMAN, United Kingdom, F. GERA, Italy

1. TUFF AND CLAY

The United States is the only country presently working on tuff. Japan is planning to start an experimental program next year. Italy has tuff formations, but no consideration has been given so far to their disposal potential.

Belgium, Italy and the United Kingdom have programs to investigate disposal of radioactive waste in clay. Switzerland is also interested in clay for the disposal of MLW. The United States has carried out disposal of waste in a shale formation by means of hydraulic fracturing.

Disposal in tuff and clay has some similarities since in both cases the formations are porous, water is present and sorption processes constitute the main barrier that restricts radionuclide migration in the far field. Matrix diffusion of radionuclides can also be an important retardation factor particularly in tuff. Significant differences also exist ; for example, the permeability of tuff is higher, the geotechnical properties of tuff are significantly better, while clays are characterized by greater self-sealing capability.

For both formations it is important to characterize the source term. For this purpose the knowledge of in-situ chemical conditions is necessary. Experiments must be carried out in such a way that no changes are induced in the chemistry of the system and that the chemical form of the mobilized species is known. In-situ migration experiments are necessary to better understand the behavior of the various barriers in representative conditions.

The verification of models should be carried out with realistic experimental data. It was recognized that there is insufficient understanding of thermomechanical and chemical phenomena associated with repository conditions in these media. This appears to be especially true for time dependent properties. It will be important to perform laboratory and field experiments to obtain the data necessary to evaluate the importance of slow changes in physical and chemical properties of the host rock. Verification of models over the very long term may only be possible by application to natural analogs.

It was pointed out that the response of tuff and clay formations to faulting at representative repository depth is not adequately known and that an effort is required in order to obtain this information through the study of appropriate geological examples.

2. ENGINEERED BARRIERS IN ALL ROCK TYPES

The group was strongly in favor of a systems approach to safety assessments of disposal techniques. The engineered barriers should be seen as integral to a complete isolation system and their functions determined accordingly, rather than having arbitrary numerical performance or arbitrary multiple redundancy requirements placed on them. Such requirements are often irrational and may be counter-productive. Both as a result of papers presented at this workshop and previous studies, rate dependent processes are seen to dominate such assessments. Simple time delay roles for engineered barriers do not appear to significantly affect their results. For the near-field the rate of release of nuclides is the important process. In this context the advantages of the anticipated distribution of canister failures over time in limiting the rate of release of nuclides to the geosphere was noted by the group. The principal role of the backfill was seen as that of providing a hydraulic barrier to limit access of water to the waste, although it may act as a groundwater conditioning agent or provide a barrier to subsequent nuclide movement. Analyses indicated that the latter process was likely to be temporally insignificant in the overall migration process for critical nuclides.

Many of the thermal effects which might adversely affect the near-field would be avoided by disposing of older, colder waste. Most countries were actively considering long-term interim storage with this in mind.

The paucity of very long-term experimental data in several fields prompted some support for embarking on such tests now, with a view to operating very simple but pertinent experiments over several decades, on representative sites.

The benefit of canisters of high integrity in guarding against accidental releases during the operational and pre-closure life of a repository was widely appreciated. However, unless extremely long-lived containers are considered, with lifetimes in excess of 10^5 years, no value is attached to the subsequent longevity of the canister. This is because models show dose rates to man to be insensitive to the time of release provided subsequent migration times are long in comparison with the half-lives of the nuclides concerned. The type of environments considered for disposal should ensure such migration times. Additionally, with apparent long leach times controlled by high-level waste form and nuclide solubilities, corrosion rates are shown to be insignificant.

Séance 7

DISCUSSIONS DES GROUPES DE TRAVAIL

GROUPE DE TRAVAIL SUR LES ROCHES RÉCEPTRICES
GRANITIQUES ET CRISTALLINES

Président : G.R. SIMMONS (Canada)
Rapporteur : P. JOHNSTON (NEA)

GROUPE DE TRAVAIL SUR LES FORMATIONS SALINES

Président : R.H. KOSTER (République fédérale d'Allemagne)
Rapporteur : H. BURKHOLDER (Etats-Unis)

GROUPE DE TRAVAIL SUR LES AUTRES MILIEUX RÉCEPTEURS ET
BARRIÈRES ARTIFICIELLES

Président : R. HEREMANS (Belgique)
Rapporteurs : N. CHAPMAN (Royaume-Uni), F. GERA (Italie)

GROUPE DE TRAVAIL SUR LES ROCHES RECEPTRICES
GRANITIQUES ET CRISTALLINES

Président : G.R. SIMMONS, AECL, Canada
Rapporteur : P. JOHNSTON, AEN

Les programmes de recherche ont pour objectif principal de faire des évaluations de l'évacuation dans des formations géologiques, sous l'angle de l'environnement et de la sûreté. Ces évaluations portent sur la voûte, la géosphère et la biosphère, qu'il faut considérer comme des parties intégrantes du système d'isolement, et non prendre séparément. Les participants ont reconnu que les recherches sur les phénomènes en champ proche n'étaient qu'un aspect de l'évaluation globale de la sûreté d'un dépôt, et que l'importance de ces phénomènes doit être appréciée dans le cadre d'une étude plus vaste. On peut avoir une idée de la place qu'il convient de leur accorder, grâce aux analyses de sensibilité des dégagements de radionucléides. Dans son exposé, M. Mano en a donné un exemple, mais il sera nécessaire de mieux comprendre les choses avant de pouvoir porter une appréciation réaliste sur l'importance de certains effets en champ proche. Au stade actuel, on n'en sait pas encore assez sur les phénomènes fondamentaux pour être en mesure de faire des analyses de sensibilité détaillées et d'attribuer des priorités aux différents aspects des recherches dont il a été rendu compte à la réunion.

La question des espèces chimiques que l'on rencontre dans les eaux souterraines présentes dans les formations granitiques a été abordée dans les contributions qui ont été présentées et au cours des débats du groupe de travail. Les participants ont considéré qu'il était indispensable de connaître la nature chimique des radionucléides dans la région d'un dépôt pour comprendre leur rétention et leur migration, et pour calculer le terme source aux fins des évaluations de sûreté. La modélisation de la sorption et de la solubilité des radionucléides est cependant très difficile en champ proche, où le comportement chimique peut être fortement modifié par des altérations thermiques de la surface des roches réceptrices ainsi que par les conteneurs, les matériaux tampons et les matériaux de remplissage.

Les participants ont estimé qu'en champ proche, il ne serait pas approprié de retenir une valeur unique de K_d relative à chaque radionucléide si l'on voulait avoir une vue réaliste des mécanismes de rétention. La modélisation précise de la migration des radionucléides n'est possible que si l'on a une bonne compréhension des mécanismes thermodynamiques qui entrent en jeu, et une caractérisation précise de l'environnement géochimique. La radiolyse, à proximité des emballages de déchets, peut également avoir une forte incidence sur les espèces chimiques et les états d'oxydation des radionucléides. En outre, les produits de corrosion provenant du conteneur et de l'emballage de déchets peuvent influer sur les espèces chimiques et sur la capacité de sorption des roches tampons et réceptrices.

Etant donné les difficultés que pose la modélisation de la rétention des radionucléides dans un environnment complexe, plusieurs participants ont proposé que l'on évite, dans la conception des dépôts et des emballages de déchets, d'utiliser des matériaux dont le comportement est imprévisible dans l'environnement hydrothermique du champ proche. A cette fin, on peut ou bien choisir les matériaux utilisés ou bien surveiller l'environnement hydrothermique, de façon à éviter les altérations des matériaux.

D'après le groupe de travail, il est possible, avec les techniques minières et de génie civil actuelles, de réaliser un dépôt de déchets dans du granite. De tels dépôts seraient construits et exploités comme les mines traditionnelles. Les débats ont porté sur l'évaluation de la sûreté, et plus particulièrement sur la croissance lente des fissures et sur le fluage des roches qui accompagnent le creusement de cavités, ainsi que sur la chaleur dégagée par les déchets. Ces phénomènes peuvent modifier la résistance de la masse rocheuse et la migration des radionucléides en provoquant des fissures. Plusieurs participants ont présenté des analyses thermo-mécaniques des dépôts, mais aucun des modèles ne comprenait d'équations fondamentales du fluage. On dispose de très peu de données expérimentales pour juger de l'importance de ce phénomène dans les roches dures.

Certains ont émis l'idée que le dépôt serait entouré d'une enveloppe, au sein de laquelle il entraînerait d'importantes modifications de la structure rocheuse, et au-delà de laquelle on n'observerait pas de perturbations sensibles. Il faudrait essayer de quantifier les modifications structurelles de grande ampleur.

Les participants ont aussi abordé la question des modèles de réponse dynamique et des critères sismiques. Certains travaux sur ce sujet sont poursuivis au Canada, et des études lui sont consacrées au Japon et aux Etats-Unis. Elles donnent à penser que les événements sismiques ne devraient pas affecter gravement la stabilité des dépôts souterrains.

Il a également été question de l'échange international de programmes d'ordinateur et de résultats expérimentaux relatifs au champ proche. La possibilité d'utiliser des filières établies, comme cela se fait dans d'autres domaines de la sûreté nucléaire et de la modélisation, a été mentionnée. Déjà, des comparaisons de programmes ont lieu aux Etats-Unis, en Suède et au Canada, et le groupe de travail a reconnu que ces activités pourront prendre de l'importance à mesure que les codes se répandront.

Le groupe de travail a convenu que la coopération internationale était indispensable en matière de recherches sur les phénomènes en champ proche. Elle pourrait, dans un premier temps, s'exercer aux niveaux technique et scientifique, et prendre par la suite un caractère plus formel. Parmi les sites actuellement disponibles pour les travaux de recherche in situ dans le granite figurent la mine de Stripa en Suède, les installations de la Climax and Colorado School of Mines aux Etats-Unis, et des structures en roches dures au Japon et au Royaume-Uni. Des centres d'évacuation dans des formations granitiques seront construits dans l'avenir proche en Suisse et au Canada. Le groupe de travail a insisté sur la nécessité d'encourager la coordination et d'utiliser efficacement des ressources qui sont limitées.

GROUPE DE TRAVAIL SUR LES FORMATIONS SALINES

Président : R.H. KOSTER, République fédérale d'Allemagne
Rapporteur : H. BURKHOLDER, ONWI, Etats-Unis

Le groupe de travail a débattu des objectifs et des sujets qui ont été soumis à son examen par le secrétaire de la réunion. Ses discussions et conclusions sont résumés ci-après.

IDENTIFICATION DES BESOINS ET PRIORITES EN MATIERE DE RECHERCHE

Le groupe a considéré que la base de données concernant les phénomènes en champ proche associés à certaines formes de déchets ainsi que les roches encaissantes était, dans l'ensemble, satisfaisante, mais que celle qui se rapportait aux conteneurs, aux matériaux de remplissage et aux matériaux de scellement devait être améliorée. Il a estimé que la forme des déchets, leur conteneur, la roche encaissante de remplissage et l'environnement géologique devaient être considérés comme un système de barrières multiples. Il s'est cependant montré sceptique quant à l'utilité des conteneurs et des roches de remplissage comme barrières dans le système d'isolement après fermeture complète du dépôt. Des données supplémentaires semblent nécessaires dans les domaines suivants : décomposition, lixiviation et solubilité des déchets de haute activité, du combustible irradié et des déchets de moyenne activité ; effets des rayonnements sur les constituants de l'enrobage, fluides interstitiels de la roche encaissante et effets thermochimiques et thermo-mécaniques sur les formes de déchets, conteneurs, et impuretés de la roche encaissante.

Dans l'ensemble, le groupe a considéré que les efforts devraient porter plus sur l'intégration du modèle de champ proche que sur la mise au point de modèles nouveaux. De meilleurs modèles sont nécessaires dans certains domaines bien précis : interactions thermochimiques entre la forme des déchets, le conteneur et la roche encaissante dans la période postérieure à la fermeture, et effets thermohydrologiques se produisant dans le dépôt pendant la période d'exploitation.

Pour le groupe, la mise au point d'un modèle de comportement du système de champ proche et l'intégration de ce modèle à un modèle de performance du système de champ éloigné est de la plus haute importance. Selon lui, il n'est pas possible, sans un tel modèle, de prendre des décisions rationnelles quant aux priorités en matière de R & D dans les différents domaines relatifs au champ proche. L'*analyse systémique* est apparue comme le seul moyen de pondérer l'importance des divers phénomènes en champ proche par rapport au comportement global du système d'évacuation et de déterminer dans quelle mesure les différents éléments de ce dernier contribuent à son comportement.

PROMOTION DE LA COOPERATION INTERNATIONALE

Le groupe a conclu que l'AEN/OCDE était la mieux placée pour promouvoir la coopération internationale en organisant d'autres réunions consacrées aux phénomènes en champ proche. Le travail en commun

sur la mise au point d'un modèle intégré de comportement du champ
proche a semblé impossible en raison de la diversité des doctrines en
matière d'évacuation ainsi que des environnements décisionnels. Il est
vrai aussi que les petits pays ne peuvent se permettre de détacher des
individus pour de telles activités. Le groupe a estimé qu'il serait
bon d'organiser une réunion consacrée aux modèles intégrés de comportement du champ proche (c'est-à-dire aux interactions des divers modèles de phénomènes se produisant dans les systèmes d'évacuation). Il
a jugé nécessaire d'accroître l'efficacité des échanges d'informations.
Pour cela, des accords portant sur l'échange d'informations prélimi-
naire paraissent indispensables.

L'IMPORTANCE DES PHENOMENES EN CHAMP PROCHE

Le groupe a estimé que les recherches consacrées aux phénomènes en champ proche devraient se poursuivre, mais a conclu que leur
importance, par rapport aux phénomènes en champ éloigné, ne pouvait
être déterminée quantitativement en l'absence de modèle de comportement
global des systèmes d'évacuation. Il semble que deux facteurs jouent
un rôle décisif : la doctrine fondamentale en matière d'évacuation (par
exemple, limite de température pour la roche encaissante) et le cadre
réglementaire (par exemple, normes de durée de vie des conteneurs) ;
des phénomènes qui, normalement, n'auraient pas d'importance, pourraient en prendre à la suite de décisions dans ces domaines.

TRAVAUX EXPERIMENTAUX ET TRAVAUX DE MODELISATION CONCERNANT LE FLUAGE ET LA FRACTURATION DES FORMATIONS SALINES

Le groupe a conclu qu'il n'était pas utile d'entreprendre
d'autres activités de R & D à condition que celles qui sont en cours
soient menées à bien.

TRAVAUX EXPERIMENTAUX ET TRAVAUX DE MODELISATION DU MOUVEMENT DE L'EAU ET DU GAZ

Le groupe a conclu que les activités de R & D en cours
devraient être poursuivies mais qu'il fallait revoir les priorités
dans ce domaine en utilisant un modèle intégré de comportement du
champ proche.

VALIDATION DES PROGRAMMES DE CALCUL

Le groupe a conclu que les programmes de calcul devraient
être validés au moyen de données expérimentales. Ces dernières pourront provenir d'expériences de laboratoire et d'essais sur le terrain.
Le choix de la méthode dépendra dans une très large mesure des phénomènes modélisés. Certains se produisent si rapidement que l'essai sur
le terrain est possible ; d'autres, en revanche, sont si lents qu'il
faut procéder à des essais accélérés. Mais ces derniers sont très dif-
ficiles à interpréter et demandent des extrapolations à long terme.

GROUPE DE TRAVAIL SUR LES AUTRES MILIEUX RECEPTEURS ET
BARRIERES ARTIFICIELLES

Président : R. HEREMANS, Belgique
Rapporteurs : N. CHAPMAN, Royaume-Uni, F. GERA, Italie

1. TUF ET ARGILE

Les Etats-Unis sont actuellement le seul pays travaillant sur le tuf. Le Japon envisage de lancer un programme expérimental l'année prochaine. On trouve des formations de tuf en Italie, mais elles n'ont pas, jusqu'à présent, retenu l'attention comme milieu d'évacuation.

La Belgique, l'Italie et le Royaume-Uni ont des programmes de recherches sur l'évacuation des déchets radioactifs dans l'argile. La Suisse s'y intéresse également pour les déchets de moyenne activité. Les Etats-Unis ont évacué des déchets dans une formation schisteuse par fracturation hydraulique.

L'évacuation dans le tuf et dans l'argile présente des similitudes : les deux formations sont poreuses, on y trouve de l'eau, et les processus de sorption constituent la principale barrière empêchant la migration des radionucléides dans le champ éloigné. La diffusion des radionucléides à travers la matrice rocheuse peut également être un important facteur de retardement, notamment dans le tuf. On observe toutefois des différences : c'est ainsi que le tuf a une plus grande perméabilité, des propriétés géotechniques nettement meilleures, tandis que les argiles se caractérisent par une plus grande capacité d'auto-scellement.

Pour les deux formations, il importe de caractériser le terme source. A cette fin, il faut connaître les conditions chimiques in situ et faire des expériences de telle façon qu'il n'y ait pas de modifications dans la chimie du système et que la forme chimique des espèces entraînées soit connue. Ces expériences sont nécessaires si l'on veut mieux comprendre le comportement des diverses barrières dans des conditions représentatives.

La vérification des modèles devrait s'effectuer à l'aide de données expérimentales réalistes. Les participants ont admis que les phénomènes thermo-mécaniques et chimiques associés aux conditions des dépôts dans ces milieux n'étaient pas assez bien connus. Cela est vrai notamment pour les propriétés liées au facteur temps ; il faudra effectuer des expériences en laboratoire et sur le terrain pour obtenir les données nécessaires à l'évaluation de l'importance des modifications lentes des propriétés physiques et chimiques de la roche réceptrice. A très long terme, il n'est possible de vérifier les modèles qu'en les appliquant à des conditions naturelles analogues.

Il a été souligné que la réaction à la fracturation des formations de tuf et d'argile à des profondeurs représentatives des dépôts était mal connue, et que pour en savoir plus il fallait étudier des cas géologiques appropriés.

2. BARRIERES ARTIFICIELLES DANS TOUS LES TYPES DE ROCHES

Le groupe s'est montré fortement partisan d'une approche systémique des évaluations de sûreté des techniques d'évacuation. Il serait préférable de considérer les barrières artificielles comme faisant partie intégrante du système complet d'isolement et de déterminer leurs fonctions en conséquence, plutôt que de leur imposer des caractéristiques numériques ou des redondances multiples arbitraires. De telles exigences sont souvent irréalistes et risquent d'aller à l'encontre du but recherché. Il ressort des documents présentés à ce séminaire et de travaux antérieurs, que les processus liés à la vitesse soient déterminants dans ces évaluations de sûreté. Le fait que les barrières techniques jouent un simple rôle retardateur ne semble pas modifier sensiblement leurs résultats. Pour les phénomènes en champ proche, la vitesse de libération des nucléides est l'élément important. Dans ce contexte, le groupe a noté que la prévision de la répartition des défaillances des conteneurs au cours du temps aurait pour avantage de limiter la vitesse de dégagement des nucléides dans la géosphère. Il a également considéré que les matériaux de remplissage avaient pour rôle principal de constituer une barrière hydraulique, limitant l'accès des eaux aux déchets, mais qu'ils pouvaient aussi, par la suite, agir sur la composition des eaux souterraines et faire obstacle au dégagement de nucléides. Les analyses montrent que cette dernière action serait probablement insignifiante du point de vue temporel dans le processus global de migration des nucléides critiques.

On pourrait éviter de nombreux effets thermiques risquant d'avoir des effets préjudiciables sur le champ proche en évacuant des déchets plus anciens et plus froids. La plupart des pays envisagent sérieusement le stockage intérimaire à long terme dans cette optique.

Le manque de données expérimentales à très long terme dans plusieurs domaines a conduit à penser qu'il serait bon d'explorer cette voie maintenant, en vue de procéder sur des sites représentatifs, à des expériences très simples mais significatives portant sur plusieurs décennies.

Les participants ont été largement d'accord sur l'avantage que présentent les conteneurs de haute intégrité, qui empêchent des rejets accidentels pendant l'exploitation et la période précédant la fermeture d'un dépôt. Toutefois, ou bien on envisage des conteneurs ayant une durée de vie extrêmement longue - c'est-à-dire dépassant 10^5 années - ou bien le problème de la longévité ne se pose pas. En effet, les modèles montrent que les débits de dose délivrés à l'homme ne dépendent pas du moment où la radioactivité est libérée, à condition que les temps de migration soient longs par rapport aux périodes des nucléides concernés. Les types d'environnements envisagés pour l'évacuation devraient garantir de tels temps de migration. En outre, les longs délais apparents de lixiviation étant régis par la forme des déchets de haute activité et par les solubilités des nucléides, les vitesses de corrosion se révèlent insignifiantes.

BELGIUM - BELGIQUE

HEREMANS, R.H., Géo-technologie, Centre d'Etude de l'Energie Nucléaire,
CEN/SCK, Boeretang 200, B-2400 Mol

CANADA

CARMICHAEL, T.J., Rock Sciences Section, Civil Research Department,
Ontario Hydro Research Division, KR 253, 800 Kipling Avenue,
Toronto, Ontario M8Z 5S4

CHEUNG, S., Atomic Energy of Canada Limited, Whiteshell Nuclear
Research Establishment, Pinawa, Manitoba ROE 1LO

GOODWIN, B.W., Atomic Energy of Canada Limited, Whiteshell Nuclear
Research Establishment, Environmental and Safety Assessment Branch,
Pinawa, Manitoba ROE 1LO

JAKUBICK, A.T., Rock Sciences Section, Civil Research Department,
Ontario Hydro Research Division, KR 258, 800 Kipling Avenue,
Toronto, Ontario M8Z 5S4

KOOPMANS, R., Rock Sciences Section, Civil Research Department,
Ontario Hydro Research Division, KR 258, 800 Kipling Avenue,
Toronto, Ontario M8Z 5S4

RADHAKRISHNA, H.S., Rock Sciences Section, Civil Research Department,
Ontario Hydro Research Division, KR 258, 800 Kipling Avenue,
Toronto, Ontario M8Z 5S4

RIGBY, G.L., Atomic Energy of Canada Limited, Whiteshell Nuclear
Research Establishment, Pinawa, Manitoba ROE 1LO

SHULTZ, K.R., Atomic Energy Control Board, P.O. Box 1046, Ottawa,
Ontario K1P 5S9

SIMMONS, G.R., Applied Geoscience Branch, Atomic Energy of Canada
Limited, Whiteshell Nuclear Research Establishment, Pinawa,
Manitoba ROE 1LO

TSUI, K.K., Geotechnical Engineering Department, Ontario Hydro,
700 University Avenue, Toronto, Ontario M5G 1X6

WILKINS, B.J.S., Atomic Energy of Canada Limited, Whiteshell Nuclear
Research Establishment, Pinawa, Manitoba ROE 1LO

DENMARK - DANEMARK

SKYTTE-JENSEN, B., Risø National Laboratory, DK-4000 Roskilde

FINLAND - FINLANDE

MIETTINEN, J.K., Department of Radiochemistry, University of Helsinki,
Unioninkatu 35, SF-00170 Helsinki 17

NIINI, H., Geological Survey of Finland, Kivimiehentie 1,
    SF-02150 Espoo 15

PELTONEN, E.K., Technical Research Centre of Finland, Nuclear
    Engineering Laboratory, P.O. Box 169, SF-00181 Helsinki 18

FRANCE

BARBREAU, A., Commissariat à l'Energie Atomique, Institut de Protec-
    tion et de Sûreté Nucléaire, CSDR, B.P. n° 6, F-92260 Fontenay-aux-
    Roses

CAMBON, J.L., Geostock, Tour Aurore, Cedex n° 5, F-92080 Paris la
    Défense

COUAIRON, M., Commissariat à l'Energie Atomique, 31-33 rue de la
    Fédération, F-75015 Paris

GOBLET, P., Ecole des Mines de Paris, Centre d'Informatique Géologique,
    35 rue Saint-Honoré, F-77305 Fontainebleau

JAOUEN, D., Société Générale des Techniques Nouvelles, Saint-Quentin-
    en-Yvelines

FEDERAL REPUBLIC OF GERMANY - REPUBLIQUE FEDERALE D'ALLEMAGNE

FISCHLE, W.R., Gesellschaft für Strahlen- und Umweltforschung mbH,
    Institut für Tieflagerung, Schachtanlage Asse, D-3346 Remlingen

JOCKWER, N., Gesellschaft für Strahlen- und Umweltforschung mbH,
    Institut für Tieflagerung, Theodor-Heuss-Strasse 4,
    D-3300 Braunschweig

KOSTER, R.H., Kernforschungszentrum Karlsruhe GmbH, Institut für
    Nukleare Entsorgungstechnik, Postfach 3640, D-7500 Karlsruhe 1

PITZ, W.K.F., Deutsche Gesellschaft zum Bau und Betreib von Endlagern
    für Abfallstoffe mbH (DBE), Postfach 1169, D-3150 Peine

STORCK, R., Institut für Kerntechnik, Technische Universität Berlin,
    Marchstrasse 18, D-1000 Berlin 10

UERPMANN, P., Gesellschaft für Strahlen- und Umweltforschung mbH,
    Institut für Tieflagerung, Schachtanlage Asse, D-3346 Remlingen

WALLNER, M., Bundesanstalt für Geowissenschaften und Rohstoffe,
    Postfach 51 01 53, D-3000 Hannover 51

ITALY - ITALIE

BOCOLA, W., Comitato Nazionale per l'Energia Nucleare, Laboratorio
    Rifiuti Radioattivi, CNEN-CSN Casaccia, CP 2400, I-00100 Roma

GERA, F., ISMES, Via T. Taramelli 14, I-00197 Roma

JAPAN - JAPON

FUKUMITSU, K., Shimizu Construction Co., 16-1 Kyobashi, 2-chome,
    Chuo-ku, Tokyo

HWANG, M.J., The Institute of Applied Energy, 33rd Mori-Building,
    3-8-21 Toranomon, Minato-ku, Tokyo

ISHII, T., Shimizu Construction Co., 16-1 Kyobashi, 2-chome, Chuo-ku,
    Tokyo

MAEKAWA, K., Mitsubishi Metal Corporation, 1-5-2, Otemachi, Cheyoda-ku,
    Tokyo

MAKINO, M., Nuclear and Process Engineering Department, JGC Corporation,
    14-1, Bessho, 1-chome, Minami-ku, Yokohama 232

MANO, T., Waste Management Office, Power Reactor and Nuclear Fuel
    Development Corporation, 1-9-13 Akasaka, Minato-ku, Tokyo

SWEDEN- SUEDE

ALLARD, B.M., Department of Nuclear Chemistry, Chalmers University
    of Technology, S-412 96 Gothenburg

GRENTHE, I.R., Department of Inorganic Chemistry, Royal Institute of
    Technology, S-100 44 Stockholm

KARLSSON, F.D.B., SKBF/KBS, Box 5864, S-102 48 Stockholm

SWITZERLAND - SUISSE

ISSLER, H., NAGRA, Parkstrasse 23, CH-5401 Baden

McCOMBIE, C., NAGRA, Parkstrasse 23, CH-5401 Baden

UNITED KINGDOM- ROYAUME-UNI

CHAPMAN, N.A., Institute of Geological Sciences, Building 151, Harwell
    Laboratory, Didcot, Oxfordshire OX11 ORA

UNITED STATES - ETATS-UNIS

BACA, R.G., Rockwell Hanford Operations, P.O. Box 800, Richland,
    Washington 99352

BLACIC, J.D., Los Alamos National Laboratory, P.O. Box 1663, Los
    Alamos, New Mexico 87545

BURKHOLDER, H.C., Office of Nuclear Waste Isolation, Battelle Memorial
    Institute, 505 King Avenue, Columbus, Ohio 43201

CRANWELL, R.M., Sandia National Laboratories, Division 4413,
    Albuquerque, New Mexico 87185

JOHNSTONE, J.K., Sandia National Laboratories, Division 4537, Albuquerque, New Mexico 87185

LESTER, D.H., Science Applications, Inc., P.O. Box 2351, La Jolla, California 92038

MATALUCCI, R.V., Sandia National Laboratories, Division 4512, Albuquerque, New Mexico 87185

MULLER, A., Sandia National Laboratories, Division 4413, Albuquerque, New Mexico 87185

PARRISH, D.K., RE/SPEC, Inc., P.O. Box 725, Rapid City, South Dakota 57709

PATRICK, W.C., Lawrence Livermore National Laboratory, L-204, P.O. Box 808, Livermore, California 94550

RAI, D., Battelle Memorial Institute, Pacific Northwest Laboratory, P.O. Box 999, Richland, Washington 99352

TODESCHINI, R.A., Research and Engineering, Bechtel Group Inc., P.O. Box 3965, San Francisco, California 94119

WILEMS, R.E., Nuclear Waste Programs, INTERA Environmental Consultants, Inc., 11999 Katy Freeway, Suite 610, Houston, Texas 77079

INTERNATIONAL ATOMIC ENERGY AGENCY
AGENCE INTERNATIONALE DE L'ENERGIE ATOMIQUE

TSYPLENKOV, V.S., International Atomic Energy Agency, P.O. Box 100, A-1400 Vienna (Austria)

SECRETARIAT

JOHNSTON, P., Division of Radiation Protection and Waste Management, Nuclear Energy Agency, 38 boulevard Suchet, F-75016 Paris (France)

SOME
NEW PUBLICATIONS
OF NEA

QUELQUES
NOUVELLES PUBLICATIONS
DE L'AEN

ACTIVITY REPORTS

RAPPORTS D'ACTIVITÉ

Activity Reports of the OECD Nuclear Energy Agency (NEA)
— 8th Activity Report (1979)
— 9th Activity Report (1980)

Rapports d'activité de l'Agence de l'OCDE pour l'Énergie Nucléaire (AEN)
— 8e Rapport d'Activité (1979)
— 9e Rapport d'Activité (1980)

Free on request — Gratuits sur demande

Annual Reports of the OECD HALDEN Reactor Project
— 19th Annual Report (1978)
— 20th Annual Report (1979)

Rapports annuels du Projet OCDE de réacteur de HALDEN
— 19e Rapport annuel (1978)
— 20e Rapport annuel (1979)

Free on request — Gratuits sur demande

• • •

INFORMATION BROCHURES

— The NEA Data Bank
— International Co-operation for Safe Nuclear Power
— NEA at a Glance
— OECD Nuclear Energy Agency: Functions and Main Activities

BROCHURES D'INFORMATION

— La Banque de Données de l'AEN
— Une coopération internationale pour une énergie nucléaire sûre
— Coup d'œil sur l'AEN
— Agence de l'OCDE pour l'Énergie Nucléaire : Rôle et principales activités

Free on request — Gratuits sur demande

• • •

SCIENTIFIC AND
TECHNICAL
PUBLICATIONS

PUBLICATIONS
SCIENTIFIQUES
ET TECHNIQUES

NUCLEAR FUEL CYCLE

LE CYCLE DU COMBUSTIBLE NUCLÉAIRE

Nuclear Fuel Cycle Requirements and Supply Considerations, Through the Long-Term (1978)

Besoins liés au cycle du combustible nucléaire et considérations sur l'approvisionnement à long terme (1978)

£4.30   US$8.75   F35,00

World Uranium Potential —
An International Evaluation (1978)

Potentiel mondial en uranium —
Une évaluation internationale (1978)

£7.80   US$16.00   F64.00

Uranium — Resources, Production and Demand (1979)

Uranium — ressources, production et demande (1979)

£8.70   US$19.50   F78,00

● ● ●

RADIATION PROTECTION

RADIOPROTECTION

Iodine-129
(Proceedings of an NEA Specialist Meeting, Paris, 1977)

Iode-129
(Compte rendu d'une réunion de spécialistes de l'AEN, Paris, 1977)

£3.40   US$7.00   F28,00

Recommendations for Ionization Chamber Smoke Detectors in Implementation of Radiation Protection Standards (1977)

Recommandations relatives aux détecteurs de fumée à chambre d'ionisation en application des normes de radioprotection (1977)

Free on request — Gratuit sur demande

Radon Monitoring
(Proceedings of the NEA Specialist Meeting, Paris, 1978)

Surveillance du radon
(Compte rendu d'une réunion de spécialistes de l'AEN, Paris, 1978)

£8.00   US$16.50   F66,00

Management, Stabilisation and Environmental Impact of Uranium Mill Tailings (Proceedings of the Albuquerque Seminar, United States, 1978)

Gestion, stabilisation et incidence sur l'environnement des résidus de traitement de l'uranium
(Compte rendu du Séminaire d'Albuquerque, États-Unis, 1978)

£9.80   US$20.00   F80,00

Exposure to Radiation from the Natural Radioactivity in Building Materials (Report by an NEA Group of Experts, 1979)

Exposition aux rayonnements due à la radioactivité naturelle des matériaux de construction
(Rapport établi par un Groupe d'experts de l'AEN, 1979)

Free on request — Gratuit sur demande

Marine Radioecology
(Proceedings of the Tokyo Seminar, 1979)

Radioécologie marine
(Compte rendu du Colloque de Tokyo, 1979)

£9.60   US$21.50   F86.00

Radiological Significance and Management of Tritium, Carbon-14, Krypton-85 and Iodine-129 arising from the Nuclear Fuel Cycle (Report by an NEA Group of Experts, 1980)

Importance radiologique et gestion des radionucléides : tritium, carbone-14, krypton-85 et iode-129, produits au cours du cycle du combustible nucléaire (Rapport établi par un Groupe d'experts de l'AEN, 1980)

£8.40   US$19.00   F76,00

The Environmental and Biological Behaviour of Plutonium and Some Other Transuranium Elements (Report by an NEA Group of Experts, 1981)

Le comportement mésologique et biologique du plutonium et de certains autres éléments transuraniens (Rapport établi par un Groupe d'experts de l'AEN, 1981)

£ 4.60   US$ 10.00   F 46,00

• • •

RADIOACTIVE WASTE MANAGEMENT

GESTION DES DÉCHETS RADIOACTIFS

Objectives, Concepts and Strategies for the Management of Radioactive Waste Arising from Nuclear Power Programmes (Report by an NEA Group of Experts, 1977)

Objectifs, concepts et stratégies en matière de gestion des déchets radioactifs résultant des programmes nucléaires de puissance (Rapport établi par un Groupe d'experts de l'AEN, 1977)

£8.50   US$17.50   F70,00

Treatment, Conditioning and Storage of Solid Alpha-Bearing Waste and Cladding Hulls (Proceedings of the NEA/IAEA Technical Seminar, Paris, 1977)

Traitement, conditionnement et stockage des déchets solides alpha et des coques de dégainage (Compte rendu du Séminaire technique AEN/AIEA, Paris, 1977)

£7.30   US$15.00   F60,00

Storage of Spent Fuel Elements (Proceedings of the Madrid Seminar, 1978)

Stockage des éléments combustibles irradiés (Compte rendu du Séminaire de Madrid, 1978)

£7.30   US$15.00   F60,00

In Situ Heating Experiments in Geological Formations (Proceedings of the Ludvika Seminar, Sweden, 1978)

Expériences de dégagement de chaleur in situ dans les formations géologiques (Compte rendu du Séminaire de Ludvika, Suède, 1978)

£8.00   US$16.50   F66,00

Migration of Long-lived Radionuclides in the Geosphere (Proceedings of the Brussels Workshop, 1979)

Migration des radionucléides à vie longue dans la géosphère (Compte rendu de la réunion de travail de Bruxelles, 1979)

£8.30   US$17.00   F68,00

Low-Flow, Low-Permeability Measurements in Largely Impermeable Rocks (Proceedings of the Paris Workshop, 1979)

Mesures des faibles écoulements et des faibles perméabilités dans des roches relativement imperméables (Compte rendu de la réunion de travail de Paris, 1979)

£7.80   US$16.00   F64,00

On-Site Management of Power Reactor Wastes (Proceedings of the Zurich Symposium, 1979)

Gestion des déchets en provenance des réacteurs de puissance sur le site de la centrale (Compte rendu du Colloque de Zurich, 1979)

£11.00   US$22.50   F90,00

Recommended Operational Procedures for Sea Dumping of Radioactive Waste (1979)

Recommandations relatives aux procédures d'exécution des opérations d'immersion de déchets radioactifs en mer (1979)

Free on request — Gratuit sur demande

Guidelines for Sea Dumping Packages of Radioactive Waste (Revised version, 1979)

Guide relatif aux conteneurs de déchets radioactifs destinés au rejet en mer (Version révisée, 1979)

Free on request — Gratuit sur demande

Use of Argillaceous Materials for the Isolation of Radioactive Waste (Proceedings of the Paris Workshop, 1979)

Utilisation des matériaux argileux pour l'isolement des déchets radioactifs (Compte rendu de la Réunion de travail de Paris, 1979)

£7.60   US$17.00   F68,00

Review of the Continued Suitability of the Dumping Site for Radioactive Waste in the North-East Atlantic (1980)

Réévaluation de la validité du site d'immersion de déchets radioactifs dans la région nord-est de l'Atlantique (1980)

Free on request — Gratuit sur demande

Decommissioning Requirements in the Design of Nuclear Facilities (Proceedings of the NEA Specialist Meeting, Paris, 1980)

Déclassement des installations nucléaires : exigences à prendre en compte au stade de la conception (Compte rendu d'une réunion de spécialistes de l'AEN, Paris, 1980)

£7.80   US$17.50   F70,00

Borehole and Shaft Plugging (Proceedings of the Columbus Workshop, United States, 1980)

Colmatage des forages et des puits (Compte rendu de la réunion de travail de Columbus, États-Unis, 1980)

£12.00   US$30.00   F120,00

Radionucleide Release Scenarios for Geologic Repositories (Proceedings of the Paris Workshop, 1980)

Scénarios de libération des radionucléides à partir de dépôts situés dans les formations géologiques (Compte rendu de la réunion de travail de Paris, 1980)

£6.00   US$15.00   F60,00

Cutting Techniques as related to Decommissioning of Nuclear Facilities (Report by an NEA Group of Experts, 1981)

Techniques de découpe utilisées au cours du déclassement d'installations nucléaires (Rapport établi par un Groupe d'experts de l'AEN, 1981)

£ 3.00   US$ 7.50   F 30.00

Decontamination Methods as related to Decommissioning of Nuclear Facilities (Report by an NEA Group of Experts, 1981)

Méthodes de décontamination relatives au déclassement des installations nucléaires (Rapport établi par un Groupe d'experts de l'AEN, 1981)

£ 2.80   US$ 7.00   F 28,00

• • •

SAFETY

Safety of Nuclear Ships
(Proceedings of the Hamburg Symposium,
1977)

£17.00   US$35.00   F140,00

Nuclear Aerosols in Reactor Safety
(A State-of-the-Art Report by a Group of
Experts, 1979)

£8.30   US$18.75   F75,00

Plate Inspection Programme
(Report from the Plate Inspection
Steering Committee — PISC — on the
Ultrasonic Examination of Three
Test Plates), 1980

£3.30   US$7.50   F30.00

Reference Seismic Ground Motions
in Nuclear Safety Assessments
(A State-of-the-Art Report by a
Group of Experts, 1980)

£7.00   US$16.00   F64,00

Nuclear Safety Research in the OECD Area.
The Response to the Three Mile Island
Accident (1980)

£3.20   US$8.00   F32,00

Safety Aspects of Fuel Behaviour in Off-
Normal and Accident Conditions
(Proceedings of the Specialist Meeting,
Espoo, Finland, 1980)

£12.60   US$28.00   F126,00

Safety of the Nuclear Fuel Cycle (A State-
of-the-Art Report by a Group of Experts,
1981)

£ 6.60   US$ 16.50   F 66,00

SÛRETÉ

Sûreté des navires nucléaires
(Compte rendu du Symposium de
Hambourg, 1977)

Les aérosols nucléaires dans la sûreté
des réacteurs
(Rapport sur l'état des connaissances
établi par un Groupe d'Experts, 1979)

Programme d'inspection des tôles
(Rapport du Comité de Direction sur
l'inspection des tôles — PISC — sur l'examen
par ultrasons de trois tôles d'essai au moyen
de la procédure «PISC» basée sur le code
ASME XI), 1980

Les mouvements sismiques de référence
du sol dans l'évaluation de la sûreté
des installations nucléaires
(Rapport sur l'état des connaissances
établi par un Groupe d'experts, 1980)

Les recherches en matière de sûreté
nucléaire dans les pays de l'OCDE. L'adaptation des programmes à la suite de l'accident de Three Mile Island (1980)

Considérations de sûreté relatives au comportement du combustible dans des conditions anormales et accidentelles
(Compte rendu de la réunion de spécialistes,
Espoo, Finlande, 1980)

Sûreté du Cycle du Combustible Nucléaire
(Rapport sur l'état des connaissances établi
par un Groupe d'Experts, 1981)

• • •

SCIENTIFIC INFORMATION

Neutron Physics and Nuclear Data for Reactors and other Applied Purposes
(Proceedings of the Harwell International Conference, 1978)

£26.80   US$55.00   F220,00

Calculation of 3-Dimensional Rating Distributions in Operating Reactors
(Proceedings of the Paris Specialists' Meeting, 1979)

£9.60   US$21.50   F86.00

Nuclear Data and Benchmarks for Reactor Shielding
(Proceedings of a Specialists' Meeting, Paris, 1980)

£9.60   US$24.00   F96,00

INFORMATION SCIENTIFIQUE

La physique neutronique et les données nucléaires pour les réacteurs et autres applications
(Compte rendu de la Conférence Internationale de Harwell, 1978)

Calcul des distributions tridimensionnelles de densité de puissance dans les réacteurs en cours d'exploitation (Compte rendu de la Réunion de spécialistes de Paris, 1979)

Données nucléaires et expériences repères en matière de protection des réacteurs (Compte rendu d'une réunion de spécialistes, Paris, 1980)

LEGAL PUBLICATIONS

PUBLICATIONS JURIDIQUES

Convention on Third Party Liability in the Field of Nuclear Energy — incorporating the provisions of Additional Protocol of January 1964

Free on request — Gratuit sur demande

Convention sur la responsabilité civile dans le domaine de l'énergie nucléaire — Texte incluant les dispositions du Protocole additionnel de janvier 1964

Nuclear Legislation, Analytical Study: "Nuclear Third Party Liability" (revised version, 1976)

£6.00   US$12.50   F50,00

Législations nucléaires, étude analytique: "Responsabilité civile nucléaire" (version révisée, 1976)

Nuclear Legislation, Analytical Study: "Regulations governing the Transport of Radioactive Materials" (1980)

£8.40   US$21.00   F84,00

Législations nucléaires, étude analytique : "Réglementation relative au transport des matières radioactives" (1980)

Nuclear Law Bulletin
(Annual Subscription — two issues and supplements)

£5.60   US$12.50   F50,00

Bulletin de Droit Nucléaire
(Abonnement annuel — deux numéros et suppléments)

Index of the first twenty five issues of the Nuclear Law Bulletin

Index des vingt-cinq premiers numéros du Bulletin de Droit Nucléaire

Description of Licensing Systems and Inspection of Nuclear Installation (1980)

£7.60   US$19.00   F76,00

Description du régime d'autorisation et d'inspection des installations nucléaires (1980)

NEA Statute

Free on request — Gratuit sur demande

Statuts de l'AEN

• • •

OECD SALES AGENTS
DÉPOSITAIRES DES PUBLICATIONS DE L'OCDE

ARGENTINA – ARGENTINE
Carlos Hirsch S.R.L., Florida 165, 4° Piso (Galería Guemes)
1333 BUENOS AIRES, Tel. 33.1787.2391 y 30.7122

AUSTRALIA – AUSTRALIE
Australia and New Zealand Book Company Pty, Ltd.,
10 Aquatic Drive, Frenchs Forest, N.S.W. 2086
P.O. Box 459, BROOKVALE, N.S.W. 2100

AUSTRIA – AUTRICHE
OECD Publications and Information Center
4 Simrockstrasse 5300 BONN. Tel. (0228) 21.60.45
Local Agent/Agent local :
Gerold and Co., Graben 31, WIEN 1. Tel. 52.22.35

BELGIUM – BELGIQUE
LCLS
35, avenue de Stalingrad, 1000 BRUXELLES. Tel. 02.512.89.74

BRAZIL – BRÉSIL
Mestre Jou S.A., Rua Guaipa 518,
Caixa Postal 24090, 05089 SAO PAULO 10. Tel. 261.1920
Rua Senador Dantas 19 s/205-6, RIO DE JANEIRO GB.
Tel. 232.07.32

CANADA
Renouf Publishing Company Limited,
2182 St. Catherine Street West,
MONTRÉAL, Que. H3H 1M7. Tel. (514)937.3519
522 West Hasting,
VANCOUVER, B.C. V6B 1L6. Tel. (604) 687.3320

DENMARK – DANEMARK
Munksgaard Export and Subscription Service
35, Nørre Søgade
DK 1370 KØBENHAVN K. Tel. +45.1.12.85.70

FINLAND – FINLANDE
Akateeminen Kirjakauppa
Keskuskatu 1, 00100 HELSINKI 10. Tel. 65.11.22

FRANCE
Bureau des Publications de l'OCDE,
2 rue André-Pascal, 75775 PARIS CEDEX 16. Tel. (1) 524.81.67
Principal correspondant :
13602 AIX-EN-PROVENCE : Librairie de l'Université.
Tel. 26.18.08

GERMANY – ALLEMAGNE
OECD Publications and Information Center
4 Simrockstrasse 5300 BONN Tel. (0228) 21.60.45

GREECE – GRÈCE
Librairie Kauffmann, 28 rue du Stade,
ATHÈNES 132. Tel. 322.21.60

HONG-KONG
Government Information Services,
Sales and Publications Office, Baskerville House, 2nd floor,
13 Duddell Street, Central. Tel. 5.214375

ICELAND – ISLANDE
Snaebjörn Jönsson and Co., h.f.,
Hafnarstraeti 4 and 9, P.O.B. 1131, REYKJAVIK.
Tel. 13133/14281/11936

INDIA – INDE
Oxford Book and Stationery Co. :
NEW DELHI-1, Scindia House. Tel. 45896
CALCUTTA 700016, 17 Park Street. Tel. 240832

INDONESIA – INDONÉSIE
PDIN-LIPI, P.O. Box 3065/JKT., JAKARTA, Tel. 583467

IRELAND – IRLANDE
TDC Publishers – Library Suppliers
12 North Frederick Street, DUBLIN 1 Tel. 744835-749677

ITALY – ITALIE
Libreria Commissionaria Sansoni :
Via Lamarmora 45, 50121 FIRENZE. Tel. 579751
Via Bartolini 29, 20155 MILANO. Tel. 365083
Sub-depositari :
Editrice e Libreria Herder,
Piazza Montecitorio 120, 00 186 ROMA. Tel. 6794628
Libreria Hoepli, Via Hoepli 5, 20121 MILANO. Tel. 865446
Libreria Lattes, Via Garibaldi 3, 10122 TORINO. Tel. 519274
La diffusione delle edizioni OCSE è inoltre assicurata dalle migliori
librerie nelle città più importanti.

JAPAN – JAPON
OECD Publications and Information Center,
Landic Akasaka Bldg., 2-3-4 Akasaka,
Minato-ku, TOKYO 107 Tel. 586.2016

KOREA – CORÉE
Pan Korea Book Corporation,
P.O. Box n° 101 Kwangwhamun, SÉOUL. Tel. 72.7369

LEBANON – LIBAN
Documenta Scientifica/Redico,
Edison Building, Bliss Street, P.O. Box 5641, BEIRUT.
Tel. 354429 – 344425

MALAYSIA – MALAISIE
and/et SINGAPORE - SINGAPOUR
University of Malaysia Co-operative Bookshop Ltd.
P.O. Box 1127, Jalan Pantai Baru
KUALA LUMPUR. Tel. 51425, 54058, 54361

THE NETHERLANDS – PAYS-BAS
Staatsuitgeverij
Verzendboekhandel Chr. Plantijnnstraat
S-GRAVENAGE. Tel. nr. 070.789911
Voor bestellingen: Tel. 070.789208

NEW ZEALAND – NOUVELLE-ZÉLANDE
Publications Section,
Government Printing Office Bookshops:
AUCKLAND: Retail Bookshop: 25 Rutland Street.
Mail Orders: 85 Beach Road, Private Bag C.P.O.
HAMILTON: Retail: Ward Street,
Mail Orders, P.O. Box 857
WELLINGTON: Retail: Mulgrave Street (Head Office),
Cubacade World Trade Centre
Mail Orders: Private Bag
CHRISTCHURCH: Retail: 159 Hereford Street,
Mail Orders: Private Bag
DUNEDIN: Retail: Princes Street
Mail Order: P.O. Box 1104

NORWAY – NORVÈGE
J.G. TANUM A/S Karl Johansgate 43
P.O. Box 1177 Sentrum OSLO 1. Tel. (02) 80.12.60

PAKISTAN
Mirza Book Agency, 65 Shahrah Quaid-E-Azam, LAHORE 3.
Tel. 66839

PHILIPPINES
National Book Store, Inc.
Library Services Division, P.O. Box 1934, MANILA.
Tel. Nos. 49.43.06 to 09, 40.53.45, 49.45.12

PORTUGAL
Livraria Portugal, Rua do Carmo 70-74,
1117 LISBOA CODEX. Tel. 360582/3

SPAIN – ESPAGNE
Mundi-Prensa Libros, S.A.
Castello 37, Apartado 1223, MADRID-1. Tel. 275.46.55
Libreria Bastinos, Pelayo 52, BARCELONA 1. Tel. 222.06.00

SWEDEN – SUÈDE
AB CE Fritzes Kungl Hovbokhandel,
Box 16 356, S 103 27 STH, Regeringsgatan 12,
DS STOCKHOLM. Tel. 08/23.89.00

SWITZERLAND – SUISSE
OECD Publications and Information Center
4 Simrockstrasse 5300 BONN. Tel. (0228) 21.60.45
Local Agents/Agents locaux
Librairie Payot, 6 rue Grenus, 1211 GENÈVE 11. Tel. 022.31.89.50
Freihofer A.G., Weinbergstr. 109, CH-8006 ZÜRICH.
Tel. 01.3634282

TAIWAN – FORMOSE
National Book Company,
84-5 Sing Sung South Rd, Sec. 3, TAIPEI 107. Tel. 321.0698

THAILAND – THAILANDE
Suksit Siam Co., Ltd., 1715 Rama IV Rd,
Samyan, BANGKOK 5. Tel. 2511630

UNITED KINGDOM – ROYAUME-UNI
H.M. Stationery Office, P.O.B. 569,
LONDON SE1   9NH. Tel. 01.928.6977, Ext. 410 or
49 High Holborn, LONDON WC1V 6 HB (personal callers)
Branches at: EDINBURGH, BIRMINGHAM, BRISTOL,
MANCHESTER, CARDIFF, BELFAST.

UNITED STATES OF AMERICA – ÉTATS-UNIS
OECD Publications and Information Center, Suite 1207,
1750 Pennsylvania Ave., N.W. WASHINGTON, D.C.20006 – 4582
Tel. (202) 724.1857

VENEZUELA
Libreria del Este, Avda. F. Miranda 52, Edificio Galipan,
CARACAS 106. Tel. 32.23.01/33.26.04/33.24.73

YUGOSLAVIA – YOUGOSLAVIE
Jugoslovenska Knjiga, Terazije 27, P.O.B. 36, BEOGRAD.
Tel. 621.992

Les commandes provenant de pays où l'OCDE n'a pas encore désigné de dépositaire peuvent être adressées à :
OCDE, Bureau des Publications, 2, rue André-Pascal, 75775 PARIS CEDEX 16.

Orders and inquiries from countries where sales agents have not yet been appointed may be sent to:
OECD, Publications Office, 2 rue André-Pascal, 75775 PARIS CEDEX 16.

64323-9-1981

PUBLICATIONS DE L'OCDE, 2, rue André-Pascal, 75775 PARIS CEDEX 16 - N° 42114 1981
IMPRIMÉ EN FRANCE
(66 81 12 3) ISBN 92-64-02236-8